Energy Harvesting

Energy Harvesting

Edited by **Nora Ayling**

CLANRYE
INTERNATIONAL

New Jersey

Published by Clanrye International,
55 Van Reypen Street,
Jersey City, NJ 07306, USA
www.clanryeinternational.com

Energy Harvesting
Edited by Nora Ayling

International Standard Book Number: 978-1-63240-209-7 (Hardback)

Contents

Preface

This book has been a concerted effort by a group of academicians, researchers and scientists, who have contributed their research works for the realization of the book. This book has materialized in the wake of emerging advancements and innovations in this field. Therefore, the need of the hour was to compile all the required researches and disseminate the knowledge to a broad spectrum of people comprising of students, researchers and specialists of the field.

The extensive processes of energy harvesting and their concepts are highlighted in this profound book. The objective of this book is to supply a comprehensive overview of the current research advancements in the design of efficient small scale energy harvesters via the contributions of internationally known researchers. The book covers a broad range of topics from the physics of energy conversion, the elaboration of electroactive materials and their technique to the conception of a complete micro generator, and is organized based upon the input energy source. It will help its readers in their research and/or provide them with new ideas in the vast area of energy harvesting.

At the end of the preface, I would like to thank the authors for their brilliant chapters and the publisher for guiding us all-through the making of the book till its final stage. Also, I would like to thank my family for providing the support and encouragement throughout my academic career and research projects.

Editor

Photonic

Advances in Photoelectrochemical Fuel Cell Research

Kai Ren and Yong X. Gan

Additional information is available at the end of the chapter

1. Introduction

Fuel cells are electrochemical devices which can convert chemical energy into electrical power. They have the advantages of quiet in operation, high efficiency and low pollutant emissions. Photoelectrochemical fuel cells (PEFCs or PECs) are special fuel cells. PEFCs are used in organic waste degradation (Patsoura A et al., 2006), solar energy utilization (Bak T et al., 2002), gaseous product decomposition (Ollis DF et al., 2000), aqueous pollutants removal (Sakthivel S et al., 2004) and photocatalytic sterilization (Fujishima A et al., 1972). A PEFC or PFC consumes fuels and utilizes luminous energy to generate electricity power when the photoanode is excited by radiation. (Lianos P et al., 2010).

Fig. 1 shows a typical two-compartment photo fuel cell separated by a silica frit (Antoniadou M et al., 2010). The electrolyte is NaOH. The anode is nanocrystalline titania. The cathode is a carbon black deposited with Pt as the catalyst. This device works under UV irradiation. The open circuit voltage was 0.88V without ethanol and 1.22 V with ethanol.

2. Mechanisms of Photoelectrochemical Fuel Cells (PEFCs)

PEFCs normally consist of a semiconductor photoanode, metal cathode and electrolyte which could be an acid, base or just water. Light excites electrons at the photoanode if the light energy is larger than the material energy band gap. The photoanode generates electrons (e^-) and holes (h^+). At the anode, production of oxygen happens. Hydrogen generates at the water/cathode interface. The reactions are shown as follows (Chang C et al., 2012):

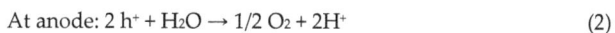

$$\text{Light energy: } 2h\nu \rightarrow 2h^+ + 2\,e^- \tag{1}$$

$$\text{At anode: } 2\,h^+ + H_2O \rightarrow 1/2\,O_2 + 2H^+ \tag{2}$$

$$\text{At cathode: } 2\ e^- + 2\ H^+ \rightarrow H_2 \tag{3}$$

$$\text{Overall reaction: } 2\ hv + H_2O \ \rightarrow 1/2\ O_2 + H_2 \tag{4}$$

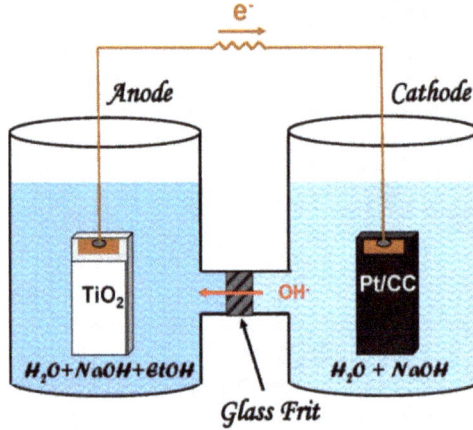

Figure 1. The sketch of a two-compartment PEFC. (Antoniadou M et al., 2010).

3. Photoanode materials

Fig. 2 shows light absorption and electron transport on a photo sensitive material. The light energy is absorbed by the photo sensitive material. Electrons and holes generate. The electrons flow to cathode. The holes decompose water to produce oxygen. Nanostructured materials may be added to substrates such as Ti, glass, copper etc.

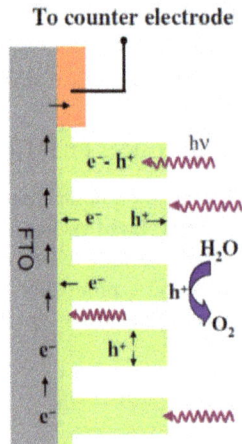

Figure 2. Schematic of a typical nanostructured photoanode. (Chakrapani V et al., 2009)

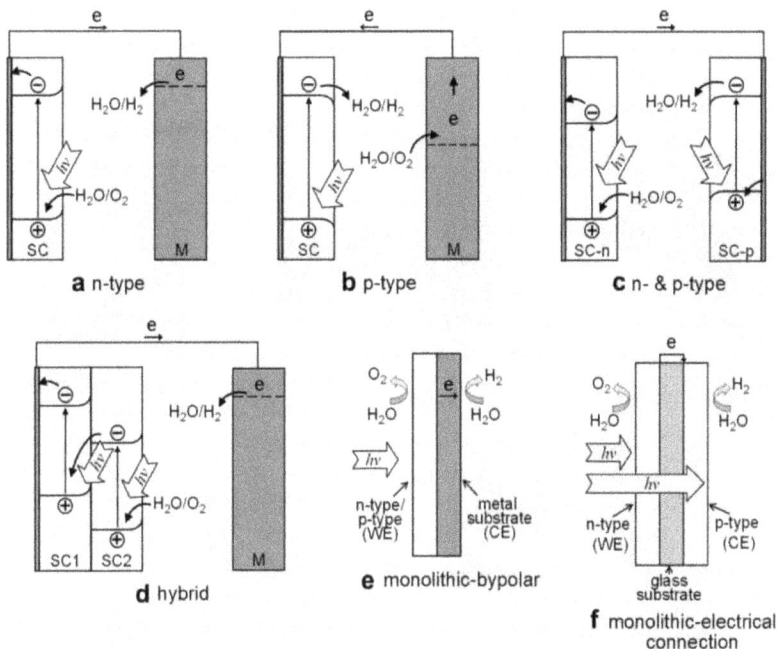

Figure 3. Some types of photoelectrode (PE) commonly used. (Minggu L et al., 2010).

Semiconductor is widely used as photoelectrode which including n-type (TiO_2), p-type (lnP) and n-p type (n-GaAs/p-lnP). They can be combined together to form multi-layered structures to tune the band gaps (Minggu L et al., 2010). In Fig. 3, SC stands for a semiconductor and M stands for a metal which is usually used as a substrate. Nanoporous materials are widely used in fuel cells. There are a number of transparent conductive oxides (TCOs) used as photoanode materials including indium-tin-oxide and fluorine-doped tin oxide. Some non-transparent conductive oxides (NTCOs) including nanocrystalline titania TiO_2, n-type semiconductor ZnO, Fe_2O_3, $SrTiO_3$ etc. can also be used as photoanode materials. Among them, TiO_2 is the most commonly used one due to its stability and high photo activity.

Fig. 4 shows the design of photoelectrode (Miller EL et al., 2003). Fig. 4a shows the first stage of design using p-type silicon. The catalyst layer is on the left side and the platinum catalyst is deposited on the right side. The arrow indicates the direction of light illumination. In Fig. 4b, the right side is coated with a Shottky barrier metal. Fig. 4c illustrates a three-junction structure consisting of Si-Ge-glass. The photo-hydrogen conversion efficiency is up to 7.8%. This design needs an external connection. Fig. 4d has no external connection, as compared with Fig. 4c. Fig. 4e is the latest integrated planar photoelectrode design. On the right side, there is a highly transparent and corrosion-resistant film to keep the high efficiency. This new design can connect single cells in series, which can generate large power.

(a) **liquid junction**   (b) **Schottky junction**   (c) **3-junction a-Si:Ge on glass/ITO with separated electrodes**

(d) **triple junction a-Si on SS**   (e) **integrated planar photoelectrode design**

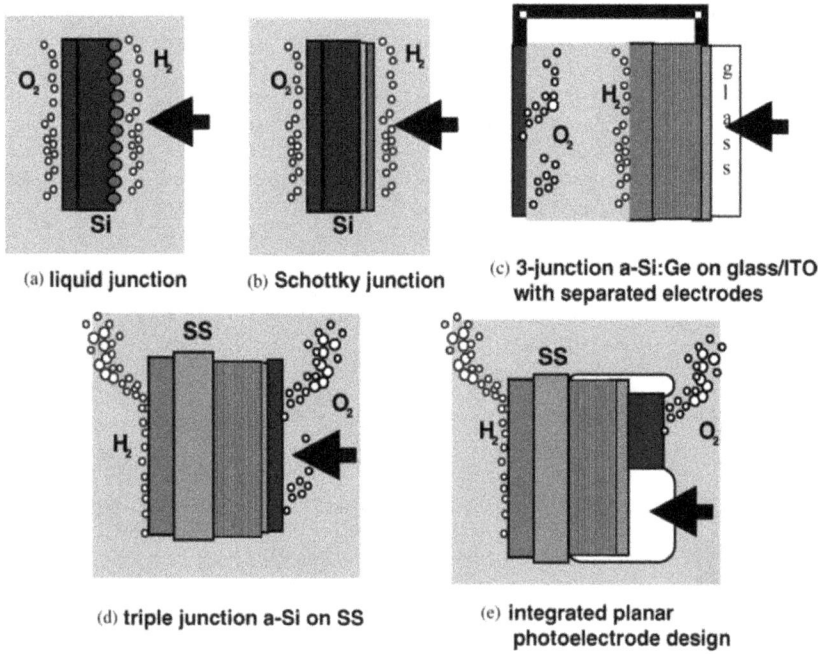

Figure 4. Photoelectrode designs. (Miller EL et al., 2003).

TiO$_2$ is an effective photocatalysis (PC). It is often used as the anode of PFC (Gratzel M et al., 2001). The reaction of TiO$_2$ under UV illumination is follows (Park KW et al., 2007):

$$TiO_2 + hv(UV) \rightarrow TiO_2 + e^- + h^+ \qquad (5)$$

This formula is applicable for any metal oxide as the anode in a photo fuel cell. When the metal oxide absorbs photons from any light sources, electron-hole pairs are produced. The photo-generated holes react with fuels.

4. Fuels

There are many types of fuels for PECs including pure water, alcohols (MeOH, EtOH, PrOH), polyols (glycerol, xylitol, sorbito, glucose, fructose, lactose), organic pollutants (urea, ammonia, triton X-100, SDS, CTAB). Alcohols have larger efficiencies than others do (Antoniadou M et al., 2009). In polyols, glycerol has the highest current density. Pure water has the lowest efficiency. Fuels are decomposed in the ways as described below.

Methanol (Lianos P, 2010):

Anode electrode in acidic media :

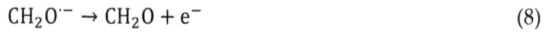

$$CH_3OH + 2h^+ \rightarrow CH_3O^{\cdot} + 2H^+ \tag{6}$$

$$CH_3O^{\cdot} \rightarrow CH_2O^{\cdot-} + H^+ \tag{7}$$

$$CH_2O^{\cdot-} \rightarrow CH_2O + e^- \tag{8}$$

Anode electrode in base media:

$$6OH^- + 6h^- \rightarrow 6OH^{\cdot} \tag{9}$$

$$CH_3OH + 6OH^{\cdot} \rightarrow CO_2 + 5H_2O \tag{10}$$

Under the photo illumination, PFC absorbs light energy and the TiO₂ is excited to release electrons. By this method, higher electric potential can be generated compared with other fuel cells. The completely reaction of TiO₂ with methanol's shown as:

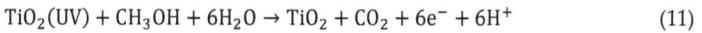

$$TiO_2(UV) + CH_3OH + 6H_2O \rightarrow TiO_2 + CO_2 + 6e^- + 6H^+ \tag{11}$$

Ethanol:

The completely reaction of TiO₂ with ethanol is as follows:

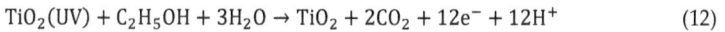

$$TiO_2(UV) + C_2H_5OH + 3H_2O \rightarrow TiO_2 + 2CO_2 + 12e^- + 12H^+ \tag{12}$$

Reber JF et al., (1984) stated that a common formula could be:

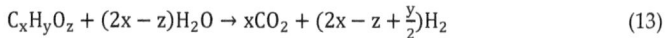

$$C_xH_yO_z + (2x - z)H_2O \rightarrow xCO_2 + (2x - z + \tfrac{y}{2})H_2 \tag{13}$$

Several types of biomass used in fuel cells are reported by Kaneko M et al., (2006), and shown in Table 1. The experimental condition is in acid solutions contain 0.1M Na₂SO₄. The anode of PEC is TiO₂ nanoporous film and the cathode is Pt black on Pt foil. The light intensity is 503 mW/cm² and ambient temperature is 25 °C. The results of open circuit voltage show that acetic acid is the best. Ammonia, glycine, phenylalanine and glutamic acid also show good performances. The short circuit current of methanol has the highest value. The fill factor (FF) as defined by the ratio of maximum obtainable power to the product of the open circuit voltage and short circuit current was calculated. Ammonia has the maximum FF of 0.63.

Liu Y et al., (2011), did similar research on various fuels with a self-organized TiO₂ nanotube array (STNA) as the photoanode of the photo fuel cell (Table 2). Multiply fuels were tested but each fuel's concentration was smaller than what Kaneko et al. used. By comparing the data in these two tables, we can see that the open circuit voltage and short circuit current obtained by Liu et al. are slightly larger, which means that they got higher efficiencies from the PFC system they built. When they varied the concentration of Na₂SO₄ from 0 to 0.5M, V_{oc} and J_{sc} reached the peak values at 0.1 M and the FF has the maximum value at 0.05 M. All the experiments were done under solar light illumination.

Fuel (conc./M)	Solvent (pH)	V_{oc}/V	Jsc/ mA cm^{-2}	FF
Methanol	None	0.54	0.8	0.23
Methanol (50 vol.%)	Water (not controlled)	0.44	0.76	0.28
Ethanol	None	0.49	0.52	0.25
Glucose (0.5)	Water (5)	0.64	0.5	0.32
Urea (5)	Water (5)	0.6	0.3	0.26
Ammonia (10)	Water (12)	0.84	0.53	0.63
Acetic acid (2 wt.%)	Water (not controlled)	0.94	0.47	0.37
Glycine (0.5)	Water (5)	0.76	0.45	0.45
Glutamic acid (0.5)	Water (1)	0.9	0.64	0.42
Tyrosine (0.5)	Water (13)	0.86	0.43	0.36
Phenylalanine (0.5)	Water (13)	0.9	0.61	0.53
Agarose (0.2 wt.%)	Water (5)	0.6	0.12	0.26
Gelatin (2 wt.%)	Water (1)	0.64	0.23	0.32
Collagen (3 mg/ml)	Water (l)	0.62	0.16	0.34
Cellulose sulfate (2 wt.%)	Water (not controlled)	0.56	0.29	0.34
Lignosulfonic acid (0.5 wt.%)	Water (not controlled)	0.57	0.02	0.51
Polyethylene glycol (2 wt.%)	Water (5)	0.6	0.28	0.27
Poly(acrylamide) (2 wt.%)	Water (5)	0.6	0.23	0.24

Table 1. PFC performances by using different fuels in 0.1M Na$_2$SO$_4$ with a TiO$_2$ photoanode and Pt/Pt black cathode. (Kaneko M et al., 2006).

	Organic compounds	V_{oc} (V)	Jsc (mA cm^{-2})	JVmax (mW cm^{-2})	FF
Model compound	Na$_2$SO$_4$ (0.1 mol/L)	1.13	0.35	0.12	0.31
	Glucose (0.05 mol/L)	1.28	0.83	0.38	0.36
	Glutamic acid (0.05 mol/L)	1.34	1.08	0.51	0.35
	Nicotinic acid (0.05 mol/L)	1.39	0.61	0.3	0.35
	Acetic acid (0.05 mol/L)	1.48	1.42	0.67	0.32
	Urea (0.05 mol/L)	1.41	0.91	0.51	0.4
	Ammonia (0.05 mol/L)	1.24	0.72	0.37	0.41
Actual wastewater	Pharmaceutical wastewater (COD =24572 mg/L)	0.88	1.36	0.43	0.36
	Petroleum exploiting wastewater (COD =19087 mg/L)	1.34	0.98	0.34	0.26
	Dying wastewater (COD =10842 mg/L)	1.53	1.21	0.5	0.27
	Chemical plant wastewater (COD =11700 mg/L)	1.11	0.99	0.3	0.27
	Original urine solution (COD =9642mg/L)	0.93	0.61	0.19	0.34

Table 2. PFC performances by using different fuels. (Liu Y et al., 2011).

5. Cathode materials

As compared with multiple choices of photoanodes, the materials for the cathode of photo fuel cells are limited. Normally a Pt wire or a Pt foil is used. Another option is to use Pt-black. The Pt black powders can be cast, sprayed or hot-pressed on the surface of a Pt (Kaneko M et al., 2006). The surface area becomes larger when the Pt-black powers were deposited onto Pt wires or foils. In addition to platinum cathodes including platinum wire, non-platinized platinum foil, platinized platinum foil, platinized SnO_2 with F, metal nanoparticles deposited on a TiO_2/SnO_2 with F doping are made into electrodes. $Pt/TiO_2/SnO_2$, $Pd/TiO_2/SnO_2$, $Au/TiO_2/SnO_2$, $Ag/TiO_2/SnO_2$, and $Ni/TiO_2/SnO_2$) are some of the examples. A platinum-loaded carbon cloth has also been used as a cathode material. The platinized SnO_2 with F electrode has better performance than others. It speaks current, voltage and efficiency are 1.15 mA/cm^2, 1340 mV and 12.3%, respectively. The platinum-loaded carbon cloth has the maximum efficiency of 32.3%. Thin layer of Si-H film photo cathode can be made by plasma assisted chemical vapor deposition (PECVD). A Si-H cathode deposited organic or inorganic protective layer or coating with catalytic platinum can enhance the stability for long time use. The best thickness of the polymer protective layer is 5 nm. The optimized thickness of Pt coating is 2 nm.

CuO is a cheap material. CuO nanoparticles and films prepared by flame spray pyrolysis (FSP) were used as photocathodes by Chiang C et al., (2011). The optical band gap was decreased from 1.68 eV to 1.44 eV with the annealing temperature increasing from room temperature to 600°C. The nanoparticle size is from 50 nm to 150 nm, as shown in Fig. 5. The best photocurrent density is 1.2 mA/cm^2 obtained from CuO particles which were annealed at 600 °C for 3 hour. The bias voltage is 0.55 V in 1M KOH. The total conversion efficiency is 1.48% and the hydrogen generation efficiency is 0.91%.

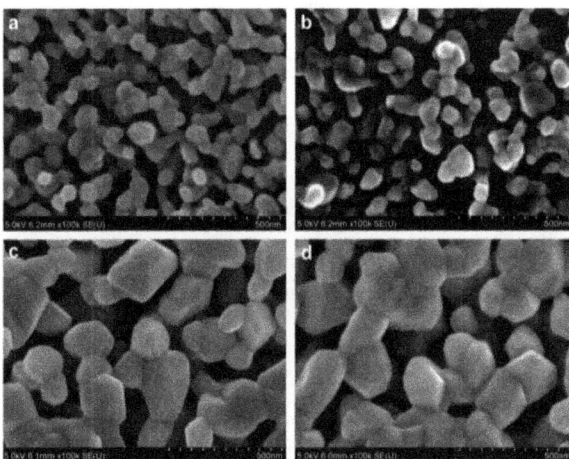

Figure 5. SEM images of CuO photo cathodes prepared under different conditions: (a) 450 °C, 1 h, (b) 450 °C, 3 h, (c) 600 °C, 1 h, (d) 600 °C, 3 h. (Chang C et al., 2011).

6. Terminologies associated with the photo fuel cells

6.1. Optical absorption coefficient for band gap determination

The optical absorption coefficient, α, is related to the wavelength, transmittance, reflectance of the light illuminating on a material. Low absorption coefficient means low photo absorption ability. The following equation holds (Pihosh Y et al., 2009)

$$\alpha(hv) = d^{-1}\ln(\frac{1-R}{T}) \tag{14}$$

where T is the transmittance, R the reflectance, and d the thickness of the material. The term hv refers to the photon energy.

The optical coefficient is used to obtain the band gap E_g following

$$constant \times (hv - E_g)^2 = \alpha(hv) \times (hv) \tag{15}$$

6.2. Roughness factor

Roughness factor is related to the surface area of an electrode. For nanotubes, the geometry roughness is calculated as (Shankar K et al., 2007)

$$G = \left[\frac{4\pi L(D+W)}{\sqrt{3(D+2W)^2}}\right] \tag{16}$$

where D is the inner diameter, W the wall thickness and L the tube length of the nanotubes. From the experiment on titania nanotubes by Isimjan TT et al., (2012), a higher surface area (roughness) was obtained at higher processing voltages. At a constant voltage, the pore size of nanotubes is dependent of distance between anode and cathode in the electrochemical process.

6.3. Photo conversion efficiency

The photo conversion efficiency is the overall efficiency of a PEC which can be defined by the following equation

$$\eta(\%) = \left[\frac{(total\ power\ output-electrical\ power\ output)}{light\ power\ input}\right] \times 100 \tag{17}$$

7. Nanostructures photoanode materials processing

7.1. TiO₂ nanotube (TNT) photoanode

TiO$_2$ nanotubes on the surface of Ti as shown in Fig. 6 demonstrate a self-organized nanostructure. The advantage of the nanobutes is the high surface/volume ratio. TiO$_2$ nanotubes have active photo catalysis characteristic, good corrosion resistance, thermal stability and good operation stability as described by Mahajan V et al., (2008). TiO$_2$ nanotubes can be made by various ways including hydro/solvothermal method (Kasuga T et al., 1998), template-assisted approach (carbon nanotube, alumina or monocrystal as the template), sol–gel method (Kasuga T et al., 1998), microwave irradiation (Zhao Q et al., 2009), and direct electrochemical anodization. The advantage of the hydro/solvothermal method is easy to operate. The disadvantage is that only disordered and twisted TiO_2 nanotubes can be obtained. For the template-assisted method, the size of the nanobutes is uniform. For the electrochemical anodic oxidation method, it has the advantage of easy to operate and the obtained nanotubes are highly ordered. Therefore, many researchers prefer the electrochemical method.

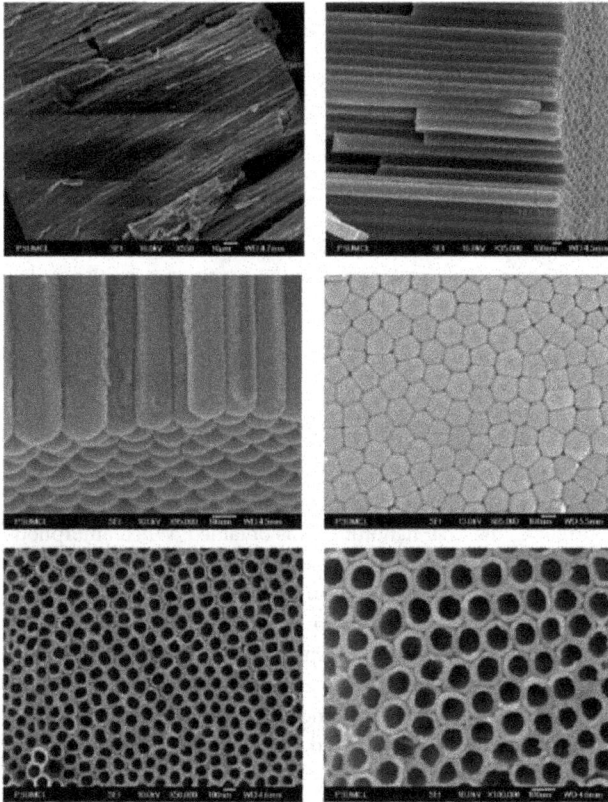

Figure 6. Self-organized TiO_2 nanotubes via anodization. (Shankar K et al., 2007).

7.1.1. Hydrothermal treatment

Hydrothermal method is one of the popular approaches to prepare TNTs. The first group having successfully fabricated TiO_2 nanotubes by hydrothermal method is Kasuga T et al., in 1998. During the process, titania nanopowders are placed in alkaline aqueous solutions held in high pressure steel vessels. The temperate should be between 50-180 °C. The process continues for 10 to 20 hours. Some post treatment can be applied, for example, washing with acid or alkaline solutions for 10 hours, drying at 80 °C and annealing at 500 °C. The reaction process is divided into four steps (Hafez H et al., 2009) i.e. (1) synthesis of TiO_2 nanotubes in alkaline aqueous solutions, (2) protons replacing alkali ions in the reaction, (3) drying, (4) acid washing (post treatment). There is controversy about the necessity of the acid washing. Some researchers (Liu S et al., 2009) think acid washing is a necessary procedure to form TNTs, but other researchers (Chen X et al., 2007) think hydrothermal is more important than the acid washing step as sketched in Fig. 7. The step of washing by acid is not even necessary to form TNTs.

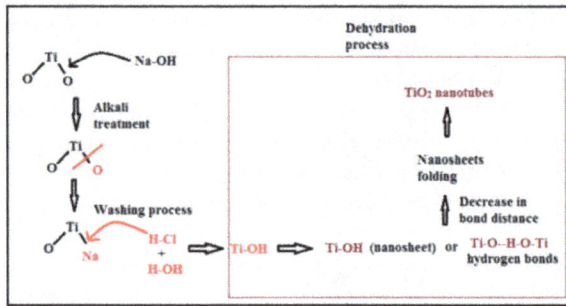

Figure 7. Hydrothermal method for fabricating TiO_2 nanotubes (Chen XB et al., 2007).

7.1.2. The effects factors of material and solution

With different raw materials and reaction solutions, the different morphology of TiO_2 was obtained by hydrothermal method (Yuan ZY et al., 2004). When crystalline TiO_2 react with NaOH under 100-160 °C, the TiO_2 nanotubes was obtained. When amorphous TiO_2 be used under same conditions, the TiO_2 nanofibers are fabricated. Either crystalline or amorphous TiO_2 can be used reaction with NaOH can result TiO_2 nanoribbons when temperature rise to 180 °C. If the solution used by KOH, the nanowires morphology is formed. The pH value of solution also plays an important role in morphology of TiO_2 nanomaterials (Xu YM et al., 2010). Fen LB et al., (2011) used anatase TiO_2 nanopowders (Aldrich 637254-50G, 99.7%) with NaOH solution fabricated TNTs. The inner diameter is 3-6 nm and wall thickness is 1.9 nm. Lan Y et al., (2005) used rutile nanopowders with 10 M NaOH solution obtained TNTs which inner diameter 2-3 nm and wall thickness is 7-8 nm, besides the length is 200-300 nm. The inner diameter is smaller but the wall thickness is larger than the TNTs made by Fen LB et al., (2011).

Hydrothermal treatment temperature and time are significant factors during the formation of TNTs. The temperature range should be from 100 °C to 180 °C and the time range should be from 1 to 24 hours. Sreekantan S et al., (2010), selected the temperatures at 90, 110, 130, 150 °C and time for 3, 6, 9, 15, 18, 24 hours. The $NaOH/TiO_2$ solution was used. At 90 °C, the TiO_2 particles form sheets. When the temperature was set at 110 °C, the sheets were transformed into nanotubes because the thermal energy increases with temperature (Seo HK et al., 2008). With the temperature increasing to 130, 150 °C, there is no change of the outer diameter (10 nm) of nanotubes but the TNTs transform to anatase phase. For the effect of reaction time, particles begin to form sheet at 3 hours. Sreekantan S et al., (2010) indicated that Ti-O-Ti bond is replaced by Ti–O–Na and Ti–OH bonds at this time. After 6 and 9 hours, more and more sheets form nanotubes (10 nm). After 15 hours, TNTs form completely. They found that 150 °C is the best temperature for making TNTs with the highest photocatalytic activity.

Seo HK et al., (2008), studied the phase transformation of TNTs at different hydrothermal temperatures. They used a 10 M NaOH solution and the temperature range was from 70 °C to 150 °C. A 0.1 M HCl solution was used for washing the TNTs. They founded that at 70 °C, the particles begun forming nanosheets. Nanosheets and nanofibers co-existed at 90 °C. At 110 °C, the nanosheets were transformed into nanotubes. This conclusion is also reported by Sreekantan S et al., (2009). Hydrothermal processing can also produce nanoribbons instead of nanotubes if the reaction temperature is higher than 180 °C.

7.2. Synthesis of self-organized TiO$_2$ nanotubes via electrochemical anodization

In 1999, Zwilling V et al. first used electrochemical anodization method for synthesis of TiO$_2$ nanotubes in the solution containing chromic acid and hydrofluoric acid. Later many researchers (e.g. Macak JM et al., 2005) showed that using different applied potentials, electrolytes, pH values (much longer nanotubes at neutral pH electrolytes) and anodization time can control the lengths, thickness, diameters and morphology of TiO$_2$ nanotubes. Zeng X et al., (2011), reported electrochemical oxidation of Ti in a 1.0 M H$_3$PO$_4$ and 0.25 M NaF solution. With the increasing in the potential, TiO$_2$ experienced three forms. When the potential was very low, Ti dissolved into the solution. With the increasing of potential, Ti was oxidized to form TiO$_2$. When the potential was less than 2.5 V, TiO$_2$ film was obtained. Between 2.5 V and 6.0 V, the TiO$_2$ porous structure formed. When potential was higher than 6, the self-organized TiO$_2$ nanotubes were obtained (Fig. 9b).

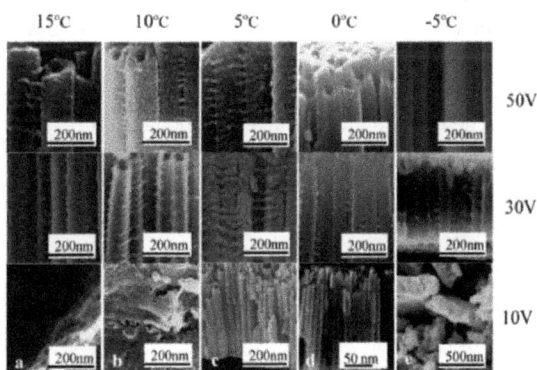

Figure 8. Morphology of self-organized anodic TiO$_2$ nanotubes formed at different temperature and voltage levels. (Liu H et al., 2011).

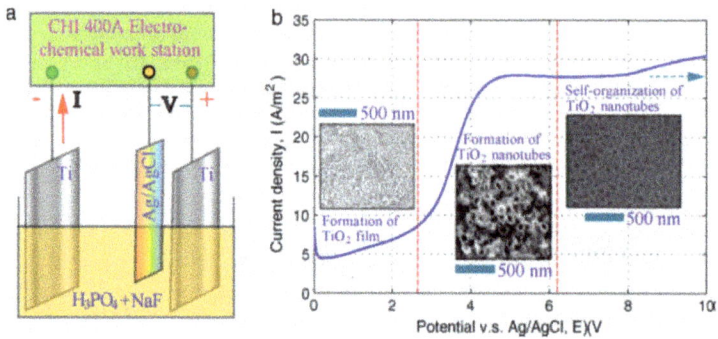

Figure 9. (a) Sketches for electrochemical oxidation of Ti. (b) effectof voltage level on the morphology of TiO₂. (Zeng X et al., 2011).

Before 2005, all of these researches were exclusively using inorganic solutions as electrolytes, such as HF (Varghese OK et al., 2003), KF, NaF (Cai QY et al., 2005). Macak JM et al., (2005), investigated TiO₂ nanotube formation in Na₂SO₄ electrolytes with NaF. The maximum length of nanotubes was up to 2.4 μm. It takes about 6 hours. But longer than this time, the irregular morphology showed up. As compared with HF, NaF can thicker the porous layers. The use of organic electrolytes is a milestone for the TiO₂ nanotubes fabrication. Liu H et al., (2011), studied the temperature effect on morphology of TiO₂ nanotubes. The specified temperatures are -5, 0, 5, 10, 15 °C and the applied potentials are 10, 30, 50V. It helped control the nanotube size and structure under the complex condition as show in Fig. 8. In summary, there are two types of electrolytes in TiO₂ andoization, one is aqueous-based electrolytes, and the other is organics-based electrolytes. Aqueous electrolytes allow the nanotubes to form more quickly because of the low electrical resistance. Besides, lower voltage is enough. However, it is hard to form longer nanotubes because of the dissolution of the nanotubes in the solutions. The organic electrolytes, for example, ethylene glycol and glycerol, have higher electrical resistances. They can slow down the ion transfer. Therefore, higher voltages and longer times are needed. In organic electrolytes, it is easier to form long nanotubes.

7.2.1. Anodization mechanisms

During anodization, a constant voltage in the range from 1V to 150V is applied. The electrolytes containing fluorides have the concentration range from 0.05 to 0.5M. The processing time ranges from a few minutes to a couple of days.

There are two main reactions with the anodization of Ti (Macak JM et al., 2005):

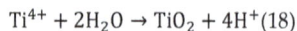

$$Ti^{4+} + 2H_2O \rightarrow TiO_2 + 4H^+ (18)$$

$$TiO_2 + 6F^- + 4H^+ \rightarrow [TiF_6]^{2-} + 2H_2O \ (19)$$

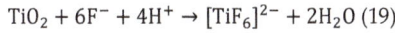

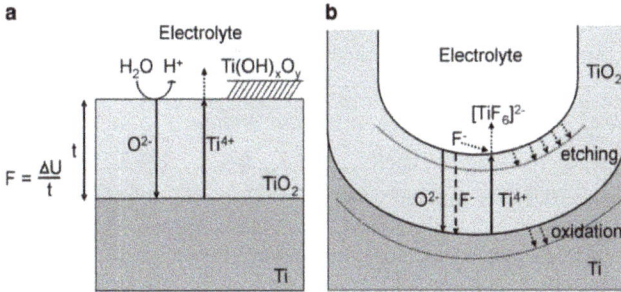

Figure 10. Sketches of Ti anodization (a) without F^-, (b) with F^-. (Macak JM et al., 2007).

First, titanium in the electrolyte produces Ti^{4+}. Then Ti^{4+} reacts with water to form TiO_2 and hydrogen ion (Eq. 18). TiO_2 becomes oxide film on the surface of the titanium as a barrier layer. Meantime, TiO_2 is etched by F^- and many holes form in the film (Eq. 19). With the processing time increasing, the holes become deeper and form nanotubes. When the anodization rate of Ti is equal to the etching rate of TiO_2, the process reaches to a steady-state. The length of nanotubes keeps unchanged.

F ion plays an important role in synthesizing TiO_2 nanotubes. Fig. 10 shows the results of NTs obtained from different solutions with and without F^-. Without F^- the TiO_2 is flat without porous structure. With F^-, reaction (Eq.19) occurs. F ion generates TiF_6^{2-} which is the driving force of etching TiO_2. H^+ can enhance the etching ability of F^-. TiF_6^{2-} ions owing the small diameter can easily move through TiO_2 crystal lattice. Comparing the electrolytes containing Cl^- and Br^- (Chen X et al., 2007), TiO_2 nanotubes arrays fabricated in electrolytes containing F^- have better quality. Fluoride concentration can affect the electrochemical characteristics (Beranek R et al., 2003). If the fluoride concentration is low (less than 0.05 wt. %), there are almost no fluoride ions. If the fluoride concentration is high (1 wt. %), no oxide formation can be observed. Ti^{4+} reacts with F^- immediately to form TiF_6^{2-}. The maximum nanotube length is about 500 nm synthesized in HF electrolytes. The maximum length is several micron meters using NaF and NH_4F electrolytes.

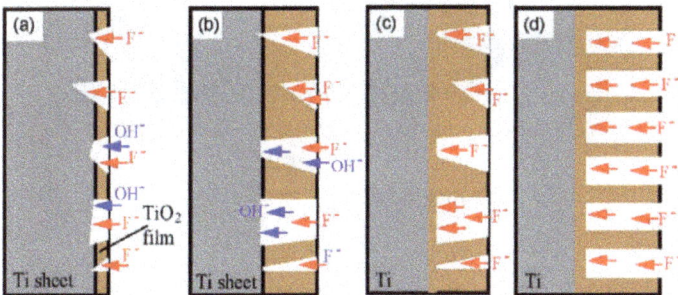

Figure 11. Self-organization of TiO_2 nanotubes in F^- containing solutions. (Gan Y et al., 2011).

The mechanism of TiO_2 growth can be shown in Fig. 11. TiO_2 grows on the Ti substrate gradually. With the TiO_2 film being thicker and thicker, TiO_2 has the function of a protecting film to slow down the Ti dissolution. With the development of F- etching TiO_2, the self-organized TiO_2 nanotubes form as illustrated in Fig.11d.

7.2.2. Synthesis of TiO₂ nanotubes using organic electrolytes

Organic electrolytes containing F- have some advantages. In 2005, Ruan CM et al. used dimethyl sulfoxide (DMSO) and ethylalcohol (1:1) as electrolytes for fabricating TiO_2 nanotubes with a length of 2.3 μm. Macak MJ et al., (2005) used glycerinum synthesized TiO_2 nanotubes with 7 μm length. The maximum length could over 1000 μm. Prakasam HE et al., (2007), using ethylene glycol with 1%-3% H_2O volume and 0.1% to 0.5 % wt of NH_4F solution, anodized Ti foil for 17 h at 20, 40, 50, 60 and 65 V. The result showed that with the increase in the voltage, the inner diameter, outer diameter and length of nanotubes were increased. The maximum values are 135 nm, 185 nm and 105 μm, respectively. The nanotubes grow rate is 15 μm/h. The important factor to affect the length of the nanotubes is the water content. The water volume content should be under 5% for obtain good quality of nanotubes. The morphology of TiO_2 nanotubes formed in organic electrolytesis more smooth and orderly. Besides, the nanotubes have higher photocatalytic efficiencies.

Non-F- electrolytes were also used (Allam N et al., 2007) for the environmental protection purpose. Pulse anodization (Chanmanee W et al., 2008) generated TiO_2 nanotubes with good photoelectrochemical property. Glancing angle deposition (GLAD) was used to obtain Ti films. The anodization of the Ti films produced nanotubes and nanorods (NRs) on a glass substrate. Even brush type nanostructures (BTNs) were obtained (Pihosh Y et al., 2009) as shown in Fig. 12. As compared with the plate counterparts, the TiO_2 NRs, NTs and BTNs have significantly higher photocatalytic activity under Vis-light and UV illumination. The NTs and BTNs have better photocatalytic activity than the NRs because of their larger surface areas. The BTNs can be obtained by andoization of NRs in base.

Figure 12. SEM images of brush type nanostructures (BTNs). (Pihosh Y et al., 2009).

7.3. Post-treatment of TiO₂ nanostructures

7.3.1. Annealing

The purpose of annealing is to change the morphology of TiO_2 nanotubes from amorphous to crystalline (anatase or rutile). Over the past 10 to 15 years, there were a large number of researchers focusing on annealing. Stem N et al., (2011), thermal treated TiO_2 at 1000 °C in wet N_2 for 2 hours, which enhances the photocatalytic performance. Wang MC et al., (2012), showed that annealing temperature affected photocatalytic capability of N-doped TiO_2 thin films. The temperature ranges from 250 °C to 550 °C. The time lasts for 1 hour. Below 350 °C, the surface roughness is low. The photocatalytic activity is the highest after the 350 °C annealing. Lin JY et al., (2011), applied the rapid thermal annealing (RTA) method. The temperatures been used that were 700, 800, 900, 1000 and 1100 °C. The temperate increasing rate was 5 °C/s. The total annealing time was 30 s in oxygen. Through the X-ray diffraction (XRD) examination, it was found that oxygen-related defects were reduced when the TiO_2 nanotubes changed from amorphous to anatase phase. Fang D et al., (2011), studied high temperate calcinations. TiO_2 nanotubes were exposed at the temperatures of 450, 600, 800, and 900°C. The results (Fig. 13) show that 450 °C helps generate a pure anatase phase. At 600 °C, a mixed phase of anatase and rutile can be got. At 800 °C, pure anatase phase grows into large crystallites. As the conclusion, 450 °C is the best calcination temperature. Bauer S et al., (2011), showed that the nanotube's size affects the crystal phase. When the nanotube's diameter is smaller than 30 nm, it is more likely to form rutile. In contrary, when the diameter is larger than 30 nm, anatase can be obtained. Not only the temperate, the type of gases used also affects the properties of the nanotubes (Sang LX et al., 2011). Annealing in air (TNT-A), nitrogen (TNT-N) and 5% hydrogen/nitrogen (TNT-H)generate the similar morphology and band gap. But the difference in the UV absorption photocurrent density exists. The maximum photocurrent density is 0.60 mA/cm^2 for the nanotubes named as TNT-H. The minimum is 0.27 mA/cm^2 for TNT-A. TNT-H has more surface defects. The more surface defects, the higher the photocurrent was generated.

On the contrary to traditional annealing process, Liu JM et al., (2011), performed vacuum annealing and multi-cycle annealing on the Nb-doped TiO_2 thin film. During the three-cycle vacuum annealing, the TiO_2 was heated up to 550 °C (0.05 Pa air pressure) for 1 hour in one cycle. This process was repeated for three times. In another experiment, the TiO_2 film was held at 550 °C at 5 Pa air pressure for 1 hour. Then the annealing was repeated for three times. These two different procedures both can improve the conductivity of the Nb-doped TiO_2 thin film. At different annealing temperatures, the TiO_2 nanotubes showed different photoelectrochemical characteristics (Tang Y et al., 2008). The treatment temperatures are in the range from 300 °C to 550 °C. Again, the sample annealed at 450 °C showed better performance under UV light. With the UV on, the nanotube electrode showed good photoelectric current stability. When the UV was off, the photocurrent quickly decreased to the initial value.

7.3.2. Ultrasonic clean

During anodization synthesis, TiO_2 nanotubes formed, but unexpected deposits may also be on the nanotubes. They can be cleaned by ultrasonic waves (Cai QY et al., 2005). Xu H et al., (2011), applied ultrasonic waves to clean the surface of TiO_2 nanotubes for different time periods. They employed the ultrasonic wave with the power of 80 W at 40 kHz. 9 min is the best treatment time for cleaning the nanotubes. When the time was extended to 40 min, the nanotubes were broken. The nanotubes were peeled off from the Ti completely at 60 min.

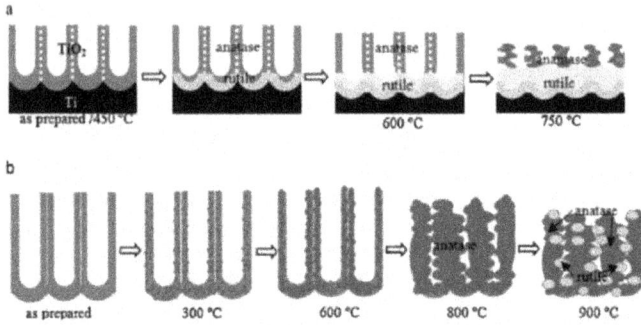

Figure 13. Annealing treatment of TiO_2 nanotubes (a) with Ti substrate, (b) free standing nanotubes. (Fang D et al., 2011).

7.3.3. Doping

There are two main limitations of pure TiO_2 nanotubes

1. Pure TiO_2 can only absorb UV light of wavelength shorter than 400 nm because the band gap of TiO_2 is 3.2 eV, which means that pure TiO_2 can only utilize 6% solar energy. The visible light has the energy band gaps from 1.8 eV to 3.1 eV.
2. High electrical resistance of pure TiO_2 at the room temperature results in very low electron transfer rate. This causes electric energy loss. The converted heat energy dissipates into ambient. At 20 °C, TiO_2 is not a conductor. Only when the temperature rises to 400 °C, the resistance of TiO_2 becomes lower.

Direct doping is one way to overcome the limitations of pure titania. Another method of doping is to stack different materials which have different band gaps. That could make hybrid photoanode (HPE) as first reported by Morisaki H et al., (1976).

Some noble metals doped to TNTs such as gold (Malwadkar SS et al., 2009), silver (Guo GM et al., 2009), platinum can improve the photocatalytic activity of the TNTs. This is because these noble metals can inhibit recombination of electron (Chan SC et al., 2005). Metal doped-TNTs for photo fuel cell applications are reported.

In addition to using noble metals, Macak JM et al., (2007) showed doping copper by electrodeposition. First, synthesis of TiO_2 nanotubes which have low conductivity especially at the bottom of the nanotubes was carried out. Second, using an aqueous electrolyte, about 1% of Ti^{4+} in the TiO_2 outer layer was converted into Ti^{3+} ($Ti^{4+} + e^- + H^+ = Ti^{3+}H^+$ and $2H^+ + 2e^- = H_2$). With Ti^{3+}, the mobility gap of TiO_2 was reduced from 3.2 eV to 2.4 eV and the bottom of the nanotubes become highly conductive. The third step is to dope the nanotubes with Cu. Since the bottom has a high conductivity with Ti^{3+} and H^-, Cu is easily be doped in the nanotubes by a current pulsing electroplating approach.

Sun L et al., (2009), fabricated Fe-doped TNTs using Fe ion containing electrolytes. This study shows that the content of Fe^{3+} is a significant factor affecting the photo catalysis capacity. They used three $Fe(NO_3)_3$ solutions for comparison. The concentrations are 0.05 M, 0.1 M and 0.2 M. The result shows that 0.1 M $Fe(NO_3)_3$ doped TNTs have the maximum photo current and photocatalytic degradation rate. 0.15 M $Fe(NO_3)_3$ doped nanotubes have the maximum absorbance under UV-Vis. Different application need different $Fe(NO_3)_3$ contents. Tu YF et al., (2010), employed template-based LPD method to dope Fe to TNTs. Redshift of the absorption was found. The best Fe content was 5.9 at %. The doped-TNTs achieved the best efficiency of photo catalysis under visible light. Wu Q et al., (2012), fabricated Fe-doped (Fe_2O_3 and Fe^{3+}) with ultrasound assisted impregnating calcinations method. Results showed that Fe_2O_3 went into the TNTs and Fe^{3+} into the TiO_2 lattice. The operation time and temperatures affect the photo responses of Fe-doped TNTs. Ultrasound treating for 5 min following by annealing at 500 °C provides NTNs the highest photo catalysis efficiency.

C and N doping are non-metal doping examples. B, P, and other nonmetallic dopants are also used. Nitrogen is the earliest, most effective and most studied doping element for TNTs. There are many methods to dope nitrogen into TNTs such as annealing TNTs in ammonia (Vitiello RP, et al., 2006), ion implantation (Ghicov A et al., 2006) etc. Asahi R et al., (2001), doped TiO_2 with nitrogen using a solution method. Vitiello RP et al., (2006), showed a simple method for making N-doped TNTs, which is treating TNTs at 300-600 °C in NH_3 atmosphere. Results showed that 500 °C- 600 °C is the best annealing temperate range at which TNTs transfer to anatase and have the most effective photoresponse. Xu JJ et al., (2010), showed difference in photo catalytic activity between N-doped and Non-doped TiO_2 nanotubes under Vis-light. The photocurrent density of N-doped nanotubes was twice as that of the non-doped nanotubes under visible light illumination. Yuan J et al., (2006), synthesized N-doped TiO_2 by heating urea with TiO_2 at 300-700 °C. The doped TiO_2 can absorb light with the wavelength up to 600 nm. The result shows that urea changes to chemisorbed N_2 and substituted N staying in the TiO_2.

Liu ZY et al., (2009), doped carbon into TNTs for solar photochemical cell hydrogen generation. Shaban YA et al., (2007), studied the fabrication time and temperature effects on grooved and non-grooved Ti metal sheet doped with carbon for photochemical catalysis. The result shows that the grooved sample has higher photocurrent density than the non-grooved one. The grooved simple with a depth of 0.005 inch has the maximum photo conversion efficiency of 11.37 % (treated at 820 °C, 18 minutes, thermal flame oxidation, tested in 5.0 M KOH, illuminated by a 150 W Xenon lamp).The non-grooved simple, 0.003 inch grooved one, and 0.001 inch groove done have the maximum photo conversion efficiency of 9.08 %, 8.68 %, 7.20 %, respectively under the same treatment condition.

Co-doping multiple elements was also applied to TiO_2 nanotubes. Tungsten and nitrogen co-doping is a typical example (Shen YF et al., 2008). Nitrogen and sulfur (Yan GT et al., 2011), fluorine and boron (Su YL et al., 2008), Pt and N (Huang LH et al., 2007) co-doping has also been studied. Liu SH et al., (2009), developed a carbon and nitrogen co-doping method by adding 5 mg polyvinyl alcohol (PVA) and 20 mg urea. Then calcination was performed in nitrogen at 600 °C. The photocurrent density is 3 times, 2 times, and 1.2 times compared with the non-doped, C-doped and N-doped TiO_2 nanotubes under solar light and 0.2 V bias-potential combined excitation. He HC et al., (2011), doped Pt-Ni into NTNs using pulsed electrodeposition method. The photo catalytic activity is better than that of only Pt-doped NTNs. Pt-Ni doped NTN is a good anode material for photo fuel cell (direct methanol type). The performance of co-doped TNT is better than that of Pt doped one. Huang LH et al., (2007), synthesized Pt-N doped TNTs by two steps. First, they obtained N-doped TNTs. Second, they used H_2PtCl_6 solution to supply Pt, resulting in Pt-N co-doping. N-doping can enhance the photo response activity and Pt-doping can strengthen the electron separation from holes. Ag can be deposited into N-doped TNTs (Zhang SS et al., 2011), via electrochemical deposition in a 0.2 g/L $AgNO_3$ solution. The result shows that the average photocurrent density of the Ag/N-doped TiO_2 nanotubes is 6 times higher.

Li XQ et al., (2011), developed the CdS nanoparticle and CuTsPc molecule co-doped TNTs. The I-V curve shows that CdS-CuTsPc has the maximum photocurrent density as compared with CdS-CuTsPc, CdS–CuPc, CdS, and CuTsPc doped TNTs. Jia FZ et al., (2012), successful processed ZnS-In_2S_3-Ag_2S doped $TiO_{2-x}S_x$ by a two-step (anodization and solvothermal) approach. Zhang X et al. (2009), doped $PW_{12}O_{40}^{3-}$ and Cr^{3+} into TNTs through the anodization and impregnation methods. The function of Cr^{3+} is narrowing the band gap of TiO_2. They have found that the synergetic factor is 1.42. Su Y et al., (2008), doped nitrogen and fluorine into TNTs. They simply used anodization (20 V) of Ti in the $C_2H_2O_4{\cdot}2H_2O$+NH_4F electrolyte through TiO_2 self-organization. After annealing at 400°C, N-F-doped TNTs showed very good photocatalytic ability and stability. The efficiency of methyl orange (MO) decomposition test is higher than 97%. This method avoids using ammonia which is hazardous. With CeO_2 nanoparticles being doped into TNTs, enhanced charge storage capacity of TNTs was achieved (Wen H et al., 2011). Wang J et al., (2012), reported a C_3N_4 doped TiO_2 nanorod. The UV-Vis absorbance ability of this modified material is as twice as that of the TiO_2.

8. Conclusions

Photoelectrochemical fuel cells have experienced fast development recently because of the progress in nanomaterials. Using various materials processing techniques, it is possible to obtain various nanostructure forms such as nanoparticles, nanorods, nanothin film and nanotubes for photo fuel cell applications. There are many ways for fabricating nanostructures including hydro/solvothermal method, template-assisted method, sol–gel method, microwave irradiation method and electrochemical direct anodizaiton method. Electrochemical anodization becomes a popular method in recent years because it is easy to control the size of nanotubes. Typical photo sensitive materials such as TiO_2, WO_3, Fe_2O_3, CuO and ZnO have been studied. These materials have different band gaps and many researchers reported how to enhance the photo response of them. Doping is a significant and efficient method for improving the photo response of nanomaterials. Metal doping and Non-metal doping are two major types. Besides, organic doping, co-doping alloys and muti-component materials also result in good performance of PEFCs. In summary, PEFCs represent promising energy conversion systems. Future studies should focus on increasing the photoelectric energy conversion efficiency.

Author details

Kai Ren and Yong X. Gan

Department of Mechanical, Industrial and Manufacturing Engineering, College of Engineering, University of Toledo, Toledo, OH, USA

9. References

Allam N, Grimes C. Formation of vertically oriented TiO_2 nanotube arrays using a fluoride free HCl aqueous electrolyte . Journal of Physical Chemistry C. 2007;111; 13028-13032.

Antoniadou M, Kondarides D , Labou D, Neophytides S , Lianos P. An efficient photoelectrochemical cell functioning in the presence of organic wastes. Solar Energy Materials & Solar Cells. 2010; 94; 592–597.

Antoniadou M, Lianos P. Photoelectrochemical oxidation of organic substances over nanocrystalline titania: Optimization of the photoelectrochemical cell. Catalysis Today. 2009; 144; 166-171.

Antoniadou M, Lianos P. Production of electricity by photoelectrochemical oxidation of ethanol in a PhotoFuelCell. Applied Catalysis B: Environmental. 2010; 99; 307-313.

Asahi R, Morikawa T, Ohwaki T, Aoki K, Taga Y. Visible-light photocatalysis in nitrogen-doped titanium oxides. Science. 2001; 293; 269-271.

Bak T, Nowotny J, Rekas M, Sorrell CC. Photo-electrochemical hydrogen generation from water using solar energy. Materials-related aspects. International Journal of Hydrogen Energy. 2002;27:991–1022.

Bauer S, Pittrof A, Tsuchiya H, Schmuki P. Size-effects in TiO_2 nanotubes: Diameter dependent anatase/rutile stabilization. Electrochemistry Communications. 2011; 13; 538–541.

Beranek R, Hildebrand H, Schmuki P. Self-organized porous titanium oxide prepared in H_2SO_4/HF electrolytes. Electrochemical and Solid-State Letters.2003; 6; B12-B14.

Cai QY, Paulose M, Varghese OK, Grimes CA. The effect of electrolyte composition on the fabrication of self-organized titanium oxide nanotube arrays by anodic oxidation. Journal of Materials Research. 2005 ,20 : 230 -236.

Chakrapani V, Thangala J, Sunkara MK. WO_3 and W_2N nanowire arrays for photoelectrochemical hydrogen production. International Journal of Hydrogen Energy. 2009; 34; 9050-9059.

Chan SC, Barteau MA. Preparation of highly uniform Ag/TiO_2 and Au/TiO_2 supported nanoparticle catalysts by photodeposition. Langmuir. 2005; 21; 5588-5595.

Chang C, Wang C, Tseng C, Cheng K, Hourng L, Tsai B. Self-oriented iron oxide nanorod array thin film for photoelectrochemical hydrogen production. International Journal of Hydrogen Energy. 2012. Article in press.

Chanmanee W, Watcharenwong A, Chenthamarakshan C R, Kajitvichyanukul P, Tacconi N, and Rajeshwar K. Formation and characterization of self-organized TiO_2 nanotube arrays by pulse anodization. Journal of the American Chemical Society. 2008; 130; 965-974.

Chen X, Schriver M, Suen T, Mao SS. Fabrication of 10 nm diameter TiO_2 nanotube arrays by titanium anodization. Thin Solid Films. 2007; 515 ;8511 -8514.

Chen XB, Mao SS. Titanium dioxide nanomaterials: synthesis, properties, modifications and applications. Chemical Reviews. 2007; 107; 2891-2959.

Chiang C, Aroh K, Franson N, Satsangi V, Dass S, Ehrman S. Copper oxide nanoparticle made by flame spray pyrolysis for photoelectrochemical water splitting -Part II. Photoelectrochemical study.International Journal of Hydrogen Energy.2011;36;15519-15529.

Fang D, Luo ZP, Huanga KL, Lagoudas DC. Effect of heat treatment on morphology, crystalline structure and photocatalysis properties of TiO_2 nanotubes on Ti substrate and freestanding membrane. Applied Surface Science. 2011; 257; 6451-6461.

Fen LB, Han TK, Nee NM, Ang BC, Johan MR. Physico-chemical properties of titania nanotubes synthesized via hydrothermal and annealing treatment. Applied Surface Science. 2011; 258;431- 435.

Fujishima A, Honda K. Electrochemical photolysis of water at a semiconductor electrode. Nature. 1972; 238: 37-38.

Gan Y, Gan B, Su L. Biophotofuel cell anode containing self-organized titanium dioxide nanotube array. Materials Science and Engineering B. 2011; 176; 1197– 1206.

Ghicov A, Aldabergerova S, Tsuchiya H, Schmuki P. TiO_2-Nb_2O_5 nanotubes with electrochemically tunable morphologies. Angewandte Chemie International Edition. 2006; 45; 6993-6996.

Ghicov A, Macak JM, Tsuchiya H, Kunze J, Haeublein V, Frey L, Schmuki P. Ion implantation and annealing for an efficient N-doping of TiO_2 nanotubes. Nano Letters. 2006;6;1080–1082.

Gratzel M. Photoelectrochemical cells. Nature. 2001; 414; 338-344.

Guo GM, Yu BB, Yu P, Chen X. Synthesis and photocatalytic applications of Ag/TiO_2-nanotubes. Talanta. 2009;79; 570–575.

He HC, Xiao P, Zhou M, Zhang YH, Jia YC, Yu SJ. Preparation of well-distributed Pt-Ni nanoparticles on/into TiO_2 NTs by pulse electrodeposition for methanol photoelectro-oxidation.Catalysis Communications. 2011; 16; 140–143.

Huang LH, Sun C, Liu YL. Pt/N-codoped TiO_2 nanotubes and its photocatalytic activity under visible light. Applied Surface Science. 2007; 253; 7029–7035.

Isimjan TT, Rohani S, Ray AK. Photoelectrochemical water splitting for hydrogen generation on highly ordered TiO_2 nanotubes fabricated by using Ti as cathode. International Journal of Hydrogen Energy. 2012;37;103-106.

Jia FZ, Yao ZP, Jiang ZH, Li CX. Preparation of carbon coated TiO_2 nanotubes film and its catalytic application for H_2 generation.Catalysis Communications. 2011; 12; 497–501.

Kaneko M, Nemoto J, Ueno H, Gokan N, Ohnuki K, Horikawa M, Saito R, Shibata T. Photoelectrochemical reaction of biomass and bio-related compounds with nanoporous

TiO$_2$ film photoanode and O$_2$ reducing cathode. Electrochemistry Communications. 2006; 8; 336-340.

Kasuga T, Hiramatsu M, Hoson A, Sekino T, Niihara K. Formation of titanium oxide nanotube, Langmuir. 1998; 14; 3160–3163.

Lan Y, Gao XP, Zhu HY, Zheng ZF, Yan TF, Wu F, Ringer SP, Song DY. Titanate nanotubes and nanorods prepared from rutile powder. Advanced Functional Materials. 2005;15; 1310-1318.

Li XQ, Cheng Y, Liu LF, Mu J. Enhanced photoelectrochemical properties of TiO$_2$ nanotubes co-sensitized with US nanoparticles and tetrasulfonated copper phthalocyanine. Colloids and Surfaces A: Physicochemical and Engineering Aspects. 2011;353; 226–231.

Lianos P. Production of electricity and hydrogen by photocatalytic degradation of organic wastes in a photoelectrochemical cell. The concept of the Photofuelcell: A review of a re-emerging research field. Journal of Hazardous Materials. 2010; 185; 575-590.

Lin JY, Chou YT, Shen JL, Yang MD, Wu CH, Chi GC, Chou WC, Ko CH. Effects of rapid thermal annealing on the structural properties of TiO$_2$ nanotubes. Applied Surface Science. 2011; 258; 530- 534.

Liu H, Tao L, Shen WZ. Optimal self-organized growth of small anodic TiO$_2$ nanotubes with "micro-annealing" effect under complex conditions via reaction-diffusion approach.Electrochimica Acta. 2011; 56 ; 3905-3913.

Liu JM, Zhao X, Duan L, Cao M, Sun H, Shao J, Chen S, Xie H, Chang X, Chen C. Influence of annealing process on conductive properties of Nb-doped TiO$_2$ polycrystalline films prepared by sol–gel method. Applied Surface Science. 2011; 257;10156– 10160.

Liu S, Yang L, Xu S, Luo S, Cai Q. Photocatalytic activities of C–N-doped TiO$_2$ nanotube array/carbon nanorod composite. Electrochemistry Communications. 2009; 11;1748–1751.

Liu SH, Yang LX, Xu SH , Luo SL, Cai QY. Photocatalytic activities of C–N-doped TiO$_2$ nanotube array/carbon nanorod composite. Electrochemistry Communications. 2009;11;1748-1751.

Liu Y, Li J, Zhou B, Li X, Chen H, Chen Q, Wang Z, Li L, Wang J, Cai W. Efficient electricity production and simultaneously wastewater treatment via a high-performance photocatalytic fuel cell. Water research. 2011; 45; 3991-3998.

Liu ZY, Pesic B, Raja KS, Rangaraju RR, Misra M. Hydrogen generation under sunlight by self ordered TiO$_2$ nanotube arrays. International Journal of Hydrogen Energy. 2009; 34; 3250 -3257.

Macak J M, Tsuchiya H, Ghicov A, Yasuda K, Hahn R, Bauer S, Schmuki P. TiO$_2$ nanotubes: Self-organized electrochemical formation, properties and applications. Current Opinion in Solid State and Materials Science. 2007; 1; 3–18.

Macak J M, Tsuchiya H., P Schmuki. High-aspect-ratio TiO$_2$ nanotubes by anodization of titanium.Angewandte Chemie International Edition. 2005; 44 ;2100 -2102.

Macak J M, Tsuchiya H., Taveira L, Aldabergerova S, Schmuki P. Smooth anodic TiO$_2$ nanotubes. Angewandte Chemie International Edition. 2005; 44 : 7463-7465.

Macak JM, Gong BG, Hueppe M, Schumk P, Filling of TiO$_2$ Nanotubes by self-Doping and Electrodeposition. Advanced Materials. 2007;19;3027-3031.

Mahajan V, Mohapatra S, Misra M. Stability of TiO$_2$ nanotube arrays in photoelectrochemical studies. International Journal of Hydrogen Energy. 2008; 33; 5369-5374.

Malwadkar SS, Gholap RS, Awate SV, Korake PV, Chaskar MG, Gupta NM. Physico-chemical, photo-catalytic and O$_2$-adsorption properties of TiO$_2$ nanotubes coated with gold nanoparticles. Journal of Photochemistry and Photobiology A: Chemistry.2009; 203; 24-31.

Miller EL, Rocheleau RE, Deng XM. Design considerations for a hybrid amorphous silicon/photoelectrochemical multijunction cell for hydrogen production. International Journal of Hydrogen Energy. 2003; 28; 615-623.

Minggu LJ, Daud WRW, Kassim MB, Cronin SB. An overview of photocells and photoreactors for photoelectrochemical water splitting. International Journal of Hydrogen Energy. 2010;35;5233-5244.

Morisaki H, Watanabe T, Iwase M, Yazawa K. Photoelectrolysis of water with TiO$_2$ covered solar-cell electrodes. Appl Phys Lett. 1976; 29; 338-340.

Ollis DF. Photocatalytic purification and remediation of contaminated air and water. Competes Rendus Del Academie Des Sciences Serie II Fascicule C-Chimie. 2000; 3; 405-411.

Park KW, Han SB, Lee JM. Photo(UV)-enhanced performance of Pt-TiO$_2$ nanostructure electrode for methanol oxidation. Electrochemistry Communications. 2007; 9; 1578–1581.

Patsoura A., Kondarides DI, Verykios XE. Enhancement of photoinduced hydrogen production from irradiated Pt/TiO$_2$ suspensions with simultaneous degradation of azo-dyes. Applied Catalysis B: Environmental. 2006; 64 ;171-179.

Pihosh Y, Turkevych I, Ye J, Goto M, Kasahara A, Kondo M, Tosa M. Photocatalytic Properties of TiO$_2$ Nanostructures Fabricated by Means of Glancing Angle Deposition and Anodization. Journal of the Electrochemical Society. 2009;156;160-165.

Prakasam HE, Shankar K, Paulose M, Varghese OK, Grimes, CA. A new benchmark for TiO$_2$ nanotube array growth by anodization. Journal of Physical Chemistry C. 2007; 111; 7235-7241.

Reber JF, Meier K. Photochemical production of hydrogen with zinc-sulfide suspensions. Journal of Physical Chemistry. 1984; 88; 5903-5913.

Ruan CM, Paulose M, Varghese OK, Mor GK,Grimes CA. Fabrication of highly ordered TiO$_2$ nanotube arrays using an organic electrolyte. Journal of Physical Chemistry B. 2005; 109; 15754-15759.

Sakthivel S, Shankar MV, Palanichamy M, Arabindoo B, Bahnemann DW, Murugesan V. Enhancement of photocatalytic activity by metal deposition: characterisation and photonic efficiency of Pt, Au and Pd deposited on TiO$_2$ catalyst. Water Research. 2004; 38; 3001-3008.

Sang LX, Zhang ZY, Ma CF. Photoelectrical and charge transfer properties of hydrogen-evolving TiO$_2$ nanotube arrays electrodes annealed in different gases. International Journal of Hydrogen Energy. 2011; 36; 4732-4738.

Seo HK, Kim GS, Ansari SG, Kim YS, Shin HS, Shim KH, Suh EK. A study on the structure/phase transformation of titanate nanotubes synthesized at various hydrothermal temperatures. Solar Energy Materials and Solar Cells. 2008; 92; 1533–1539.

Shaban YA, Khan SUM. Surface grooved visible light active carbon modified (CM)-n-TiO_2 thin films for efficient photoelectrochemical splitting of water. Chemical Physics. 2007; 339; 73-85.

Shankar K, Mor GK, Prakasam HE, Yoriya S, Paulose M, Varghese OK, Grimes CA. Highly-ordered TiO_2 nanotube arrays up to 220 μm in length: use in water photoelectrolysis and dye-sensitized solarcells. Nanotechnology. 2007; 18; 065707.

Shen YF, Xiong TY, Li TF, Yang K. Tungsten and nitrogen co-doped TiO_2 nano-powders with strong visible light response. Applied Catalysis B: Environmental. 2008; 83; 177-185.

Sreekantan S, Wei LC. Study on the formation and photocatalytic activity of titanate nanotubes synthesized via hydrothermal method.Journal of Alloys and Compounds. 2010; 490 ; 436–442.

Stem N, Chinaglia EF, dos Santos Filho SG. Microscale meshes of Ti_3O_5 nano- and microfibers prepared via annealing of C-doped TiO_2 thin films. Materials Science and Engineering B. 2011; 176; 1190-1196.

Su Y, Chen SO, Quan X, Zhao HM, Zhang YB. A silicon-doped TiO_2 nanotube arrays electrode with enhanced photoelectrocatalytic activity. Applied Surface Science. 2008; 255;2167-2172.

Su YL, Han S, Zhang XW, Chen XQ, Lei LC. Preparation and visible-light-driven photoelectrocatalytic properties of boron-doped TiO_2 nanotubes.Materials Chemistry and Physics. 2008; 110; 239–246.

Sun L, Li J, Wang CL, Li F, Chen HB, LinCJ. An electrochemical strategy of doping Fe^{3+} into TiO_2 nanotube array films for enhancement in photocatalytic activity. Solar Energy Materials and Solar Cells.2009; 93; 1875-1880.

Tang Y, Tao J, Zhang Y, Wu T, Tao H, Bao Z, Preparetion and Characterization of TiO_2 Nanotube Arrays via Anodization of Titanium Films Deposited on FTO Conducting Glass at Room Temperature. ACTA Physico-Chimica Sinica. 2008;24;2191-2197.

Tu YF, Huang SY, Sang JP, Zou XW. Preparation of Fe-doped TiO_2 nanotube arrays and their photocatalytic activities under visible light. Materials Research Bulletin. 2010; 45; 224-229.

Varghese OK, Gong DW, Paulose M, Grimes CA, Dickey EC. Crystallization and high-temperature structural stability of titanium oxide nanotube arrays. Journal of Materials Research .2003; 18; 156 -165.

Vitiello RP, Macak JM, Ghicov A, Tsuchiya H, Dick LFP, Schmuki P. N-Doping of anodic TiO_2 nanotubes using heat treatment in ammonia Electrochemistry Communications. 2006; 8; 544–548.

Wang J, Zhang W. Modification of TiO_2 nanorod arrays by graphite-like C_3N_4 with high visible light photoelectrochemical activity. Electrochimica Acta. 2012;71; 10-16.

Wang MC, Lin HJ, Wang CH, Wu HC. Effects of annealing temperature on the photocatalytic activity of N-doped TiO_2 thin films. Ceramics International. 2012;38; 195-200.

Wen H, Liu Z, Yang Q, Li Y, Jerry Yu J. Synthesis and electrochemical properties of CeO_2 nanoparticle modified TiO_2 nanotube arrays. Electrochimica Acta. 2011; 56; 2914–2918.

Wu Q, Ouyang JJ, Xie KP, Sun L, Wang MY, Lin CJ. Ultrasound-assisted synthesis and visible-light-driven photocatalytic activity of Fe-incorporated TiO_2 nanotube array photocatalysts.Journal of Hazardous Materials. 2012; 199; 410-417.

Xu H, Zhang Q, Zheng CL, Yan W, Chu W. Application of ultrasonic wave to clean the surface of the TiO_2 nanotubes prepared by the electrochemical anodization. Applied Surface Science. 2011; 257; 8478– 8480.

Xu JJ, Ao YH, Chen MD, Fu DG. Photoelectrochemical property and photocatalytic activity of N-doped TiO_2 nanotube arrays. Applied Surface Science. 2010; 256; 4397–4401.

Xu YM, Fang XM, Xiong JA, Zhang ZG. Hydrothermal transformation of titanate nanotubes into single-crystalline TiO_2 nanomaterials with controlled phase composition and morphology. Materials Research Bulletin. 2010; 45; 799-804.

Yan GT, Zhang M, Hou J, Yang JJ. Photoelectrochemical and photocatalytic properties of N plus S co-doped TiO_2 nanotube array films under visible light irradiation. Materials Chemistry and Physics. 2011; 129; 553-557.

Yuan J, Chen MX, Shi JW , Shangguan WF. Preparations and photocatalytic hydrogen evolution of N-doped TiO_2 from urea and titanium tetrachloride. International Journal of Hydrogen Energy. 2006; 31; 1326-1331.

Yuan ZY, Su BL. Titanium oxide nanotubes, nanofibers and nanowires.Colloids and Surfaces A: Physicochemical and Engineering Aspects. 2004; 241; 173-183.

Zeng X, Gan Y, Clark E, Su L. Amphiphilic and photocatalytic behaviors of TiO_2 nanotube arrays on Ti prepared via electrochemical oxidation. Electrochemistry Communications 2009;11; 1748–1751.

Zhang SS, Peng F, Wang HJ, Yu H, Zhang SQ, Yang J, Zhao HJ. Electrodeposition preparation of Ag loaded N-doped TiO_2 nanotube arrays with enhanced visible light photocatalytic performance. Catalysis Communications. 2011; 12; 689–693.

Zhang X, Lei L, Zhang J, Chen Q, Bao J, Fang B. Preparation of $PW_{12}O_{40}{}^{3-}$/Cr–TiO_2 nanotubes photocatalysts with the high visible light activity. Separation and Purification Technology. 2009; 67; 50–57.

Zhang XW, Lei LC, Zhang JL, Chen QX, Bao JG, Fang B. A novel CdS/S-TiO_2nanotubes photocatalyst with high visible light activity. Separation and Purification Technology. 2009; 68; 433-433.

Zhao Q, Li M, Chu JY, Jiang TS, Yin HB. Preparation, characterization of Au (or Pt)-loaded titania nanotubes and their photocatalytic activities for degradation of methyl orange.Applied Surface Science. 2009; 255; 3773-3778.

Zwilling V, Darque-Ceretti E, Boutry-Forveille A , David D, Perrin MY, Aucouturier M. Structure and physicochemistry of anodic oxide films on titanium and TA6V alloy. Surf Interface Anal. 1999 ; 27; 629-637.

Thermal

Chapter 2

Thermal Energy Harvesting Using Fluorinated Terpolymers

Hongying Zhu, Sébastien Pruvost, Pierre-Jean Cottinet and Daniel Guyomar

Additional information is available at the end of the chapter

1. Introduction

Energy supply has always been a crucial issue in designing battery-powered wireless sensor networks because the lifetime and utility of the systems are limited by how long the batteries are able to sustain the operation. Therefore, harvesting energy from the environment has been proposed to supplement or completely replace battery supplies to enhance system lifetime and reduce the maintenance cost of replacing batteries periodically. Of the available ambient energy sources, which include, for instance, light energy, mechanical energy, and thermal energy, most innovative solutions deal with energy harvesting from vibration, but very few concern harvesting with temperature variation.

In the last several decades, thermoelectric devices have received significant attention[1]. They make use of the Seebeck effect to directly convert a steady-state temperature difference to electricity at the junction of two dissimilar metals or semiconductors. However, the precondition of steady-state heat flow greatly restricts the practical applications. On the contrary, pyroelectric energy conversion directly converting time-dependent temperature variations to electricity could be promising and by applying a proper thermodynamic cycle, a high conversion efficiency would be obtained. Presently, it is found that fluorinated polymers and especially terpolymers have enormous potential for energy harvesting from heat due to its large pyroelectric activity. In 1985, Olsen et al. reported the first pyroelectric conversion cycle for the copolymer P(VDF-TrFE), the output electrical energy density was 30 mJ/cm^3, which is 15 times larger than any other polymer previously measured[2]. In 2010, Nguyen et al. developed the pyroelectric energy converter using copolymer 60/40 P(VDF-TrFE) and based on Ericsson cycle[3]. A maximum energy density of 130 mJ/cm^3 was achieved at 0.061 Hz frequency with temperature oscillating between 69.3 and 87.6°C ($\Delta T=18.3°$). In 2011, Navid et al. improved this energy converter by changing the mode of temperature change[4]. They implemented the Ericsson cycle by successively dipping the films in cold and hot silicone oil baths at 25 and 110°C, and compared three different copolymer samples: commercial purified, and porous films. A maximum energy density of 521 mJ/cm^3 and 426 mJ/cm^3 per cycle were produced by commercial and purified films, respectively, under applied electric fields ranging between 20 and 50 kV/mm, and a maximum energy density of 188 mJ/cm^3 per cycle under electric fields between 20 and 40 kV/mm.

With the development of miniaturization of integration, the aim of energy harvesting is to power smaller and smaller devices, and electrostatic energy conversion is of significant interest and so is suitable for micro-scale electrical power generation. Meninger et al. presented a MEMS (Micro-Electro-Mechanical Systems) vibration to an electricity converter based on electrostatic energy harvesting by a variable capacitor. Mechanical vibrations exert an electrostatic force changing the distance between two electrodes (leading to a variable capacitance)[5]. With an appropriate thermodynamic cycle, Stirling cycle or Ericsson cycle, electrical energy can thus be harvested. In order to avoid the external mechanical force, electrostatic energy harvesting can be performed by nonlinear capacitance variation under temperature variation.

Having taken into account the statement above, a variable capacitance is the critical element for electrostatic energy harvesting. To our best knowledge, the dielectric material can be considered as a capacitor by sandwiched between metallic electrodes, and its dielectric constant is the function of temperature, based on above, we proposed that electrostatic energy harvesting on a relaxor ferroelectric P(VDF-TrFE-CFE) terpolymer, the variable capacitance could be achieved by the dielectric permittivity variation with temperature just below the dipolar transition.

2. Material consideration

2.1. PVDF-based terpolymer

From the basic ferroelectric response consideration, the defect structure modification of the ferroelectric properties can be realized by introducing randomly in the polymer chain a third monomer, which is bulkier than VDF and TrFE monomers. Compared with the high-energy electron irradiation, the terpolymer approach to modify the copolymer from ferroelectric to relaxor is more attractive since it reduces the manufacture cost and significantly simplifies the processing steps. In addition, it greatly reduces the undesirable side effects introduced by the irradiation to the polymers.

Zhang et al. developed a method of converting the polymer to a relaxor ferroelectric by introducing defects into the P(VDF-TrFE) copolymers, i.e., terpolymer containing the chlorofluoroethylene (CFE,-CH$_2$-CFCl-) or chlorotrifluoroethylene) (CTFE) as the ter-monomer[6,7]. The introduction of the third monomer into the polymer chain serves to break up the ferroelectric domains into local nano-polar regions surrounded by an amorphous matrix, thereby reducing their size. The resulting nano-polar regions are more mobile and increase the polarization response and overall permittivity. By co-polymerization of CFE or CTFE with the P(VDF-TrFE), the polymer could be converted to a relaxor ferroelectric which eliminated the polarization hysteresis (dielectric heating) at room temperature associated with the change of polarization [8]. These relaxor-ferroelectric terpolymers P(VDF-TrFE-CFE) and P(VDF-TrFE-CTFE) exhibit a room temperature dielectric constant greater than 50. These values are among the highest ones reported in the literature, and make the terpolymers suitable in our study for energy harvesting by nonlinear capacitance variation.

2.1.1. Terpolymer P(VDF-TrFE-CTFE)

Xu et al. synthesized and evaluated terpolymer P(VDF-TrFE-CTFE) using the bulk polymerization method[9]. As the bulky and less polar termonomer CTFE is randomly introduced into P(VDF-TrFE) normal ferroelectric crystals, there are three main features in the dielectric data of the polymer due to the addition of CTFE observed: (1) the original FP transition peak of the copolymer is moved to room temperature; (2) the peak becomes much broader and its position shifts progressively with frequency towards higher temperature; (3) there is no thermal hysteresis in the dielectric data, i.e., the broad dielectric peak stays at the same temperature when measured in the heating and cooling cycles. They also measured the polarization hysteresis loops at room temperature and -40°C [9]. The terpolymer exhibits a slim polarization loop at room temperature, and as the temperature is lowered, the coercive field increases and the polarization at maximum electric field decrease. All these features are remarkably reminiscent of ferroelectric relaxor behaviour. In addition, the polarization reduction when the temperature decreased proved that, at lower temperature range (temperature lower than transition), the dielectric permittivity variation shows overwhelming advantage, while the pyroelectric effect becomes negligible. It gives a strong proof for our study in energy harvesting from capacitance variations by lowering the temperature.

2.1.2. Terpolymer P(VDF-TrFE-CFE)

There were also experimental results on terpolymer P(VDF-TrFE-CFE). Zhang et. al compared the terpolymer P(VDF-TrFE-CFE) with P(VDF-TrFE-CTFE), and concluded that the CFE is more effective on improving their electromechanical response compared with CTFE, shown in the following aspects: (1) In the containing CFE, 4-5 mol% of CFE seems to be adequate to nearly eliminate the polarization hysteresis, while in the terpolymers containing CTFE, nearly 10 mol% is required. (2) P(VDF-TrFE-CFE) 62/38/4 mol% terpolymer exhibits a much higher elastic modulus in comparison with P(VDF-TrFE-CTFE) 65/35/10 mol% terpolymer, resulting in much higher elastic energy density and electromechanical coupling factor in the P(VDF-TrFE-CFE) terpolymer[6].

Klein et al. revealed that the addition of CFE leads to two types of crystalline regions within the polymer, where a non-polar phase coexists with a polar phase. At high CFE mol% (>8.5 mol%) in the compositions range of VDF/TrFE mole ratio between 64/36 to 75/25, the polar phase region is no longer detectable, indicating a complete conversion to the relaxor ferroelectric phase and the terpolymer exhibits the highest strain level with very little hysteresis. They compared the measured strain as a function of the applied field for two compositions of 65/35/8.6 mol% and 75/25/5.3 mol%[10]. The presence of the polar-phase in the terpolymer of 75/25/5.3 mol% greatly reduces the electrostrictive strain in the polymer. On the other hand, increasing CFE content causes reduction in crystallinity, which will affect the elastic modulus and the induced polarization level of the polymer [10]. There competing effects determine the desired terpolymer compositions for given application.

In 2008, Neese et al. studied the other important property on terpolymer P(VDF-TrFE-CFE): electrocaloric effect (ECE)[11]. Electrocaloric effect corresponds to a reversible change of temperature induced by an electric field under adiabatic conditions. A large ECE requires a large entropy change associated with polarization change, and the dielectric material must be capable of generating large polarization changes. In regard to it, ferroelectric polymers are better than ceramics due to their high dielectric strength. By this way, a large entropy change can be achieved. The relaxor ferroelectric polymers have the potential for processing attractive ECE properties due to their large polarizability, small polarization hysteresis, and dielectric peak near room temperature. Neese et al. concluded that with an electric field of 307 MV/m at 55°C applied on terpolymer P(VDF-TrFE-CFE) 59.2/33.6/7.2 mol%, temperature variation $\Delta T = 12°$ and entropy change $\Delta S=55$ J/(kgK) were obtained. Especially, the entropy change is about 7 times greater than that in the ferroelectric ceramics[11]. Their study indicates that the large entropy change can be achieved associated with the electric field-induced dipole ordering – disordering (O-D) processes at temperature near O-D transitions. This transition becomes the critical technique during the process of energy harvesting on terpolymer that we will discuss it in the following part.

In view of the favourable properties above, low ferroelectric hysteresis, the temperature of the dielectric constant peak shifts to room temperature, large polarization variation induced by the O-D transition, high elastic energy density of PVDF-based terpolymer make it attractive for energy harvesting by temperature variation. Moreover, Ren et al.[12] also reported energy harvesting using a ferroelectric polymer, possessing high electromechanical response and elastic energy density, which make it possible to generate high electrical energy density and attractive for the active energy harvesting scheme. This study shows that combining the active energy harvesting scheme and high electromechanical response of the polymer yields a harvested electric energy density of ~40 mJ/cm³ with a 10% efficiency at very low frequency.

The outstanding properties of electroactive terpolymer, such as high room temperature dielectric constant, high polarization variation induced by temperature, and high electrostriction etc., make it seem to be the promising candidate for active harvesting of energy, no matter from thermal or mechanical energy sources. The investigation of polymers opens up novel possibilities for multi-source energy harvesting technique. In view of thermal energy harvesting, the use of copolymer P(VDF-TrFE) to harvest waste heat by pyroelectricity has been explored both experimentally and theoretically[2-4]. In this part, we are interested in energy harvesting from nonlinear capacitance variation by temperature on terpolymer P(VDF-TrFE-CFE)[13].

3. Energy harvesting by nonlinear capacitance variation for a relaxor ferroelectric P(VDF-TrFE-CFE) terpolymer

3.1. Introduction

In this work, the variable capacitance is realized by the relaxor poly(vinylidene fluoride-trifluoroethylene-chlorofluoroethylene) P(VDF-TrFE-CFE) terpolymer as it exhibits a large nonlinear dielectric constant variation around the dipolar ordering-disordering transition. Benefiting from the application of the Ericsson cycle and the transition at a constant electric field, the harvested energy can be greatly improved. Accordingly, the terpolymer P(VDF-TrFE-CFE) 61.3/29.7/9 mol% is chosen to be the active material of this work.

3.2. Preparation of terpolymer P(VDF-TrFE-CFE)

The terpolymer films were prepared through a solution casting method. The terpolymer P(VDF-TrFE-CFE) 61.3/29.7/9 mol%, supplied by PIEZOTECH, France, was firstly dissolved in N-dimethylformamide by stirring at room temperature for one day. The uniform solution was put in the vacuum for removing the bubbles and then deposited onto a plat of glass by the solution casting method using a doctor blade applicator (Elcometer 3700). Subsequently, the film was placed into the oven at 60°C for one day and was annealed at 85°C for 6 h in a vacuum oven to remove residual solvent. Gold electrodes were sputtered on the two surfaces of the 20-μm films which were stuck to a steel substrate as shown in Fig. 1, for accelerating heat exchange during the temperature variation step of energy harvesting cycle. The results presented below were obtained at least on five samples.

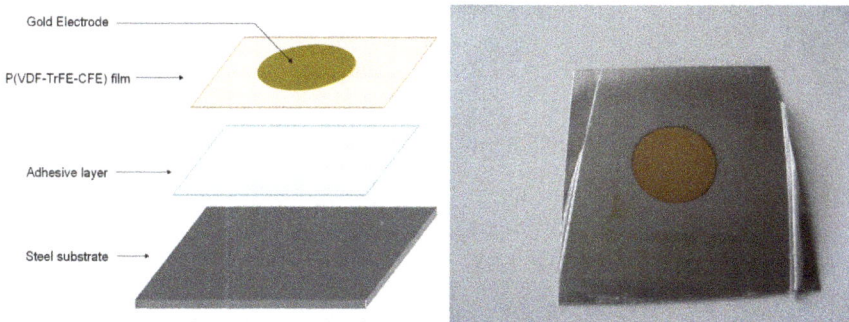

Figure 1. Sample used in this study: P(VDF-TrFE-CFE) terpolymer film stuck to steel substrate.

3.3. Principle of electrostatic energy conversion

Mitcheson et al. has presented generally the electrostatic energy conversion and its two possible conversion cycles: charge constrained and voltage constrained cycles[14]. In this part, we focus on the voltage constrained cycle which was used in our work. We put the scheme of cycles below in order to quantitatively discuss the capability of energy harvesting on P(VDF-TrFE-CFE) 61.3/29.7/9 mol% terpolymers.

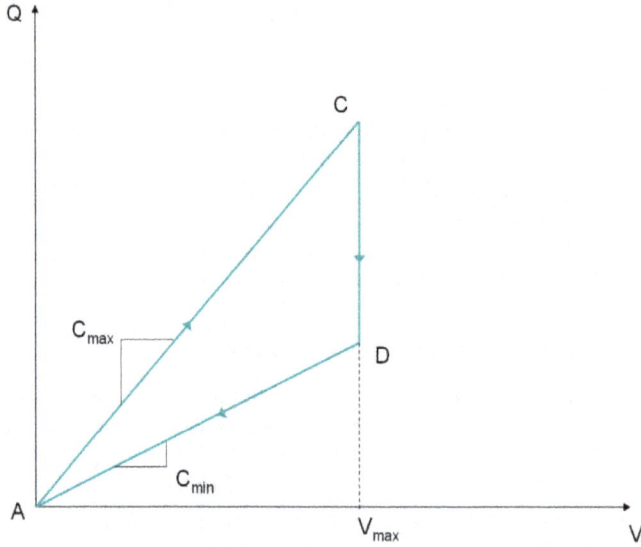

Figure 2. Scheme of electrostatic energy conversion cycles.

In this work, the voltage constrained case was chosen , cycle analogous to the other thermodynamic cycle-Ericsson cycle, which consists of two isothermal and two constant electric field processes[15]. The cycle starts when the capacitor is charged up to V_{max} from reservoir. This is done when the capacitance is at maximum. The injected energy corresponding to the AC segment of the cycle is calculated as

$$W_{inject} = \frac{1}{2}C_{max}V_{max}^2 \tag{1}$$

During this time, the value of capacitor is taken to be constant, and so the segment of A-C is a straight line. This is a valid assumption since the charge-up time to traverse path A-C (and discharge path D-A) is transient, while the segment C-D, which corresponds to the capacitance variation, is relative longer (in our case, the time is equal to that of temperature variation around 20 seconds). It is evident from the Fig. 2.Error! **Reference source not found.** that during this step in the conversion process, the voltage is held constant (hence the name voltage contained conversion.). As the capacitance decreases, path segment C-D is traversed, where the capacitance is at a minimum. In our case, the capacitance is changed by decreasing the temperature of terpolymer, the electrostatic force does work by causing charge to move from the capacitor back into the reservoir. The charge remaining on the plates is then recovered while capacitance of terpolymer reducing to minimum C_{min} following path D-A, the terpolymer is discharged (D-A). Since the segments C-D and D-A, the energy is harvested from heat to electric power, $W_{harvest}$, is the area ACD in Fig. 2.

$$W_{harvest} = -\frac{1}{2}(C_{max} - C_{min})V_{max}^2 \qquad (2)$$

This method sets a maximum limit on the conversion process. The major obstacle for this approach is that some method must be employed to hold the voltage across the capacitor of the device during the conversion process, which would require another source of value V_{max}. This is an additional source to that of the conversion charge reservoir, which is of a lower voltage and is also used to power the control electronics.

3.4. Characterization of terpolymer P(VDF-TrFE-CFE)

As mentioned earlier, the discovery of high electromechanical performance in P(VDF-TrFE-CFE) based terpolymer opens a new avenue for developing high performance electroactive polymers. Also, by introducing the termonomer, for example CFE, to form P(VDF-TrFE) based terpolymer, the normal ferroelectric P(VDF-TrFE) could be converted into a ferroelectric relaxor with high room-temperature dielectric constant peak and a very slightly polarization hysteresis. In a consequence, the terpolymer P(VDF-TrFE-CFE) 61.3/29.7/9 mol% was processed, and the characterization would be investigated in detail in this part.

3.4.1. Dielectric and ferroelectric characterization

The weak field dielectric properties were measured as a function of temperature by an impedance/GAIN-phase Analyzer (HP4194A). The terpolymer films (20μm-thickness) stuck onto the steel substrate was put into a controlled temperature chamber and connected with an impedance/GAIN-phase Analyzer. Fig. 3 shows the temperature- and frequency-dependent dielectric constant and loss tangent of terpolymer. As expected, due to the introduction of defect structure, the sharp dielectric constant peak in copolymer turns into a broad peak in terpolymer, and the O-D transition peak is moved to around room temperature. Especially, the broad dielectric peak position of dielectric constant shifts slightly with the frequency towards higher temperature[6], such a behaviour is a typical feature to all the known relaxor ferroelectric materials.

As seen in Fig. 3, it indicates that the dielectric constant shows a maximum at nearly room temperature, and decreases more rapidly at the lower temperature range than that at the higher temperature range, the large dielectric constant variation can result in the large electrical displacement variation. On the other hand, Xu et al. have confirmed that the pyroelectric effect could be ignored at this temperature range[9]. Consequently, the high nonlinearity of dielectric permittivity (capacitance) together with the negligible pyroelectric effect could be an important favourable factor for harvesting energy in our study.

In order to investigate the temperature effect of basic ferroelectric characterization on terpolymer, we measured the ferroelectric hysteresis loops at different temperature as shown in Fig. 4. D-E loops were carried out every 5°C between 20°C and -20°C in descending order, and the unipolar sine electric field was applied with the average slope of $dE/dT=40000$ kV mm^{-1}s^{-1} which is the same slope with the bipolar cycle at 100 Hz. This frequency was chosen due to its lower hysteresis loss than other frequency.

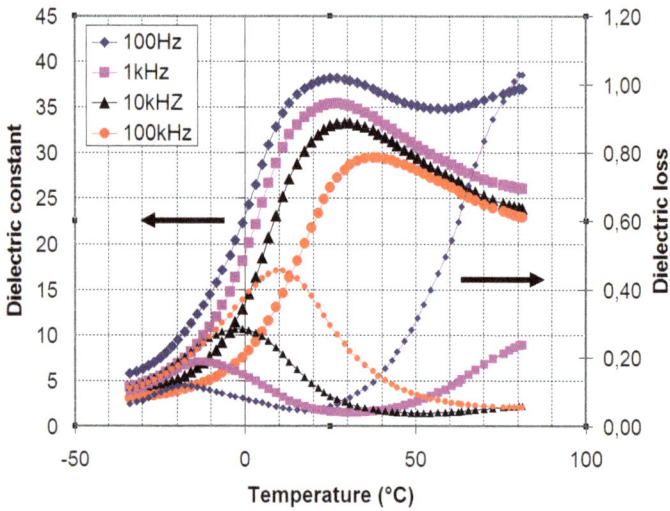

Figure 3. Temperature dependence of weak field dielectric properties of terpolymer P(VDF-TrFE-CFE) measured at different frequencies.

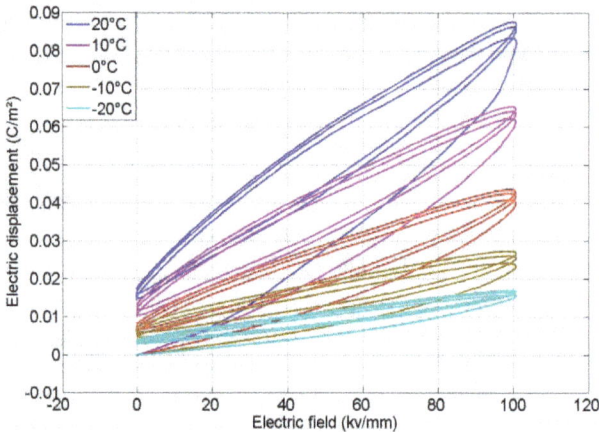

Figure 4. Unipolar D-E loops at different temperature under 100 kV mm[-1].

Fig. 4 presents that the electric displacement displays an obvious reduction with the decrease of temperature under the maximum electric field 100 kV mm[-1], it is consistent with reduction of the dielectric constant in Fig. 3. As we know, there exists two kinds of polarization mechanism in the relaxor ferroelectrics: thermally activated flips of the nanopolar regions at high temperature and the breathing of frozen nanopolar regions at lower temperature[16]. In the relaxor ferroelectric terpolymers, since the temperature lowered from high temperature to the temperature of the dielectric constant peak, the population of the nanopolar regions increases. The polarization response under high electric field is mainly obtained from the relaxation polarization which is associated with the thermally activated flips of the nanopolar regions. Around the temperature of the dielectric constant maximum, the polarization behavior reaches the maximum due to the largest quantity of the thermally active nanopolar regions. Further lowering the temperature, partial nanopolar regions become frozen, polarization behaviour is mainly determined by the breathing of the frozen nanopolar regions. The large electric displacement variation is achieved due to the nanopolar regions transition from thermally activated flip behaviour to breathing behaviour. Take the two extreme cases for instance in our measurement, at 20°C, the thermally activated flips of the nanopolar regions are predominant, so the large polarization response could be observed, as shown in Fig. 3, the electric displacement comes up to 0.086 C/m^2. Lowering the temperature down to -20°C, the breathing of frozen nanopolar regions is predominant, although still the same electric field, the electric displacement is just 0.017 C/m^2. The variation of electric displacement is satisfactory for harvesting energy.

3.5. Electrostatic energy harvesting by nonlinear capacitance variation for a relaxor ferroelectric P(VDF-TrFE-CFE)

Based on the characterization mentioned above, we indicate that, as relaxor ferroelectric terpolymer P(VDF-TrFE-CFE) 61.3/29.7/9 mol% is a promising candidate for electrostatic energy harvesting on Ericsson cycle by nonlinear capacitance variation, due to the high non-linear property of the dielectric constant in the vicinity of the polarization mechanism transformation.

3.5.1. Experiment setup

A waveform generator (Agilent 33220A), a high voltage amplifier (TREK Model 10/10B) and a current preamplifier (STANDFORD Model 570) were used to determine the dielectric constant under a DC electric field and to carry out the Ericsson cycle. An elaborate sample holder with closed protecting-cell is used to fix the sample, protect the polymer during the temperature change and applying the external electric field.

3.5.2. Theoretical model

An Ericsson cycle can be used for harvesting energy[3,15,17]. It consists of two isothermal processes (charge at θ_1 and discharge at θ_2) and a process for cooling the sample from θ_1 to θ_2 under a constant electric field. Assuming that there is a linear relationship between the electric field and the dielectric displacement:

$$D = \varepsilon_{33}(\theta)E \qquad (3)$$

Where $\varepsilon_{33}(\theta)$ and θ, denote respectively the permittivity and temperature. The injected energy during the charging process (applied electric field from 0 to E_1) is given by

$$W_{inject} = \int_0^{E_1} EdD \tag{1}$$

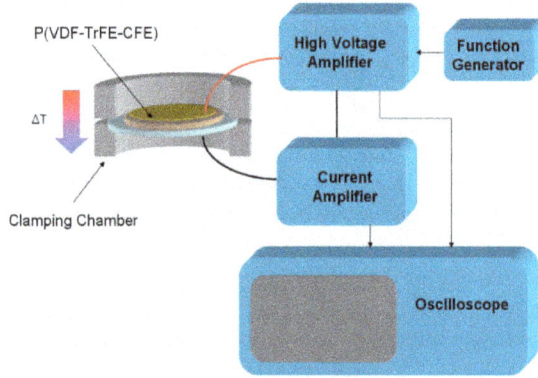

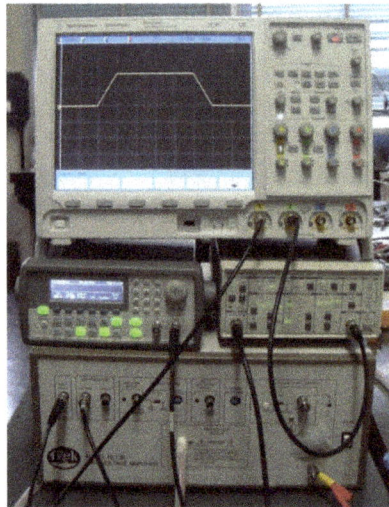

Figure 5. A schematic (upper) and a picture (lower) of the experimental setup for measuring the energy harvesting system in the terpolymers.

After the discharging process, the electric displacement returns to zero. Here, the other constant electric field process of the Ericsson cycle no longer exists since there is no remnant polarization for relaxor ferroelectrics. The whole cycle is described in a clockwise path leading to harvested energy which is equal to the area of the cycle:

$$W_{harvest} = \oint EdD \tag{5}$$

When changing the electric field and the temperature of a relaxor material, electrocaloric[18] (change in temperature induced by the application of an electric field) and pyroelectric (change in electric polarization induced by a temperature variation) effects[19] inevitably occur, but, in the present study, both were negligible. The electrocaloric effect was too small to change the temperature of the terpolymer in a significant manner (maximum 1K for the electric field applied in this study)[11]. Similarly, the pyroelectric effect, during cooling from 25°C to 0°C, was too low to lead to a significant increase of the electric displacement as described in the work performed by Xu et al.[9] where the electric displacement was remarkably decreased for a relaxor terpolymer when going from room temperature to a lower temperature.

Assuming that, theoretically, the permittivity is constant during the charging and discharging processes, the injected and harvested energy can be expressed as follows:

$$W_{inject} = \frac{1}{2}\varepsilon_1 E_1^{\;2} \tag{6}$$

$$W_{harvest} = -\frac{1}{2}(\varepsilon_1 - \varepsilon_2)E_1^{\;2} \tag{7}$$

where ε_1 and ε_2 are the permittivity during the charging and discharging processes, respectively.

3.5.3. Results and discussion

The theoretical analysis indicates that the harvested energy mainly depends on the capacitance variation, which in turn depends on the temperature. Fig. 6(a) shows the weak field dielectric properties (measured at 100 Hz) as a function of temperature, which has been verified on five samples. The permittivity demonstrated a maximum at the transition temperature (25°C), and then decreased rapidly in the lower temperature range. Benefiting from this transition, a large capacitance variation was obtained. According to these results, the temperatures of 25°C and 0°C were chosen as the charging and discharging temperatures, respectively, for the Ericsson cycle. The time for charging and discharging the sample corresponds to a frequency of 100 Hz in order to compare simulation and experiment without frequency effect.

The terpolymer was under a DC electric field during the temperature change in the Ericsson cycle. Lu et al. showed that the permittivity of PVDF-based polymers (poly vinylidene-fluoride) exhibits a tunability under a DC electric field[20]. This tunability is a function of temperature, especially in the vicinity of the dielectric peak where the polymer has the largest tunability. Therefore, the dielectric constant (relative permittivity) was measured under different DC electric fields (E_{DC}) with a small AC field (E_{AC}) at 100 Hz and at 25°C and 0°C, as shown in Fig. 6(b). The experimental uncertainties correspond to a 95% confidence interval. Two main points are emphasized: (1) it existed a tunability of the permittivity under a DC electric field (the tunability was larger at 25°C than at 0°C) and (2) this tunability presented a non linear behavior with temperature.

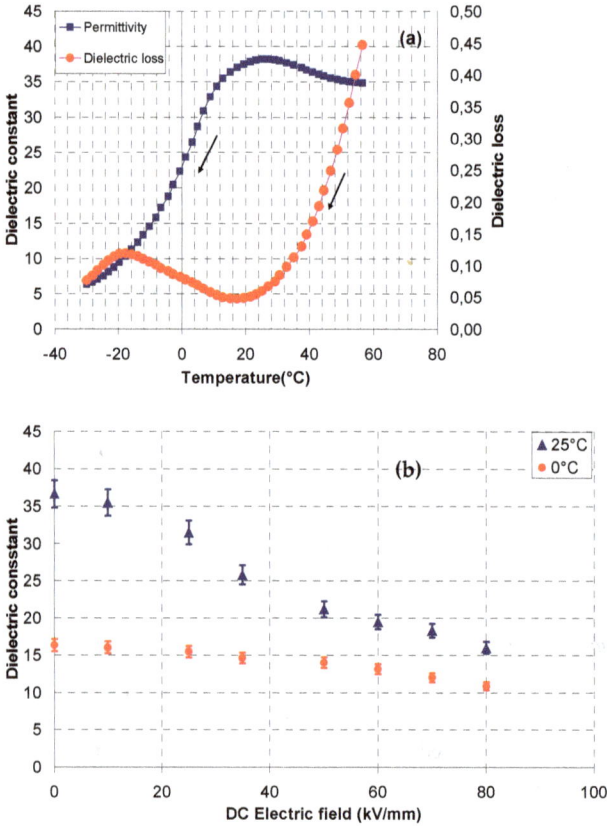

(a) dielectric constant (black squares)and dielectric loss (red circles) as a function of temperature under a weak field, (b) dielectric constant under a DC electric field at 25°C (blue triangles) and 0°C (red circles).

Figure 6. Dielectric property on P(VDF-TrFE-CFE) 61.3/29.7/9 mol%.

As the DC electric field increased from zero to 80 kV/mm, the capacitance variation between 25°C and 0°C decreased significantly (the dielectric constant variation was 21 at E_{DC}=0, and 5 at E_{DC}= 80 kV/mm, corresponding to a reduction by 75%). In physical terms, this can be understood as there existing no macroscopic domain for relaxor ferroelectrics and the polarization mainly depending on the thermally activated flips of the nanopolar regions[21]. An external electric field cannot produce a large domain reorientation, just an alignment of nanopolar regions along the field. At the dielectric peak temperature of 25°C, the quantity of the thermally active nanopolar regions was at a maximum. Nevertheless, at the lower temperature, i.e., 0°C, a majority of the nanopolar regions became frozen[16]; the dielectric behavior thus mainly received contributions from the shape change of the frozen polar regions due to the external DC electric field. Consequently, the tunability was larger at 25°C.

The Ericsson cycle was simulated from the measurements of permittivity presented in Fig. 6 and by using Eq. (7). Fig. 7 shows the harvested energy versus the DC electric field considering or not the tunability of the dielectric constant with the electric field. Measurements were undertaken on five samples and the discrepancy is around 10%. The two simulation results almost coincided in the lower electric field range, i.e., at 10 and 25 kV/mm, but at higher electric field, the difference became increasingly pronounced, reaching 46% at 80 kV/mm. These simulations underlined that the tunability needed to be taken into account in order to avoid an overestimation of the harvested energy, especially for high DC electric fields. The harvested energy increased with the maximum DC electric field, and from simulation, a maximum energy of 240 mJ/cm³ could be harvested at 80 kV/mm. However, electrical conduction and the electric field breakdown (lower with electrical conduction) are two inevitable factors, which restricted the experimental harvested energy, need to be resolved.

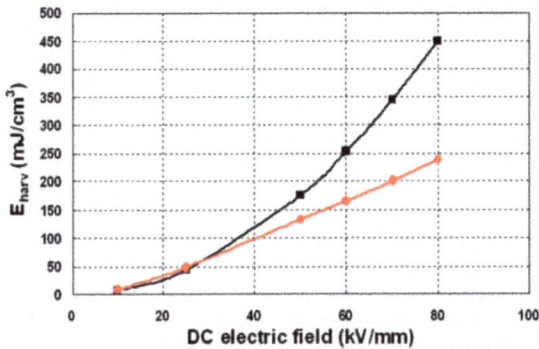

Figure 7. Harvested energy as a function of the DC electric field by a simulated Ericsson cycle for the terpolymer P(VDF-TrFE-CFE) 61.3/29.7/9. Harvested energy simulated by permittivity under a DC field (red circles) and permittivity without a DC field (black squares).

In order to confirm the simulation experimentally, an experimental Ericsson cycle was also undertaken. The energy in each segment could be obtained by the electric field integral with respect to current. Fig. 8 presents the simulated and experimental Ericsson cycles at 25 kV/mm. The direct measurement of the Ericsson cycle was carried out with a slope of dE/dt=25kV/mm.s for the charging and discharging processes. The experimental measurement showed the same type of closed Ericsson cycle as the simulation considering the tunability of the dielectric constant. For the experimental cycle, the conduction of the terpolymer is observed while the DC electric field was applied during the temperature change. An Arrhenius law was used to remove the conduction from the experimental data. Fig. 8 presents the experimental Ericsson cycle after eliminating the conduction. The harvested energy was computed by Eq. (12), and was found to be 50 mJ/cm³, which was consistent with the simulation (48 mJ/cm³). The experimental Ericsson cycle was performed over several cycles on five samples and gave nearly the same harvested energy with a deviation of 10%, as well as simulation, it is the same deviation for harvested energy due to that on dielectric constant, the most representative cycle was chosen corresponding to the average of measured cycles.

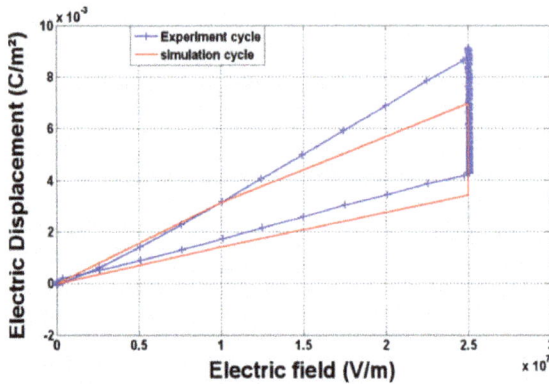

Figure 8. Simulated Ericsson cycle using a dielectric constant (red circles) and its experimental counterpart (blue crosses) for P(VDF-TrFE-CFE) 61.3/29.7/9 mol%.

4. Conclusion

PVDF-based polymer was introduced in detail in this chapter, especially for terpolymer P(VDF-TrFE-CFE). By copolymerizing the P(VDF-TrFE) with a third monomer to form a terpolymer, the defects introduce inhomogeneity in the ferroelectric phase which broadens the transition region and reduce or eliminate the hysteresis. As a result, the normal ferroelectric is transformed into a relaxor ferroelectric polymer. Due to the high room-temperature dielectric peak and slightly polarization hysteresis, terpolymer P(VDF-TrFE-CFE) 61.3/29.7/9 mol% was chosen as the active material for electrostatic energy harvesting. In addition, the basic electrostatic energy harvesting cycles -- voltage constrained cycle, was studied. The main work concentrated on Ericsson Energy Harvesting by Nonlinear Capacitance Variation between 25°C and 0°C.

In this part, we presented our work on thermal energy harvesting by electrostatic technique. It was realized by utilizing the nonlinear capacitance variation of relaxor terpolymer P(VDF-TrFE-CFE). We prepared the terpolymer through the solution casting method, and then, we investigated their characterization in detail, including dielectric properties and the temperature effect of basic ferroelectric properties. In order to simulate the harvested energy accurately, we also studied the tunability of the dielectric constant of the terpolymer.

The weak field dielectric properties were measured in order to determine the working temperature during Ericsson cycle. The largest variations in capacitance were obtained between 25°C and 0°C due to a dipolar ordering-disordering transition. In order to evaluate the harvested energy by simulation, the dielectric permittivity under DC electric field was also measured and exhibited tunability of dielectric constant as a function of temperature. Especially in the vicinity of the dielectric peak, the tunability expressed more obviously. This phenomenon was analyzed and explained perfectly from the viewpoint of nanopolar regions.

By characterizing the tunability of the dielectric constant under DC electric field, the simulated harvested energy, between 25°C and 0°C under 80 kV/mm, was equal to 240 mJ/cm^3. The direct measurement of Ericsson cycle was also carried out with a maximum electric field of 25 kV/mm. When subtracting the conduction, the harvested energy was equal to 50 mJ/cm^3 which was consistent with the simulation (48 mJ/cm^3). It proved the reliability of our theoretical evaluation and experimental feasibility in practice.

Electric conduction and the electric field breakdown are two inevitable negative factors, which restricted the experimental harvested energy, need to be resolved. The next work aims to improve the quality of polymer so as to overcome these restrictions of the experiment. Previous work dealt with the purification of a P(VDF-TrFE) copolymer and the effect of the porosity of the film. It appears that the presence of residual solvent or pores inside the film reduces its resistivity by one order of magnitude[22]. Fujisaki et al. stated that the conduction emerging at a 'defect' part (TrFE and CFE) was much smaller in amorphous regions as compared to in crystal grains[23]. Therefore, it can be reduced by optimizing the proportion of the crystal grains and amorphous regions in the mixture and their respective distribution in the film. In addition, the interface between the film and the electrode also plays an important role for the film's conduction. The improvement of the process of the film is in progress.

Author details

Hongying Zhu, Pierre-Jean Cottinet and Daniel Guyomar
INSA de Lyon, LGEF Laboratoire de Génie Electrique et Ferroélectricité EA 682, Université de Lyon, Bâtiment Gustave FERRIE, 8 rue de la Physique, F-69621 Villeurbanne Cedex, France

Sébastien Pruvost
Université de Lyon, INSA-Lyon, Ingénierie des Matériaux Polymères (IMP) UMR CNRS 5223, 69621 Villeurbanne Cedex, France

5. References

[1] S.B. Riffat and Xiaoli Ma, "Thermoelectrics: a review of present and potential applications," Applied Thermal Engineering 23, 913–935 (2003).

[2] R.B. Olsen, D.A. Bruno, J.M. Briscoe et al., "Pyroelectric conversion cycle of vinylidene fluoride-trifluoroethylene copolymer," Journal of Applied Physics 57, 5036-5042 (1985).

[3] H. Nguyen, A. Navid, and L. Pilon, "Pyroelectric energy converter using co-polymer P(VDF-TrFE) and Olsen cycle for waste heat energy harvesting " Applied Thermal Engineering 30, 2127-2137 (2010).

[4] A. Navid and L. Pilon, "Pyroelectric energy harvesting using Olsen cycles in purified and porous poly(vinylidene fluoride-trifluoroethylene) [P(VDF-TrFE)] thin films," Smart Materials and Structures 20, 025012 (2011).

[5] S. Meninger, J. O. Mur-Miranda, R. Amirtharajah et al., "Vibration-to-Electric energy conversion," IEEE Transactions on Very Large Scale Integration (VLSI) Systems 9, 64-76 (2001).

[6] Q. M. Zhang, F. Xia, Z.-Y. Cheng et al., "Poly(vinylidene fluoroethylene-trifluoroethylene) based high performance electroactive polymers," 11th International Syniposium on Electrets IEEE, 181-190 (2002).

[7] F. Xia, Z. Cheng, H. Xu et al., "High electromechanical responses in a poly(vinylidene fluoride-trifluoroethylene-chlorofluoroethylene) terpolymer," Advanced Materials 14, 1574-1577 (2002).

[8] F. Bauer, "Relaxor fluorinated polymers: novel applications and recent developments," Dielectrics and Electrical Insulation, IEEE Transactions on 17, 1106 - 1112 (2010).

[9] H. Xu, Z.-Y. Cheng, D. Olson et al., "Ferroelectric and electromechanical properties of poly(vinylidene-fluoride–trifluoroethylene–chlorotrifluoroethylene)   terpolymer," Applied Physics Letters 78, 2360-2362 (2001).

[10] R. J. Klein, F. Xia, Q. M. Zhang et al., "Influence of composition on relaxor ferroelectric and electromechanical properties of poly(vinylidene fluoride-trifluoroethylene-chlorofluoroethylene) " Journal of Applied Physics 97, 094105 (2005).

[11] B. Neese, B. Chu, S. Lu et al., "Large electrocaloric effect in ferroelectric polymers near room temperature," Science 321, 821-823 (2008).

[12] K. Ren, Y. Liu, H. Hofmann et al., "An active energy harvesting scheme with an electroactive polymer," Applied Physics Letters 91, 132910 (2007).

[13] H. Zhu, S. Pruvost, P.J. Cottinet et al., "Energy harvesting by nonlinear capacitance variation for a relaxor ferroelectric poly(vinylidene fluoridetrifluoroethylene-chlorofluoroethylene) terpolymer," Applied Physics Letters 98, 222901 (2011).

[14] P.D. Mitcheson, T. Sterken, M. Kiziroglou C. He et al., "Electrostatic Microgenerators," MEAS CONTROL-UK 41 (4), 114 - 119 (2008).

[15] H. Y. Zhu, S. Pruvost, D. Guyomar et al., "Thermal energy harvesting from $Pb(Zn_{1/3}Nb_{2/3})_{0.955}Ti_{0.045}O_3$ single crystals phase transitions," Journal of Applied Physics 106, 124102 (2009).

[16] Z.-Y. Cheng, R. S. Katiyar, X. Yao et al., "Temperature dependence of the dielectric constant of relaxor ferroelectrics," Physical Review B 57, 8166-8177 (1998).

[17] Daniel Guyomar, Sebastien Pruvost, and Gael Sebald, "Energy Harvesting Based on FE-FE Transition in Ferroelectric Single Crystals," IEEE transactions on ultrasonics, ferroelectrics, and frequency control 55, 279-285 (2008).

[18] G. Akcay, S. P. Alpay, J. V. Mantese et al., "Magnitude of the intrinsic electrocaloric effect in ferroelectric perovskite thin films at high electric fields " Applied Physics Letters 90, 252909 (2007).

[19] G. Sebald, S. Pruvost, and D. Guyomar, "Energy harvesting based on ericsson pyroelectric cycles in a relaxor ferroelectric ceramic," Smart Materials and Structures 17, 015012 (2008).

[20] S. G. Lu, B. Neese, B. J. Chu et al., "Large electric tunability in poly(vinylidene fluoride-trifluoroethylene) based polymers " Applied Physics Letters 93, 042905 (2008).

[21] V. Bobnar, B. Vodopivec, A. Levstik et al., "Dielectric properties of relaxor-like vinylidene fluoride-trifluoroethylene-based electroactive polymers," Macromolecules 36, 4436-4442 (2003).

[22] A Navid, C S Lynch, and L Pilon, "Purified and porous poly(vinylidene fluoride-trifluoroethylene) thin films for pyroelectric infrared sensing and energy harvesting " Smart Materials and Structures 19, 055006 (2010).

[23] Sumiko Fujisaki, Hiroshi Ishiwara, and Yoshihisa Fujisaki, "Low-voltage operation of ferroelectric poly(vinylidene fluoride-trifluoroethylene) copolymer capacitors and metal-ferroelectric-insulator-semiconductor diodes " Applied Physics Letters 90, 162902 (2007).

Three Dimensional TCAD Simulation of a Thermoelectric Module Suitable for Use in a Thermoelectric Energy Harvesting System

Chris Gould and Noel Shammas

Additional information is available at the end of the chapter

1. Introduction

Thermoelectric technology can be used to generate electrical power from heat, temperature differences and temperature gradients, and is ideally suited to generate low levels of electrical power in energy harvesting systems. This chapter aims to describe the main elements of a thermoelectric energy harvesting system, highlighting the limitations in performance of current thermoelectric generators, and how these problems can be overcome by using external electronic components and circuitry, in order to produce a thermoelectric energy harvesting system that is capable of providing sufficient electrical power to operate other low power electronic systems, electronic sensors, microcontrollers, and replace or recharge batteries in several applications. The chapter then discusses a novel approach to improving the thermoelectric properties and efficiency of thermoelectric generators, by creating a 3D simulation model of a three couple thermoelectric module, using the Synopsys Technology Computer Aided Design (TCAD) semiconductor simulation software package. Existing published work in the area of thermoelectric module modelling and simulation has emphasised the use of ANSYS, COMSOL and Spice compatible software. The motivation of this work is to use the TCAD semiconductor simulation environment in order to conduct a more detailed thermal and electrical simulation of a thermoelectric module, than has previously been published using computer based simulation software packages. The successful modelling and simulation of a thermoelectric module in TCAD will provide a base for further research into thermoelectric effects, new material structures, module design, and the improvement of thermoelectric efficiency and technology. The aim of the work presented in this chapter is to investigate the basic principle of thermoelectric power generation in the TCAD simulation environment. The initial model, and simulation results presented, successfully demonstrate the fundamental thermoelectric effects, and the concept of thermoelectric power generation. Future work will build on this initial model, and further analysis of the thermal and electrical simulation results will be published.

This chapter begins with a short background review of thermoelectric technology, followed by an overview of a typical thermoelectric module's construction, highlighting the main elements, material structure, and connection details for thermoelectric power generation.

The chapter then discuses a generic design of a thermoelectric energy harvesting system that incorporates a thermoelectric module with a boost converter, low power DC to DC converter, and a supercapacitor. The 3D modelling of a thermoelectric module is then presented, including the simulation results obtained for the thermal and electrical characteristics of the device when it is connected as a thermoelectric generator. Different thermoelectric couple and module designs have been investigated, and the simulation results have been discussed with reference to fundamental thermoelectric theory. The chapter draws conclusions on the application of thermoelectric technology for energy harvesting, and the validity and effectiveness of the 3D TCAD thermoelectric module simulation model for thermoelectric power generation.

2. Thermoelectric technology

Themoelectricity utilises the Seebeck, Peltier and Thomson effects that were first observed between 1821 and 1851 [1]. Practical thermoelectric devices emerged in the 1960's and have developed significantly since then with a number of manufacturers now marketing thermoelectric modules for power generation, heating and cooling applications [2]. Ongoing research and advances in thermoelectric materials and manufacturing techniques, enables the technology to make an increasing contribution to address the growing requirement for low power energy sources typically used in energy harvesting and scavenging systems [3]. Commercial thermoelectric modules can be used to generate a small amount of electrical power, typically in the mW or µW range, if a temperature difference is maintained between two terminals of a thermoelectric module. Alternatively, a thermoelectric module can operate as a heat pump, providing heating or cooling of an object connected to one side of a thermoelectric module if a DC current is applied to the module's input terminals [2].

2.1. Thermoelectric module construction

A single thermoelectric couple is constructed from two 'pellets' of semiconductor material usually made from Bismuth Telluride (Bi_2Te_3). One of these pellets is doped with acceptor impurity to create a P-type pellet, the other is doped with donor impurity to produce an N-type pellet. The two pellets are physically linked together on one side, usually with a small strip of copper, and mounted between two ceramic outer plates that provide electrical isolation and structural integrity. For thermoelectric power generation, if a temperature difference is maintained between two sides of the thermoelectric couple, thermal energy will move through the device with this heat and an electrical voltage, called the Seebeck voltage, will be created. If a resistive load is connected across the thermoelectric couple's output terminals, electrical current will flow in the load and a voltage will be generated at the load [4]. Practical thermoelectric modules are constructed with several of these thermoelectric couples connected electrically in series and thermally in parallel. Standard thermoelectric modules typically contain a minimum of three couples, rising to one hundred and twenty seven couples for larger devices [2]. A schematic diagram of a single thermoelectric couple connected for thermoelectric power generation, and a side view of a thermoelectric module is shown in Figure 1.

For thermoelectric power generation, a small amount of electrical power can be generated from a thermoelectric module if a temperature difference is maintained between two sides of the module. Normally, one side of the module is attached to a heat source and is referred to as the 'hot' side or 'TH'. The other side of the module is usually attached to a heat sink and is called the 'cold' side or 'TC'. The heat sink is used to create a temperature difference between the hot and cold sides of the module. If a resistive load (RL) is connected across the module's output terminals, electrical power will be generated at the load when a temperature difference exists between the hot and cold sides of the module due to the Seebeck effect [3].

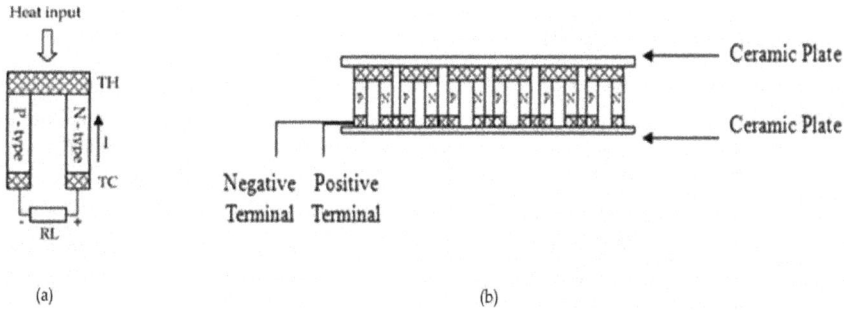

(a)   (b)

Figure 1. A schematic diagram of a single thermoelectric couple connected for thermoelectric power generation (a), and a side view of a thermoelectric module (b) [5]

A schematic diagram of a thermoelectric module, operating as a thermoelectric power generator, is shown in Figure 2.

The efficiency of a thermoelectric module for power generation can be found by:

$$\eta = \frac{Energy\ supplied\ to\ the\ load}{Heat\ energy\ absorbed\ at\ the\ hot\ junction} \tag{1}$$

In thermoelectricity, efficiency is normally expressed as a function of the temperature over which the device is operated, referred to as the dimensionless thermoelectric figure-of-merit ZT, and can be found by:

$$ZT = \frac{\alpha^2 \sigma}{\lambda} \tag{2}$$

where α is the Seebeck coefficient, σ is the electrical conductivity, and λ is the total thermal conductivity. The best thermoelectric materials used in commercial thermoelectric devices, Bi_2Te_3-Sb_2Te_3 alloys, operating around room temperature, have typical values of $\alpha=225\mu V/K$, $\sigma = 10^5/\Omega m$, and $\lambda = 1.5$ W/mK, which results in ZT $\approx$ 1 [6].

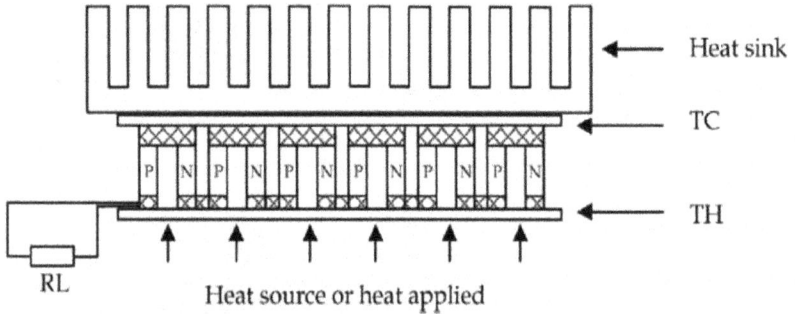

Figure 2. A schematic diagram of a thermoelectric module configured for thermoelectric power generation [5]

3. Thermoelectric energy harvesting

Although the thermoelectric output voltage, current, and electrical power generated by a standard thermoelectric module is relatively small, the thermoelectric output voltage can be boosted to a useful and stable level by using a boost converter and low power DC to DC converter. If the electrical power output from the DC to DC converter is then accumulated and stored for future use in a supercapacitor, it is possible to increase the potential output current of the system, and hence the overall electrical power output of the thermoelectric energy harvesting system. A simplified block diagram of a thermoelectric energy harvesting system is shown in Figure 3. It is not always necessary to use a boost converter, although in many applications, the output voltage from a single thermoelectric module is too low to directly operate a DC to DC converter. The output of the DC to DC converter can also be connected directly to an electrical load in order to power other low power electronic systems, to recharge a battery, or as shown - connected to a supercapacitor for electrical storage purposes. The energy stored in the supercapacitor can then be accumulated over time, and released to the load when required [3]. The addition of the supercapacitor in the system enables much higher levels of current to be drawn by a load, if only for a short period of time, and makes the system more versatile. Commercially available boost converters and low power DC to DC converters can operate from very low thermoelectric output voltages of 20mV, outputting a DC output voltage of between 2.2V to 5V [3].

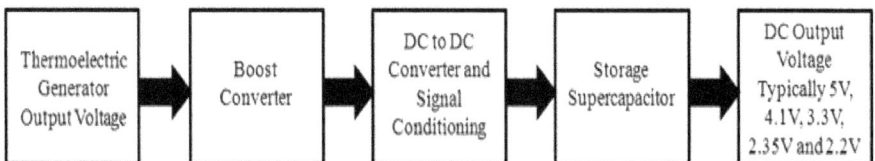

Figure 3. A generic thermoelectric energy harvesting system block diagram [3]

4. Technology Computer Aided Design (TCAD)

The Synopsys TCAD semiconductor simulation package has been chosen for this work as it is widely used in the semiconductor industry to simulate semiconductor device behaviour, and has the capability to simulate the semiconductor manufacturing process in addition to device simulation. Existing published work into thermoelectric modelling and simulation has emphasised the use of ANSYS, COMSOL and Spice compatible software. It is anticipated that modelling a thermoelectric module in TCAD will allow a more detailed analysis of the thermal properties and electrical characteristics of a device than has been published in previous studies. TCAD comprises of a suite of programs that can be executed independently, or together in the form of a Workbench project, in order to simulate the electrical characteristics and thermal properties of a device. Specific TCAD tools have been added to this workbench project in order to create a working simulation. Sentaurus Structure Editor is executed first, and the 3D thermoelectric module is created within this environment, and then meshed using Sentaurus Mesh, followed by device simulation in Sentaurus Device. Tecplot and Inspect have then been used to visualise the results [7].

5. 3D TCAD simulation model of a thermoelectric module containing three thermoelectric couples using Sentaurus Structure Editor

A three couple thermoelectric module has been modelled in Sentaurus Structure Editor, and is shown in Figure 4. The P-type pellets have been simulated using Silicon as the base material, heavily doped with Boron with a constant doping profile and initial concentration of $1e^+16cm^{-3}$. The N-type pellets are similarly constructed, using Silicon as the base material, heavily doped with Phosphorus at $1e^+16cm^{-3}$. The seven copper interconnects are labelled 'Copper Connect 1' through to 'Copper Connect 7' respectively. An electrode contact was made on the face of Copper 2 and Copper 7 to simulate the negative and positive electrical connections to the couple, and is shown in Figure 5. A thermal contact was made on the faces of Copper 1 through to Copper 7 respectively, in order to allow the temperature of each contact to be specified or calculated, and the dimension of the 3D device in the Z-direction is 1100 micron metres. Although most commercial thermoelectric modules use Bismuth Telluride as the base material, as this exhibits the most pronounced thermoelectric effects at room temperature, this work has used Silicon as the base material for simulation. TCAD's physical device equations that describe the carrier distribution and conduction mechanisms, materials database and parameter list is comprehensive for Silicon. Once the basic thermoelectric properties have been successfully demonstrated using Silicon, even though this may be at a reduced level than could be seen with state-of-the-art materials, it will be possible to alter the material structure and move to Bismuth Telluride and other material structures with increased confidence.

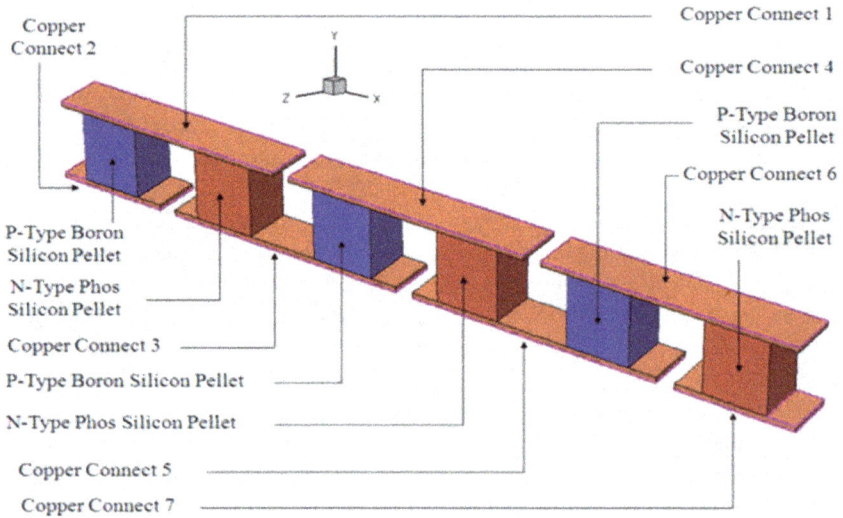

Figure 4. 3D three couple thermoelectric module modelled in Sentaurus Structure Editor [8]

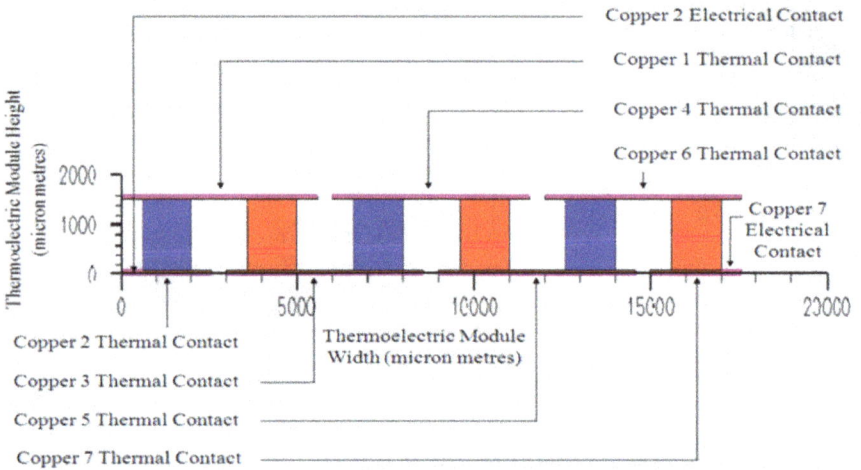

Figure 5. A cut-through in the Z direction highlighting the thermal and electrical connections [8]

The 3D thermoelectric module model was simulated as a TCAD 'Mixed Mode Simulation' rather than a 'Single Device Simulation', as it is possible to add external components and circuitry to the 3D device structure designed in Sentaurus Structure Editor. In this case a load resistor (RL) was connected between the electrical output terminals 'Copper 2' and 'Copper 7' of the device, as shown in Figure 6, in order to calculate the electrical power generated at the load. A three couple thermoelectric module with ceramic outer plates has also been simulated, and is shown in Figure 7 and Figure 8. The top and bottom face of the two ceramic plates have been used as the thermal contacts of the device, and are labelled 'Ceramic top' and 'Ceramic bottom' respectively. Otherwise, the construction of the device is the same as shown earlier for a three couple thermoelectric module without ceramic outer plates [8].

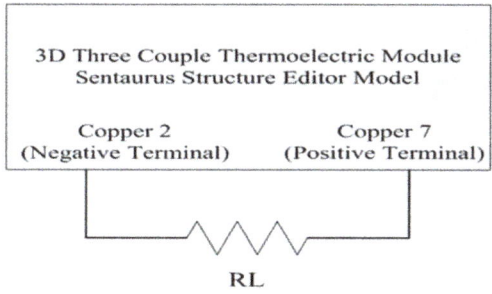

Figure 6. A schematic representation of a TCAD Mixed Mode simulation of a thermoelectric module with a load resistance RL connected between the thermoelectric model output terminals [8]

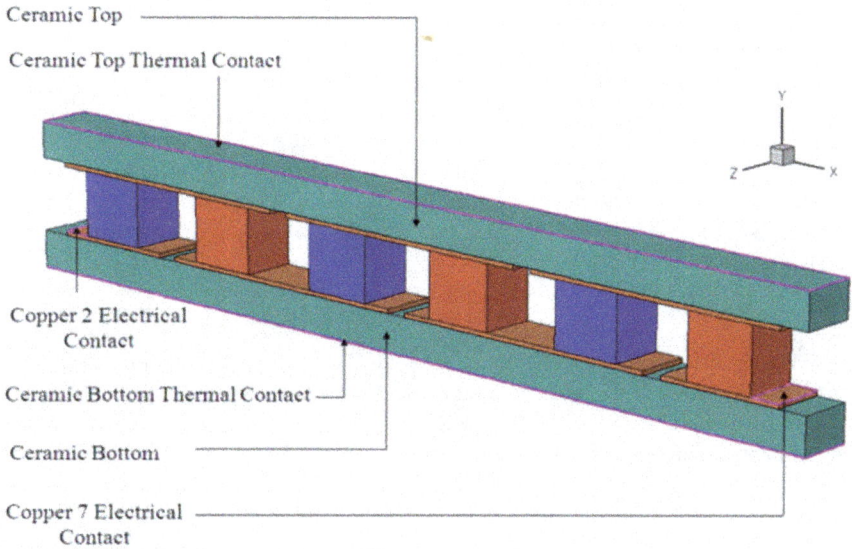

Figure 7. A 3D three couple thermoelectric module with ceramic outer plates modelled in Sentaurus Structure Editor [8]

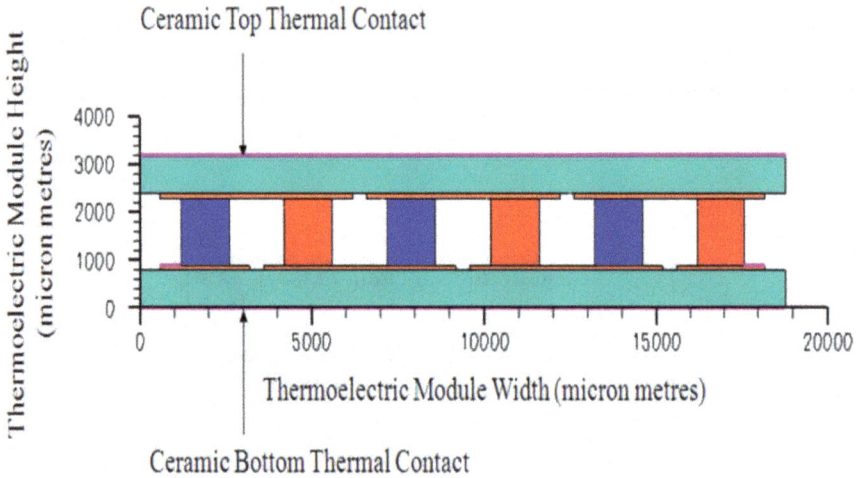

Figure 8. A cut-through in the Z direction highlighting the thermal connections [8]

6. Simulation methodology

The three couple thermoelectric module has been modelled in Sentaurus Structure Editor, connected to a load resistor RL, and tested using a Mixed Mode simulation for thermoelectric power generation. The temperature of the thermal contacts; Copper 1; Copper 4; and Copper 6; was increased from steady-state conditions of 300 Kelvin to 301 Kelvin. The temperature of the other four thermal contacts; Copper 2; Copper 3; Copper 5; and Copper 7; were kept at 300 Kelvin. This creates a 1 Kelvin temperature difference between both sides of the module. The load resistance RL was increased from 10 ohms through to 150 ohms, in 10 ohm steps, in order to establish where maximum power is generated at the load. The voltage across the load resistor, and the load current, was recorded using the simulation program, and the electrical power generated at the load calculated using:

$$P = V \times I \text{ measured in Watts} \tag{3}$$

where V is the electrical voltage measured across the load resistor RL, and I is the electrical current flowing through the load resistor RL. The P-type and N-type doping concentration was altered from $1e^{+}16cm^{-3}$ to $1e^{+}15cm^{-3}$ and $1e^{+}17cm^{-3}$ in order to establish if the doping concentration has any effect on the electrical power generated by the thermoelectric module. The temperature of the thermal contacts; Copper 1; Copper 4; and Copper 6; was then increased from 301 Kelvin to 325 Kelvin; 350 Kelvin; 375 Kelvin; and 400 Kelvin. The temperature of the other four thermal contacts; Copper 2; Copper 3; Copper 5; and Copper 7; were kept at 300 Kelvin. This creates a temperature difference between both sides of the module of 25 Kelvin; 50 Kelvin; 75 Kelvin; and 100 Kelvin respectively. The simulation was then repeated using the model of a three couple thermoelectric module with ceramic outer plates for comparison [8].

7. Simulation results

For thermoelectric power generation, the simulation results successfully demonstrate that if the thermoelectric module is subjected to a temperature gradient from one side of the device to the other, electrical power is generated at the load resistor RL connected between the device output terminals. This is in agreement with the fundamental thermoelectric theory discussed earlier. With an initial doping concentration of $1e^+16cm^{-3}$ for the P-type and N-type silicon pellets, and a temperature gradient of 1 Kelvin across the device, the lattice temperature of the module is shown in Figure 9, and the electrical power generated at the load shown in Figure 10. The maximum power generated at the load occurs with a load resistance of 50 ohms, and a peak power at the load of 0.1 micro-watts. Further tests have been conducted with a modified P-type and N-type doping concentration of $1e^+15cm^{-3}$, $1e^+16cm^{-3}$, and $1e^+17cm^{-3}$, with the results shown in Figure 11. Changing the doping concentration significantly alters the amount of electrical power generated at the load, and the resistance of the device where maximum power is observed. The doping concentration can be optimised to achieve maximum power generation, and a full set of test results will be published. Increasing the thermal gradient on both sides of the device, by increasing the temperature of the thermal contacts at Copper 1, Copper 4 and Copper 6, results in an increase in electrical power generated at the load, as shown in Figure 12. This is as expected as the Seebeck effect is temperature dependent, and the electrical power generated by a thermoelectric module is related to the temperature gradient between two sides of the device [2]. The lattice temperature of the thermoelectric module, with an applied 100 Kelvin temperature gradient between both sides of the device, is shown in Figure 13 and demonstrates that the temperature gradient within each individual thermoelectric P-type and N-type pellet, is now significantly higher than was obtained with a much lower temperature gradient of 1 Kelvin applied to the device in Figure 9 [8].

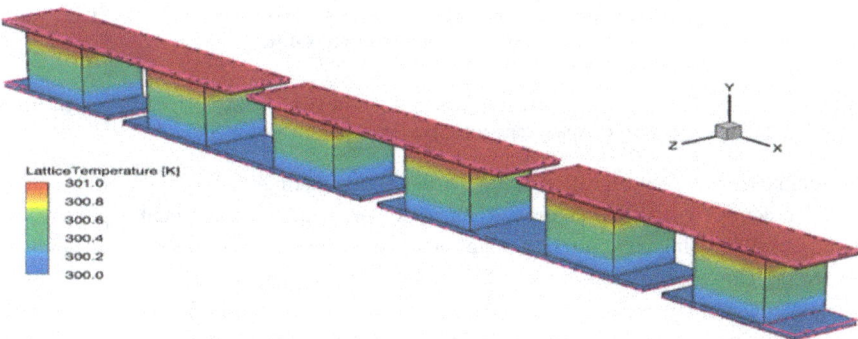

Figure 9. The lattice temperature of the thermoelectric module with an applied 1 Kelvin temperature gradient between both sides of the module [8]

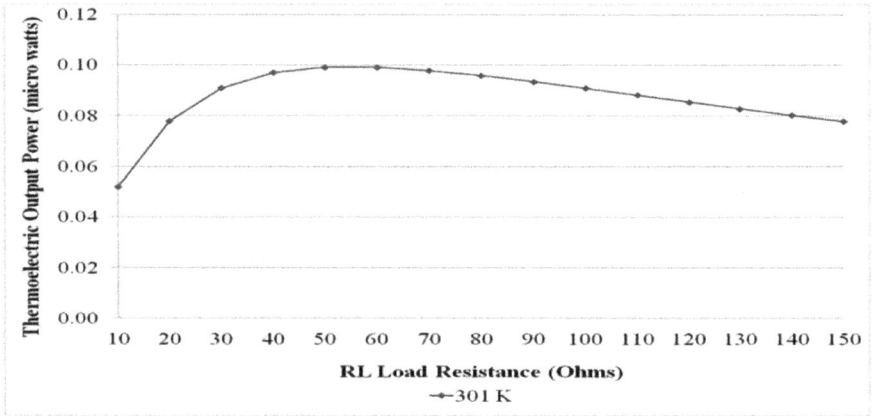

Figure 10. The electrical power generated at the load resistor (RL) with an applied 1 Kelvin temperature gradient between both sides of the module [8]

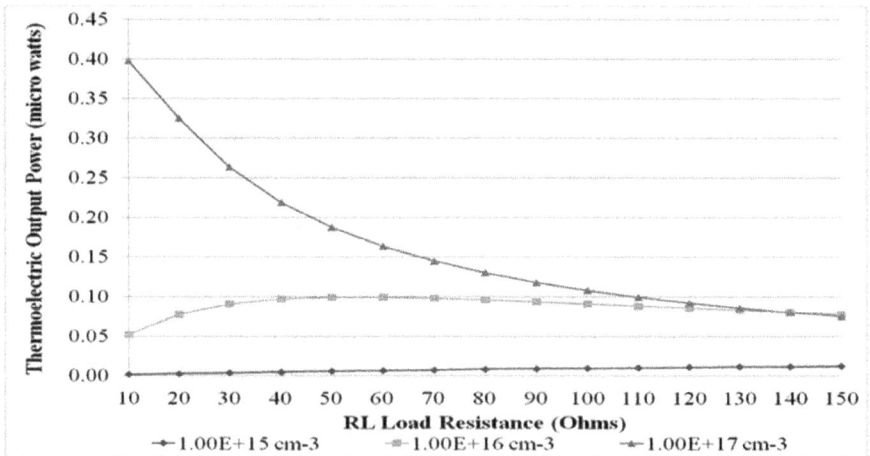

Figure 11. The electrical power generated at the load resistor (RL) with a 1 Kelvin temperature gradient and different P-type and N-type doping concentrations [8]

The simulation has been repeated on the thermoelectric module with ceramic outer plates, shown earlier in Figure 7 and Figure 8. With a 1 Kelvin temperature gradient applied to the module, and a doping concentration of $1e^{+}16cm^{-3}$, the ceramic outer plates absorb some of the applied temperature gradient, and the temperature gradient within the thermoelectric pellets is now more uniform than observed earlier, shown in Figure 14. This has the effect of reducing the electrical power generated at the load, shown in Figure 15. However, the ceramic plates are necessary in practical thermoelectric devices in order to create electrical isolation and provide a foundation to mount the thermoelectric couples. The thermal

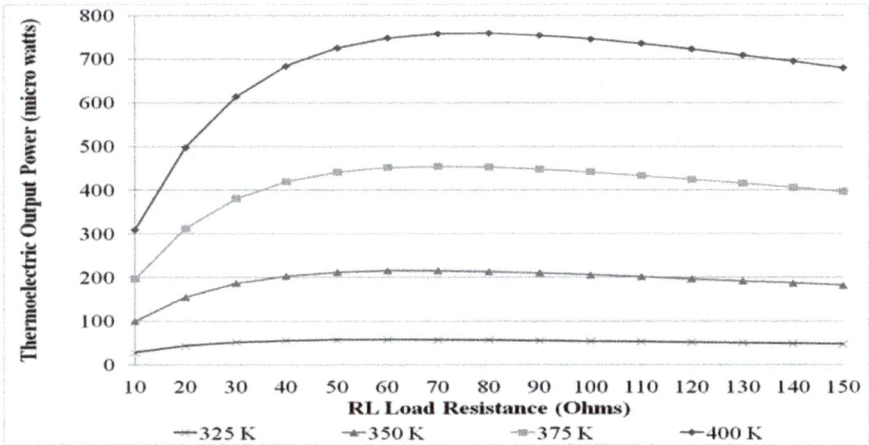

Figure 12. The electrical power generated at the load resistor (RL) with a doping concentration of
1e+16cm-3 and different temperature settings applied to the thermal contacts Copper 1, Copper 4 and
Copper 6 [8]

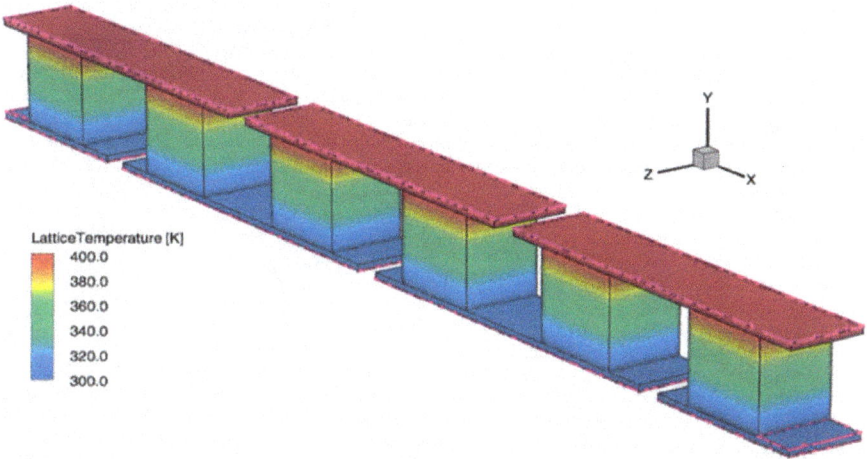

Figure 13. The lattice temperature of the thermoelectric module with an applied 100 Kelvin
temperature gradient between both sides of the module [8]

conductivity of the ceramic used in the simulation model is 0.167 [W/ cm K]. Practical
thermoelectric modules optimise the thermal conductivity of the ceramic used in the
construction of the outer plates, and are typically constructed using Alumina ceramics [9].
Optimising the material properties of the ceramic outer plates used in the simulation model,
by increasing their thermal conductivity, should improve the electrical power generated by
the thermoelectric module.

The TCAD simulation results demonstrate the basic principle of thermoelectric power generation. The use of Silicon as the base material is sufficient to demonstrate the fundamental concepts, although the output power of the thermoelectric simulation model is much lower than would be expected from a practical thermoelectric module that was manufactured with Bismuth Telluride. This is not unexpected, as Silicon has a far lower Seebeck coefficient than Bismuth Telluride. Future work will investigate different material structures, novel module design and technology, and the results will be published.

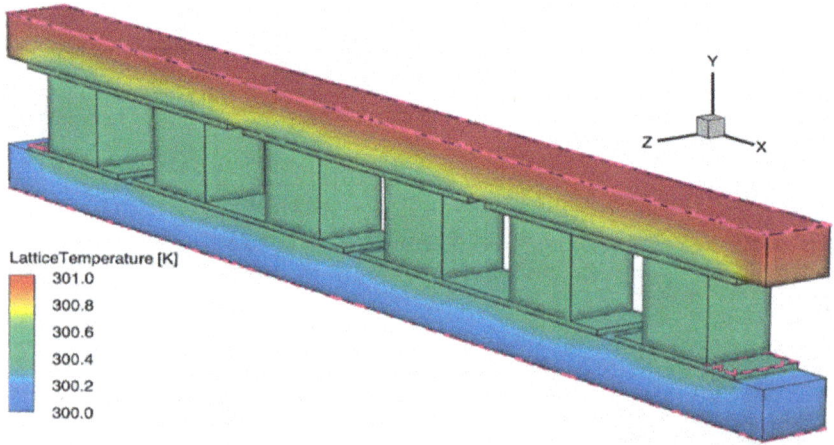

Figure 14. The lattice temperature of the thermoelectric module with ceramic outer plates and an applied 1 Kelvin temperature gradient between both sides of the module [8]

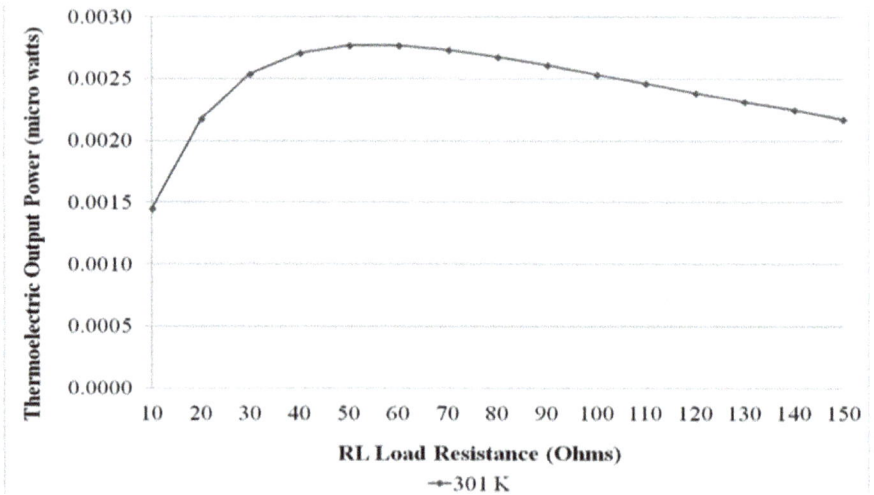

Figure 15. The electrical power generated at the load with an applied 1 Kelvin temperature gradient between both sides of the module [8]

8. Conclusions

Thermoelectric technology is ideally suited as a low power energy source for thermal energy harvesting systems, and with the addition of a boost converter and low power DC to DC conversion, coupled with electrical energy storage in supercapacitors, it is possible to construct a thermoelectric energy harvesting system capable of supplying electrical power to other low power electronic systems, and replace or recharge batteries in several applications. The 3D simulation of a three couple thermoelectric module in TCAD has been successfully achieved, and the simulation results demonstrate the basic principle of thermoelectric power generation. The use of Silicon as the base material is sufficient to demonstrate the basic concepts, and the TCAD thermoelectric simulation model can be used for further analysis into thermoelectric effects, material structure, module design and technology.

Author details

Chris Gould and Noel Shammas

Faculty of Computing, Engineering and Technology, Staffordshire University, Stafford, United Kingdom

9. References

[1] G. S. Nolas, J. Sharp, H. J. Goldsmid, *Thermoelectrics – Basic Principles and New Materials Developments*, Springer-Verlag, 2001, pp. 1-5

[2] D. M. Rowe, "General Principles and Basic Considerations", in *Thermoelectric Handbook – Macro to Nano*, D. M. Rowe (Ed.), CRC Taylor & Francis Group, 2006, pp. 1-10

[3] C. A. Gould, N. Y. A. Shammas, S. Grainger, I. Taylor, "Thermoelectric power generation: Properties, application and novel TCAD simulation", *14th IEEE European Conference on Power Electronics and Applications (EPE2011)*, Aug 30th to 1st Sept 2011, Birmingham, UK, pp. 1-10

[4] C. M. Bhandari, "Thermoelectric Transport Theory", in *CRC Handbook of Thermoelectrics*, D. M. Rowe (Ed), CRC Taylor & Francis Group, 1995, pp. 27-42

[5] Chris Gould, Noel Shammas, "A Review of Thermoelectric MEMS Devices for Micro-power Generation, Heating and Cooling Systems", in *Micro Electronic and Mechanical Systems*, Kenichi Takahata (Ed.), INTECH, 2009, pp. 15 – 24, ISBN 978-953-307-027-8

[6] B. C. Sales, "Critical review of recent approaches to improved thermoelectric materials", *International Journal of Applied Ceramic Technology*, vol. 4, no. 4, August 2007, pp. 291-296

[7] C. A. Gould, N. Y. A. Shammas, S. Grainger, I. Taylor, "A Novel 2D TCAD Simulation of a Thermoelectric Couple", *Proc. of ECT2010 – 8th European Conference on Thermoelectrics*, Sept 22nd to 24th 2010, Como, Italy, pp. 239-243

[8] C. A. Gould, N. Y. A. Shammas, S. Grainger, I. Taylor, "A 3D TCAD Thermal and Electrical Simulation of a Thermoelectric Module configured for Thermoelectric Power

Generation", *9th IEEE International Microtherm Conference*, Lodz, Poland, 28th June – 1st July 2011, pp. 1 – 6

[9] R. Marlow, E. Burke, "Module Design and Fabrication", in *CRC Handbook of Thermoelectrics*, D. M. Rowe (Ed.), CRC Taylor & Francis Group, 1995, pp. 597 - 607

Vibrations: Conversion Mechanisms

Chapter 4

High Energy Density Capacitance Microgenerators

Igor L. Baginsky and Edward G. Kostsov

Additional information is available at the end of the chapter

1. Introduction

The problem of continuous production of energy sufficient for modern microcircuits with an almost unlimited service life should be related to searching for power sources in the ambient medium. The comparison of these sources shows that only solar energy and energy of mechanical vibrations of surfaces of various solids can be used for generation of electrical energy in the milliwatt or microwatt range, which is enough for powering these microcirquits.

A typical feature for most modern sources of mechanical vibrations (surfaces of solids) is moderate amplitudes ranging from 0.1 to 2.0 μm; the analysis of the frequency distribution of amplitudes shows that low frequencies (1–100 Hz) have the most power [1]. Examples are vibrations of various building structures: supports, bridges, roadbeds, building walls, etc.

There are numerous recent publications that describe the development of microgenerators of electrical energy, including microelectromechanical systems (MEMS generators) capable of converting mechanical energy from the ambient medium to electrical energy. A new term, "energy harvesting," was accepted [2–11]. MEMS generators can be fabricated in a single technological cycle with fabrication of the basic microcircuit. The problem of powering MEMS devices is recognized as one of the most important issues in modern microelectronics.

Electrostatic energy microgenerators seem to be the most suitable for this task, because fuel or chemical elements need to be refined or renewed, solar or thermo- elements are not suitable for all situations of MEMS operation, electromagnetic generators of energy are ineffective in the range of low-amplitude vibrations and small sizes of the transducers, while piezoelectric generators are ineffective at low frequencies of vibrations [12-14].

Electrostatic generators have been known for a long time. Their operation principle is based on the work of mechanical forces transferring an electrical charge against electrostatic forces of attraction of unlike charges [15]. Depending on the method of generation and transportation of this charge, generators can be divided into two classes. In the first class, the charge generated by some external action, for instance, by an electric arc or friction, is transferred by a transporter: a belt (Van de Graaf generators) [15] or a disk (friction machines). In the second class, the charged plate of the capacitor moves. Depending on the presence or absence of a built-in charge in this capacitor such devices are classified as either electret [16,17] or capacitance generators , e.g., Toepler or Felichi machines [15].

For electrostatic capacitance machine (Fig.1a) the separation of the plates (vertical, i.e., out-of-plane, or lateral, i.e., in-plane) of the capacitor $C(t)$ initially charged from the voltage source V_0 up to the value $Q_0 = C_{max} V_0$ (where C_{max} is initial maximal value of capacitance) in the conditions of open circuit results in growth of voltage on the capacitor up to the value

$$V_{max} = V_{min} \, C_{max}/C_{min}. \tag{1}$$

Here for the case under consideration $V_{min} = V_0$. And, respectively, the energy of capacitor is changed from $W_{min} = C_{max} V_0^2/2$ to

$$W_{max} = \frac{C_{max}}{C_{min}} W_{min} = \eta W_{min} = \frac{Q_0^2}{2C_{min}} \, , \tag{2}$$

where $\eta = C_{max}/C_{min}$ is capacitance modulation depth.

The produced electric energy $W_{max} - W_{min}$ is transferred to load R. After that, the capacitor plates return to the initial position and are charged by the voltage source, and the energy conversion process is repeated. The power developed by such a generator is $P = (W_{max} - W_{min})f$ (where f is the repetition frequency of conversion cycles), and the efficiency of energy conversion, i.e. the ratio of energy produced by the generator during the conversion period to energy losses during the same period, is η-1.

Drawbacks of capacitance machine are the necessity of powering the generator by voltage source V_0 once at each cycle of energy conversion to charge the capacitor $C(t)$ and also the need in use of the key, synchronized with the phase of $C(t)$ alteration, switching the capacitor to the voltage source V_0, in the open circuit, and to the load R.

For these reasons, the circuit shown in Fig. 1a had limited applications: only for generation of high voltages (up to several hundreds of kilovolts) in solving special engineering problems.

This type of generators is used in many electrostatic MEMS generators [1,12,18-23]. Their specific power is low not exceeding 1-10 $\mu W/cm^2$ because of high value of minimum interelectrode gap at low η. It is possible to increase the generator power by decreasing the gap between the electrodes, but it results in the rising probability of an electrical breakdown in the gap. Low value of specific power is the main drawback of these generators; another problem is the necessity of using DC voltage sources.

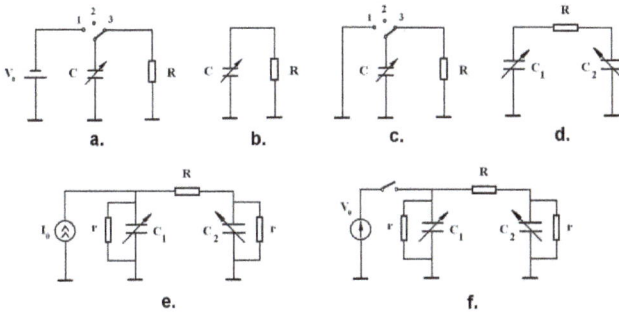

Figure 1. Various circuits of capacitance electrostatic generators: (a) capacitance machine; (b) electret current generator; (c) electret voltage generator; (d) ideal two-capacitor generator; (e) two-capacitor generator with loss compensation by a current source; (f) two-capacitor generator with loss compensation by a voltage source; V_0 is the voltage source, I_0 is the current source, R is the load resistance, and r is the leakage resistance.

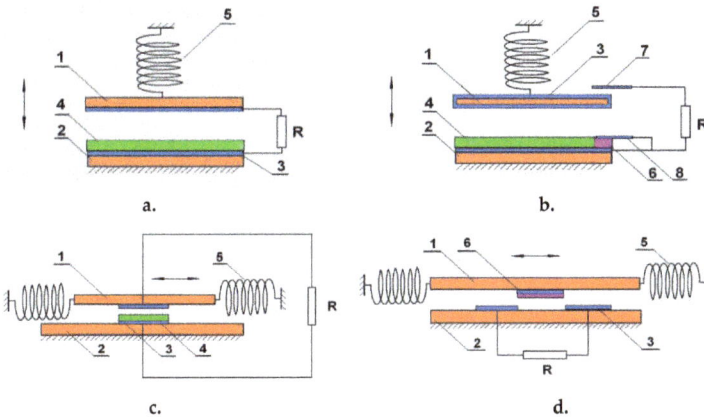

Figure 2. The schematic representation of the designs of capacitance generators, corresponding to the circuits presented on Fig.1: (a) – electret current generator, out-of-plane fabrication (see Fig.1b), (b) – electret voltage generator, out-of-plane plane fabrication (see Fig.1c), (c) – electret current generator, in-plane fabrication (see Fig.1b), (d) – two-capacitor generator, in-plane fabrication (see Fig.1d). 1 – moving substrate, 2 – stationary substrate, 3 – metal electrode, 4 – electret, 5 – spring, 6 – dielectric layer, 7 – contact 3 (Fig.1c), 8 – contact 1 (Fig.1c).

In electret generators [16,17,24-36] (see the circuits in Fig.1b,c and schematic designs in Fig.2a-c) the dielectric layer with built-in charge is formed on the inner surface of fixed plate of capacitor $C(t)$, and the charge losses are compensated by electrostatic induction of charge on the surface of moving plate, that is a considerable advantage of this way of energy transformation. These generators are further divided into current generators in which the capacitor $C(t)$ is directly connected to the load R (see Fig.1b and Fig.2a,c) and voltage generators in which the capacitor

$C(t)$ is switched just like in capacitance machine but without the voltage source: $V=0$, Fig.1c and Fig.2b. Compared to voltage generators the current generators have the advantage in the simplicity of the circuit of energy transformation. The drawback is lack of the effect of generated voltage amplification, and correspondingly of the output power, proportional to capacitance modulation depth η. It should be noted that the "in-plane" constructions of electret generators (with lateral shift of the generator plate) having a comb structure of electrodes are being actively developed now. These devices are simple in production, and the technology earlier developed for smart sensors is used for their fabrication. For these devices the specific power of order 100 $\mu W/cm^2$ was reached [36]. The further increase of specific power is impeded by large interelectrode gaps used here, of order 20 μm.

The two-capacitor mode of capacitance energy transformation is described in [5,37,38], see Fig.1d and schematic design in Fig.2d. The electric energy is generated by means of capacitance alteration in antiphase of two capacitors ($C_i(t)$, $i=1,2$), initially charged to potential V_0, under the action of the force on their moving plates. In this case there is no need to feed the capacitors by switching on the voltage source on each cycle of energy transformation, because both capacitors C_i alternate in playing this role. The electric power is generated in the load R as the current flows from the recharging capacitors. If the generator has initial charge distributed between capacitances C_1 and C_2 then in idealized case with no leakage currents the circuit could operate for unlimitedly long time producing the energy under periodical action of mechanic force.

This approach has been proposed first in [37] as an idea. The evaluations of the efficiency of energy transformation were done in [5] taking into account the compensation of charge losses in the capacitors C_i by current source I_0, Fig.1e. However the total analysis of the generator operation at all possible loads and frequencies of generation, and at various ways of capacitance modulation and compensation of charge losses by connecting the current (Fig.1e) or voltage (Fig.1f) source have not been done.

The present work is aimed at performing the analysis of specific features of operation of electrostatic capacitance generators that do not need the electrical energy sources to compensate for charge leakages permanently, at each cycle of energy transformation. For the sake of generality only electric part of the generator will be analyzed under the modulation of capacitances of generating capacitors both by means of changes of interelectrode gaps and also by lateral shift of capacitors plates. A partial experimental verification of the results of the model proposed will be done.

2. High energy density electret generators

According to the common definition the ability of dielectric (or ferroelectric) to retain the charge produced in the bulk or on the surface of the layer by external action is called the electret effect. So we assume that this charge is constant in the time of generator action.

We will analyze first the vibration mode of this generator when the area of capacitor pates overlapping is constant and the distance between them is varied, or so called out-of-plane vibration mode.

The circuit of the generator considered is shown in Fig. 1, b,c and the schematic representation of their designs is represented in Fig.2 a,b. Its structure includes four thin layers: an electrode, a dielectric (ferroelectric) of thickness d, having a space charge $\varrho(x,t)$, an air gap with a variable thickness $d_1(t)$ changing in time under the action of mechanical forces, and a moving electrode performing oscillatory motions with respect to the dielectric surface. In the general case, the operation principle of such an energy generator is the following. At $t=0$ in each layer of the structure including the metallic electrodes the initial distribution of charge and corresponding distribution of electric field $E(x,0)$ is set. The value of $E(x, t)$ is determined by the charge distribution in the dielectric layer and the voltage on the structure $V(t)$; the field in the electrodes is screened by the charge formed in them (Fig. 3).

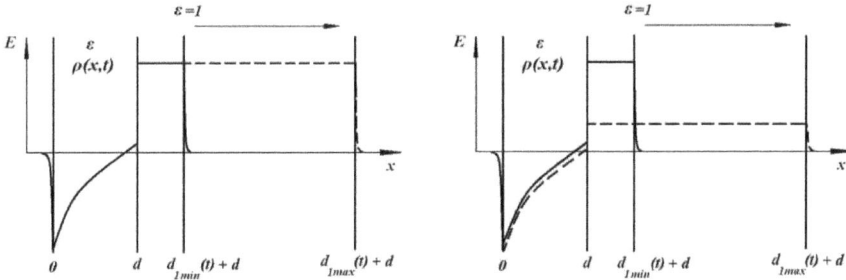

Figure 3. The schematic of the field distribution in the structure metal – dielectric with built-in charge – air gap – moving electrode at the initial state (solid lines) and in the phase of maximal plates shift (dashed lines): (a) – at the conditions of open circuit (voltage generator, fig.1c), and (b) – closed circuit (current generator, fig. 1b).

When the initial state is violated, i.e., the value of $d_1(t)$ is changed by mechanical forces, a redistribution of $E(x, t)$ occurs in each layer, accompanied by the current flow in the load circuit connecting the electrodes. When the current passes through the resistance R, the energy characterizing the energy parameters of the generator is released.

At the moment, the class of generators of electrical energy, which operate on the basis of the above mentioned principle, is well known; they are implemented in practice and are called electret generators [16]. The dielectric used in such generators belongs to a large group of materials that can retain the surface charge for a long time. The electrets [17] differ in the method of generation of this charge and in the form of the distribution of the stored charge and its sign. It can be either uniformly or nonuniformly distributed over the layer thickness and can have either an identical sign (monoelectret) or two different signs (geteroelectret). The electrets used in energy generators are sufficiently thick layers with the minimum thickness of 5–10 μm, and the information on their fabrication by microelectronics technologies is insufficient.

2.1. Model. Basic equations that describe the effect of energy generation in electrostatic machines with a dielectric containing an embedded charge

To describe the general features of operation of the generator considered, we assume that the dielectric contains a charge $\rho(x, t)$, which can change with time, with a surface density $Q_P(t) = \bar{\rho}(t)\bar{x}(t)$, where $\bar{\rho}(t)$ is the density of the space charge $\rho(x, t)$ averaged over the dielectric thickness and $\bar{x}(t)$ is its centroid.

Analyzing the behavior of the total current in this structure with variations of $d_1(t)$ for given values of d, $\bar{\rho}(t)$, $\bar{x}(t)$ we use a system of the classical one-dimensional equations consisting of the expressions for the total current $j(t)$ and conductivity current $j_c(x, t)$, equation of continuity, Poisson's equation, and expression that determines the potential V(t) between the electrodes at each instant of time. Taking into account that for electrets the space charge is constant for all time of the process, we have:

$$j(t) = \varepsilon\varepsilon_0 \frac{\partial E(x,t)}{\partial t} \tag{3}$$

$$\frac{\partial E(x,t)}{\partial x} = -\frac{\rho(x)}{\varepsilon\varepsilon_0} \tag{4}$$

$$\int_0^{d_1(t)+d} E(x,t)dx = -V(t) \ , \tag{5}$$

Integrating both parts of equation (4) with respect to the coordinate x, taking into account that the field increases in a jumplike manner by a factor of ε on the free boundary of the dielectric, i.e.,

$$\varepsilon E(d_-,t) = E(d_+,t) , \tag{6}$$

and substituting the formula derived for $E(x, t)$ into equation (5), we obtain expressions for the field on the boundaries $x = 0$ and $x = d_1(t) + d$: $E_c(t)$ and $E_A(t)$, and, correspondingly, for the specific charge induced on these boundaries:

$$Q_c(t) = -\varepsilon\varepsilon_0 E_c(t) = C(t)\left(V(t) + V_p\right) , \tag{7}$$

$$Q_s(t) = \varepsilon_0 E_A(t) = -C(t)\left(V(t) + V_p\right) , \tag{8}$$

$$C(t) = \frac{\varepsilon\varepsilon_0}{d + \varepsilon d_1(t)} \ , \tag{9}$$

$$V_p = \frac{Q_P}{C_F} \ , \tag{10}$$

where $C_F = \varepsilon \varepsilon_0 / d$ is the specific capacitance of the dielectric layer.

Therefore, according to equations (3) and (8):

$$j(t) = \frac{dQ_S(t)}{dt} = -\frac{d}{dt}\left(C(t)(V(t) + V_p)\right) \qquad (11)$$

When the electrodes are connected via the load R and in the case of open circuit (as it is shown in Fig. 1c with the switch in position 3 and with the switch in position 2), the voltage behavior in time is described by following equations:

$$\frac{d}{dt}\left[C(t)\left(V(t) + V_p\right)\right] = -\frac{V(t)}{R} \qquad (12)$$

$$\frac{d}{dt}\left[C(t)\left(V(t) + V_p\right)\right] = 0 \qquad (13)$$

These equations with the corresponding initial conditions describe all possible regimes of operation of the electrostatic generator shown in Fig. 1 b,c.

To study specific features of its operation, we choose (without loss of generality) a sine law of variation of the gap size:

$$d_1(t) = d_{10}(1 + \alpha + \sin(\omega t)) \ , \qquad (14)$$

$\omega = 2\pi f$, $f = 1/T$, T is the conversion cycle duration.

As has been shown in Introduction two types of generator construction, depending on the method of commutation of the switch (see Fig. 1 b,c), are possible in the case of motion of the moving electrode in the field of the space charge (or polarization) Q_P located in the dielectric. They have been called the voltage generator and the current generator.

2.2. Voltage generator in vibration mode

In such a generator, the output voltage is amplified compared to the case of current generator, which will be analized below in the section 2.3, in the following manner. At the initial state the capacitor electrodes are short-circuited by commutation of the switch to position 1 at the instant when the minimum distance between the surfaces of the moving electrode and dielectric is reached (see Fig. 1c and Fig.2b, the capacitance $C(t)$ has the maximum value at this instant). At the beginning of the process of electrodes separation the switch is turned to position 2; at the instant when the maximum value of $d_1(t)$ is reached, the switch is turned to position 3 and the energy worked out during the cycle is transferred to the load R.

Let us analyze the effect of voltage amplification in more detail. Under conditions of electrode motion with a non-closed circuit ($j(t) = 0$) and according to equation (13), we have

$$C(t)\left(V(t) + V_p\right) = const \ , \qquad (15)$$

then

$$C_{max}\left(V_{min}+V_p\right)=C_{min}\left(V_{max}+V_p\right). \tag{16}$$

Therefore, we obtain

$$V_{max}=\frac{C_{max}}{C_{min}}\left(V_{min}+V_p\right)-V_p \tag{17}$$

Let the process of vibrations begin from the phase of the maximum convergence of the surfaces; then, in accordance with the initial condition $V(0)=V_{min}(0)=0$, after the first displacement of the moving electrode to the maximum distance under the condition of open circuit (switch in position 2), we have

$$V_{max}=V_p\left(\frac{C_{max}}{C_{min}}-1\right) \tag{18}$$

In this case, the following amount of energy is produced:

$$W=\frac{C_{min}V_{max}^2}{2}=\frac{C_{min}V_p^2}{2}\left(\frac{C_{max}}{C_{min}}-1\right)^2\approx\frac{V_p^2}{2}\frac{C_{max}^2}{C_{min}} \tag{19}$$

The principle of mechanical energy conversion to electrical energy with electrode motion under the condition of an open circuit is illustrated by the distribution of the electrical field in the structure (Fig. 3a) both in the initial phase of motion $d_1=d_{1min}=\alpha d_{10}$, and in the phase of the maximum distance of the electrode (anode) $d_1=d_{1max}=2d_{10}+d_{1min}$. As the electrode moves, the field in the gap remains constant, the total current equals zero, and the electrical field energy increases in accordance with the increase in the area under the curve $E^2(x)$, which is manifested as an increase in the difference in the potentials V between the electrodes (Fig. 4). The energy is transferred to the load R when the switch is turned to position 3. This transfer is efficient if R does not exceed the value $1/(2\bar{f}\bar{C}S)$, where $\bar{C}$ is the cycle-averaged specific capacitance of the structure, S is the electrode area, and f is the frequency of vibrations.

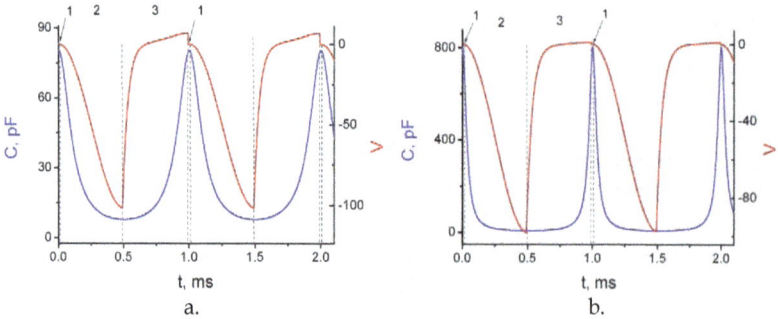

Figure 4. Behavior of $C(t)$ and $V(t)$ in a voltage electret generator based on (**a**) – dielectric: $Q_P=10^{-3}$ C/m², $\varepsilon=10$, and (**b**) – ferroelectric: $\varepsilon=1000$ and $Q_P=10^{-2}$ C/m². $d=1$ μm, $d_{10}=0.5$ μm, $d_{1min}=10$ nm, $S=1$ mm². $R=10$ MΩ

In contrast to the capacitance machine where the energy transfer to the load is finalized at the end of the cycle by complete discharge of the capacitor $C(t)$ to $V(t) = 0$, the charge induced in the electrodes and screening the field of the space charge in the dielectric flows between the electrodes through the load R during the electrode motion in the generator considered here. This process is unsteady and is determined by several constants: instantaneous value of $RC(t)S$, time of the Debye screening of the charge in the metal (which is the smallest time), and time of motion of the moving electrode during the half-period. At certain times, depending on the relation between these time constants, the total current can change its direction, when the voltage also changes its sign at the instant of the maximum approaching of the surfaces ($d_1(t) \rightarrow d_{1min}$) (see Fig. 4). To eliminate this effect, the switch (Fig.1c) is turned to position 1 at the beginning of each next cycle; then, the initial voltage $V_{min} = 0$ is recovered, and the process is repeated completely.

Despite a principally different method of capacitance recharging, this circuit of energy conversion is similar to the capacitance machine (Fig. 1a), except for the fact that the field of the built-in charge inducing the voltage V_P serves here as the voltage source.

An universal program that takes into account all parameters of the structure and energy generation modes in the cases of one- and two-capacitor generator was developed for the numerical analysis of the problem. A difference scheme with automatic choice of the time step was used; this scheme ensured solution stability and specified accuracy.

One example of such a solution, which illustrates voltage generation between the electrodes for a particular dielectric with a low value of ε, is shown in Fig. 4a. The numbers on the axis t characterize the position of the switch (see Fig. 1c). It is seen from the figure that the role of this switch is the recovery of the initial state of the system ($V_{min} = 0$) in the phase when $C(t)=C_{max}$ in each cycle of energy generation. Such synchronization allows us to obtain the voltage amplification proportional to the capacitance modulation depth in accordance with (18); the power increases thereby in accordance with (19).

If a dielectric with a high value of ε is used in the generator, the capacitance modulation C_{max}/C_{min} increases, but not in proportion to the increase in ε (when the air gap modulation depth is constant), because the value of C_{min} changes only slightly (it is determined by the maximum value of the air gap $2d_{10} + d_{1min}$). In this case, with a fixed polarization Q_P, the value of V_P decreases inversely proportional to ε (10); the value of V_{max} (18) and also the energy generated in one cycle (19) decrease accordingly.

In particular, for the structure parameters used to construct the graphs in Fig. 4a, but with the value of ε increased by a factor of 100, the value of V_P decreases by a factor of 100, but the values of V_{max} and W decreases only by a factor of 10. Therefore, to reach the output voltage and the generated energy comparable with the case of the classical electret generator (with parameters corresponding to those in Fig. 4a), the polarization in the ferroelectric should be increased by ten times (up to 10^{-2} C/m^2) (Fig. 4b). Note that such values of polarization are not critical for a number of known ferroelectrics; therefore, it is possible to increase the amount of energy generated during one conversion cycle by the ferroelectric-based generator; this increase is limited only by the voltage of the breakdown in the gap between the electrodes. Moreover, to increase the energy production, it is possible to use ferroelectrics with low values of ε and high values of spontaneous polarization, for instance, lithium niobate and tantalate (Q_P up to 0.5–0.8 C/m² with $\varepsilon \approx 40$) [39].

2.3. Current generator in vibration out-of-plane mode

The operation principle of the current generator is shown in Fig.1b and the example of its design is presented in Fig.2a, this scheme is the simplest among the other possible ones. In the case of structure capacitance modulation, the generator operates without voltage amplification [16].

There are publications on particular cases of the current generator, for instance, electret microphones [17] in which either the load resistance is small or the amplitude of electrode vibrations is small as compared with the air gap thickness. In this case, in accordance with the analysis performed above, we have $V \ll V_P$, and the current in the circuit is described as

$$j(t) = V_p \frac{dC(t)}{dt} \tag{20}$$

Let us consider the general solution of the problem of current generator operation with an arbitrary load, using equation (12) with the initial condition $V(0) = 0$ and with variation of the gap size in accordance with equation (14).

Equation (12) with provision for (14) is written in dimensionless form as

$$\frac{d\varphi(\tau)}{d\tau} = -\left(\frac{1 + \alpha' + \sin\tau}{RC_1 S\omega} - \frac{\cos\tau}{1 + \alpha' + \sin\tau} \right) \varphi(\tau) + \frac{\cos\tau}{1 + \alpha' + \sin\tau} , \tag{21}$$

where $\varphi = V/V_P$, $\tau = \omega t$, $\alpha' = \alpha + d/\varepsilon d_{10}$, and

$$C_1 = \varepsilon_0/d_{10} \tag{22}$$

is the specific capacitance at average value of the air gap. Therefore, the problem is determined only by two dimensionless parameters: α' and $RC_1 S\omega$.

Note that with a sufficiently large air gap modulation depth, the parameter α' is inversely proportional to the structure capacitance modulation depth.

One example of the numerical solution of equation (21) is shown in Fig. 5. The initial increase in voltage during the first displacement of the moving electrode is similar to its increase in the voltage generator; it is determined by capacitance modulation and by the value of V_P (18). In subsequent periods of electrode motion, a quasi-steady screening charge is formed (it is described above), and the voltage amplitude V_m decreases; this amplitude becomes sign-variable and tends to $\pm V_P$ or to a smaller value, depending on the load resistance R. The time constant of the decrease in V_m is determined by the value of $R\overline{C}S$, where $\overline{C} \leq C_1$ is the capacitance of the structure $C(t)$ (see the inset in Fig. 5) averaged over the period of vibrations.

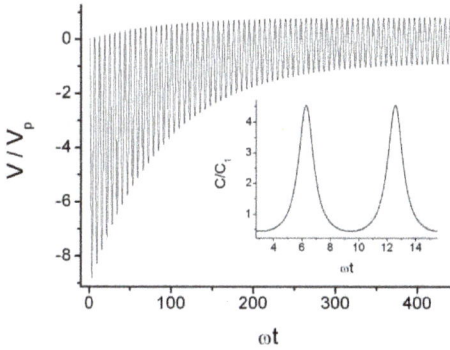

Figure 5. Transient of setting a steady state for the current generator at $\alpha' = 0.11$ and $\omega RC_1S = 1.11 \cdot 10^2$ ($Q_P = 10^{-3}$ C/m^2, $\varepsilon = 10$, $d = 1$ μm, $d_{10} = 1$ μm, $d_{1min} = 10$mn, $S = 1$ mm^2, $R = 20$ GΩ, and $f = 100$ Hz).

The specific features of the behavior of $V(t)$ in the steady regime are shown in Fig. 6. The change in the gap size $d_1(t)$ accompanied by the corresponding change in the capacitance $C(t)$ induces variations of the charge on the electrodes in time $Q_S(t)$; the maximum of this dependence in the general case can be shifted with respect to the peak value of the capacitance $C(t)$. This shift of the peaks of $Q_S(t)$ and $C(t)$ as compared with the classical case of the current generator considered in [16] (for which $V(t) = 0$ and equation (20) is valid) is caused by the delay in redistribution of the screening charge $Q_S(t)$ between the electrodes during current flow through the load R because of the finite value of $RC(t)S$.

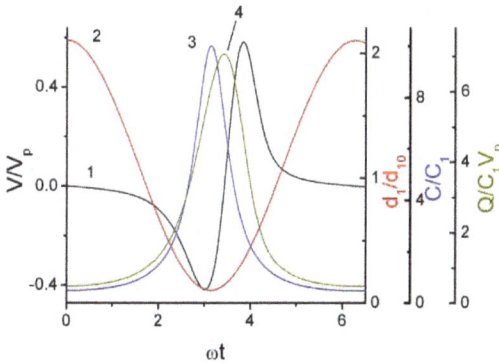

Figure 6. Example of the numerical solution of the equation that describes current generator operation in a steady regime with sinusoidal variations of the gap size d_1: V/V_P (1), d_1/d_{10} (2), C/C_1 (3), and Q_S/C_1V_P (4); $\alpha' = 0.1$ and $\omega RC_1S = 6.25 \cdot 10^{-4}$.

The generator considered is qualitatively different both from the capacitance machine and from the voltage generator in one more operation principle: its operation is determined by changes in the conditions of screening of the electric field in metallic electrodes during the motion of the moving electrode. The electrode recharging current in the circuit of the load R tends to return the system to the equilibrium state with $V = 0$.

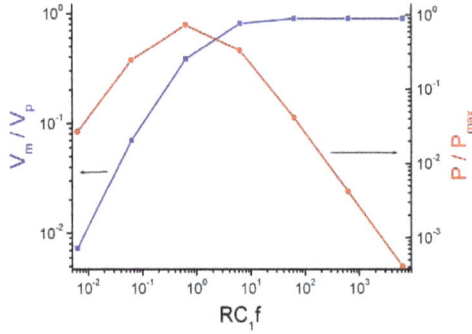

Figure 7. Frequency-load dependences of the produced power and output voltage for a current electret generator at $\alpha' = 0.1$.

If the structure capacitance modulation depth is sufficiently large, the amplitude of the voltage V_m produced by the generator tends (in the case of an optimal load) to the limiting value $\pm V_P$, and the power has the maximum value at the frequency

$$f_1 = 1/RC_1S, \tag{23}$$

as is shown in Fig. 7. Note that the law of current oscillations approaches the sine law as the parameter RC_1Sf increases to values of the order of unity and greater, in contrast to Fig. 6 where $RC_1Sf \ll 1$. Under these conditions, the power produced by the generator is $V_m^2/2R$. In the case of generation of the maximum power P_{max} at sufficiently large η, the value of V_m is close to V_P; thus, we obtain

$$P_{max} = \frac{1}{2}\frac{V_p^2}{R_m} = f\frac{C_1SV_p^2}{2}, \tag{24}$$

where $R_m = 1/(C_1Sf)$ is the load resistance at the generation of P_{max}. Correspondingly, the energy generated during the conversion period is expressed by

$$W_m = \frac{C_1SV_p^2}{2} \tag{25}$$

Thus, in the case of a sufficiently large depth of structure capacitance modulation, the amplitudes of power and the voltage of the current generator are almost independent of the values of C_{max} and C_{min}; they are determined only by the mean capacitance of the gap C_1.

Note that if $C_{max}/C_{min} > 5$ and $C_1/C_F < 0.1$ maximal output voltage approaches to V_P, and the maximum generated energy is given by (25).

2.4. Electret generator operation at lateral displacement of capacitor plates

The case of lateral displacement of capacitor plates under the operation of electret generator in the current mode (see the scematic design in Fig.2c) should be emphasized particularly, because it is realized in practice, see, e.g., so-called „in-plane gap-closing" constructions [32-36] and the rotational systems [29]. This operation mode is described by following equations (see equation (12)):

$$\frac{d}{dt}\left[CS(t)\left(V(t)+V_p\right)\right]=-\frac{V(t)}{R} ,$$ (26)

where the area of the capacitor plates overlap is described as:

$$S(t) = S_{10}(1+\beta+\sin\omega t) ,$$ (27)

and C is a specific capacitance, calculated by eqn. (9), which is constant in this case.

The equations (26) and (27) were solved numerically. The solution is represented in Fig. 8, where P_m is maximum value of power and P_{max} value is calculated using (24). This solution is qualitatively different from one for current out-of-plane electret generator, described above, by the presence of the pronounced dependence of output power on the modulation depth of structure capacitance, see Fig.8b, whereas in the previous case this dependence is practically not observed. Point is that in the case of lateral shift of capacitor plates the total value of polarization decreases along with the capacitance whereas in the previous case of vibration mode electret generator, for which the polarization is constant. The polarization decrease reduces in this time interval the influence of parazitic induced charge on the current, thus resulting in the effect of voltage amplification, as in the case of electret generator operation in the voltage vibratory mode. However, because the influence of the parazitic charge is not totally excluded, the value of maximum power P_m now is not proportional to capacitance modulation depth η (as for the case of capacitance machine) but P_m depends on η according to the logarithmic law, see Fig. 8b.

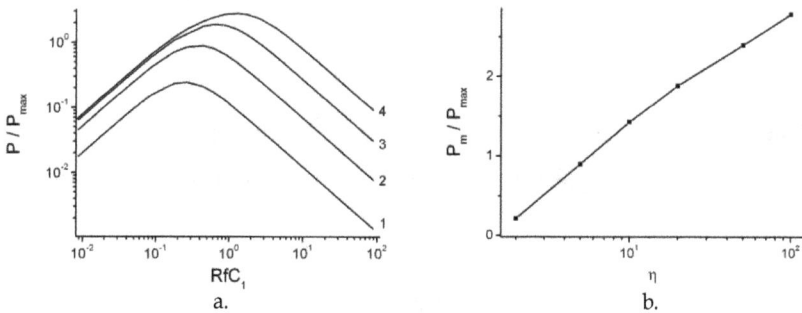

Figure 8. Electret generator with lateral shift of capacitor plates: (a) - frequency-load dependences of the output power for η = 2 (1), 5 (2), 20 (3), 100 (4) and (b) –the dependence of maximum power on capacitance modulation depth.

The drawback of this type of generator is the residual influence of parasitic induced charge mentioned above resulting in the reduction of the efficiency of the generation per unit area at equal parameters compared to out-of-plane generator in a current mode. In particular, the maximal output power of this generator becomes larger than that of its out-of-plane analog only for η>5.6, see Fig.8b. Inability to reach the small values of interelectrode gaps at high enough areas of generator plates should be marked as another drawback. These defects do not permit to reach high values of specific powers at η increase.

Note, that under transition in voltage generator mode the effects similar to those discussed in section **2.2** are observed. Therefore, in this mode the value of P_m/P_{max} will be proportional to capacitance modulation depth, and output power could be increased considerably.

3. Analysis of possible versions of implementation of two-capacitor generator circuits

Various versions of implementation of two-capacitor generators are based on the ideal generator circuit shown in Fig. 1d. They differ only in the method of compensation of charge losses caused by leakage currents in the generator capacitors (Fig. 1e,f).

Such compensation can be provided by an external source of electric energy. In the first case, a current source, i.e., a source that provides constant current and has a sufficiently high (in the ideal case, infinite) internal resistance, is connected to one of the generator capacitors (see Fig. 1e).

In the case of compensation of losses from the voltage source (whose internal resistance is low), it is necessary to use a switch connecting the source to one of the capacitors for a certain time sufficient for capacitor recharging (Fig. 1f).

Moreover, charge losses can be compensated by using an additional low-power generator connected to the input of the basic generator and consuming a minor portion of mechanical energy of the system, for instance, by using the electret effect [16].

It is also possible to compensate charge losses by organizing a feedback for transferring some part of energy from the generator output to its input for charging one of the capacitors, as shown in [10].

3.1. Compensation of charge losses with the use of a current source

The circuit based on this principle is shown in Fig. 1e. Operation of this generator is described by the system of differential equations

$$I(t) = (V_2(t) - V_1(t)) / R;$$

$$I(t) = -I_0 + \frac{d(V_1(t)C_1(t))}{dt} + \sigma_1(t)V_1(t); \qquad (28)$$

$$I(t) = -\frac{d(V_2(t)C_2(t))}{dt} - \sigma_2(t)V_2(t);$$

where $V_1(t)$ and $V_2(t)$ are the drops of voltage on capacitors C_1 and C_2, respectively; $\sigma_1(t)$ and $\sigma_2(t)$ are the conductivities arising owing to leakages in these capacitors (in the general case, they are time-dependent).

Two operation modes of two-capacitor generators are possible: with lateral shift of the plates (with variations of the electrode overlapping area, a particular case is the rotor-type generator) and with vertical out-of plane vibrations of the plates (with variations of the interelectrode gap, vibrational mode). To obtain the maximum efficiency of energy generation, the capacitor plates move in the opposite phases in both cases.

3.1.1. Two-capacitor generator with lateral shift of the capacitors plates

This operation mode is demonstrated schematically in Fig.2d. In the case with capacitances changes in the opposite phases and with lateral shift of the capacitors plates, their total capacitance is constant:

$$C_1(t)+C_2(t)=C_0 \tag{29}$$

In a particular case, when the charge leakage is proportional to the electrode overlapping area, i.e., with similar changes in the conductivities, we have

$$\sigma_1(t)/C_1(t)=\sigma_2(t)/C_2(t) \tag{30}$$

In the case of constant leakages, we have

$$\sigma_1 = \sigma_2 = 1/r \,, \tag{31}$$

and for both cases

$$\sigma_1(t)+\sigma_2(t)=\sigma_0 \tag{32}$$

System (28) was solved numerically. In the dimensionless form system (28) was formulated in [5], and its solution is determined only by two dimensionless parameters characterizing the load properties of the system (fRC_0) and the charge losses due to leakage currents (fC_0/σ_0). By solving this system, we determined the voltages V_1 and V_2 on the capacitors C_1 and C_2, and also the corresponding charges $Q_1(t)$ and $Q_2(t)$, and the total charge $Q_\Sigma(t)$, the current $I(t)$ flowing through the load resistance R and the power released in this load resistance P, averaged over the time of the cycle of energy transformation T.

The system of equations that describes operation of the two-capacitor generator was analyzed in [5], where numerical solutions were obtained for the energy generation efficiency for various methods of excitation of shift vibrations of electrode grates. The output power generated by the generator, however, was not analyzed, and no analytical estimates were obtained.

Let us estimate the value of the maximum energy generated by this generator during one conversion cycle and, correspondingly, the power. As the first approximation, we consider the ideal two-capacitor generator (see Fig. 1d), which ensures minor leakages; therefore, recharging of the capacitors (e.g., from a current source) is not needed. Let the generator capacitors be initially charged to a voltage V_0. Taking into account equation (29), we obtain

$$Q_0 = C_0 V_0 = \text{const}, \tag{33}$$

where $Q_0 = Q_\Sigma(0)$ is the total initial charge accumulated on the capacitors. As there are no charge leakages in this case, the charge is retained during the entire time of generator operation.

Under the conditions described above, there is an initial segment of current relaxation with the characteristic time constant of the order of RC_0; during this time a dynamically equilibrium mode of generation is established owing to charge redistribution on the capacitors. The behavior of voltages on the capacitors depends in this case on the initial phases of $C_1(t)$ and $C_2(t)$. Other conditions are also possible, for instance, a gradual smooth increase in the amplitude and frequency of capacitance oscillations, which is closer to reality. We do not analyze this case in detail here, because the same dynamically equilibrium mode is established for all initial conditions.

Taking into account equations (1), (29) and (33) and also that $Q_0 = C_{min}V_{max} + C_{max}V_{min}$, it is easy to get:

$$\frac{V_{min}}{V_0} = \frac{1 + 1/\eta}{2}, \tag{34}$$

As both capacitors participate here in energy conversion, we can easily show that the energy W_2 produced during one conversion cycle is

$$\frac{W_2}{W_0} = \frac{P}{P_0} = \frac{\eta^2 - 1}{2\eta}, \tag{35}$$

$$W_0 = \frac{C_0 V_0^2}{2} \tag{36}$$

is initial energy accumulated in the capacitors and P is the power of the two-capacitor generator and $P_0 = W_0 f$. A comparison with one-capacitor generator (Fig.1a, energy W_1) results in the following expression

$$\frac{W_2}{W_1} = \frac{1}{2}\left(\frac{1}{\eta} + 1\right)^2 \tag{37}$$

For $\eta \gg 1$ we have

$$W_2 / W_1 \approx 1/2 \tag{38}$$

Thus, at identical initial voltages, the power provided by single-capacitor generator is twice as high as the maximum power of the two-capacitor generator. However, for the case of one-capacitor generator the charge is completely consumed in each cycle of energy conversion, and it should be renewed, which makes this method of energy conversion more difficult in many cases. The power of the two-capacitor generator (with identical initial voltages) is lower because of the non-optimum incomplete charging of the capacitors C_i under the initial conditions mentioned above.

Solving system (28) with condition (29) numerically in the absence of charge leakages, we found the voltages on the capacitors V_i and then determined the current in the load resistance, the charge on the capacitors, and the generator power P. The dependence of P/P_0 on fRC_0, where $P_0 = fW_0$, is shown in Fig. 9.

The dependence of the maximum power P_{max}/P_0 on the modulation factor η (Fig. 10, curve 1) for the ideal generator is almost linear in the interval $\eta \gg 1$ and is adequately described by (35) and curve 3 in Fig. 10. At small values of η, $1 < \eta < 3$, there are significant deviations from equation (35), because $V_{max}/V_{min} < \eta$ in this case and equation (1) is invalid.

With our method of normalization used here the curves in Fig. 9 depend only on the capacitance modulation factor. The curves in Fig. 10 are independent on the absolute parameters of the model, i.e., they have a universal character and describe all possible solutions of system (28) for the ideal generator case. From this viewpoint, we called them the "characteristic" curves.

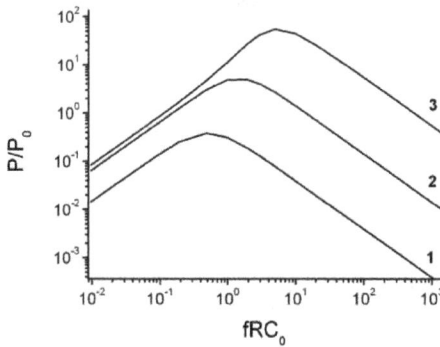

Figure 9. Characteristic curves of the generated power for the ideal generator: $\eta = 2$ (1), 10 (2), and 100 (3).

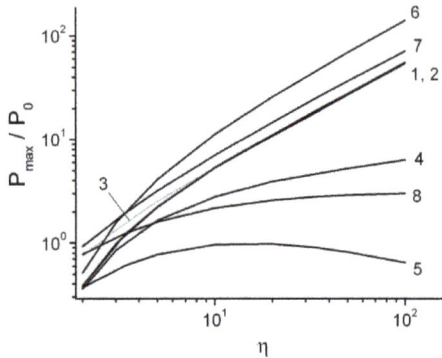

Figure 10. Characteristic curves of the maximum generated power P_{max}/P_0 versus the capacitance modulation factor for the following operation modes: ideal generator with lateral shift of the plates (1), lateral shift of the plates with modulation of charge leakages synchronized with capacitance modulation (2), lateral shift of the plates with constant leakages (4), out-of-plane antiphase vibrations of the plates (vibration generator) (5), lateral shift of the plates with recharging the capacitor C_1 from a voltage source under conditions $C_1(0) = C_{max}$ (6), out-of-plane antiphase vibrations of the plates for the ideal generator at $C_1(0) = C_{max}$ (7) and at $C_1(0)=C_2(0)$ (8), and analytical estimate for the ideal generator from equation (13) - (3).

An example of the numerical solution of (28) illustrating the operation of the two-capacitor generator in the absence of charge losses and with a set of parameters corresponding to its peak power is shown in Fig. 11.

Sinusoidal antiphase oscillations of the capacitors C_1 and C_2 lead to anharmonic oscillations of the charges Q_i potentials V_i, and also of the current I in the load resistance R. The greater the amplitude of oscillations of the values of V_i, Q_i, and I, the greater the generator power or its normalized value P/P_0 determined as W/W_0, i.e., the ratio of the energy produced by the generator during the period of oscillations to the initial energy accumulated on the capacitors:

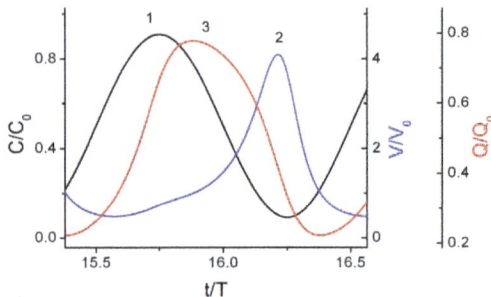

Figure 11. Time evolution of the capacitance (1), voltage (2), and charge (3) for one of the generator capacitors at the vicinity of optimum power ($\eta = 10$ and $fRC_0 = 1.9$).

$$\frac{P}{P_0} = \frac{W}{W_0} \, ,\tag{39}$$

$$W = \int_{t_1}^{t_1+T} I^2(t)R\,dt \, ,\tag{40}$$

$P = Wf$ is the power.

Note that the results of the numerical analysis support the above-formulated approximation (1), from which it follows that the charges Q_i at the maximum and minimum values of the capacitance are equal, i.e., $C_{min}V_{max} = C_{max}V_{min}$ (see Fig. 11, curve 3).

At sufficiently high frequencies ($f \gg 1/RC_0$), the capacitors C_1 and C_2 do not have enough time to exchange the charge during one cycle of energy conversion, which reduces the charge modulation factor on each capacitor in the dynamically equilibrium mode. Therefore, the energy W generated during the cycle becomes smaller than the limiting value W_2. On the other hand, at low frequencies ($f \ll 1/RC_0$), energy conversion is also ineffective, because the charge passes to the second capacitor under these conditions faster than the capacitance of the generating capacitor reaches the minimum value. The charge on the capacitors "tracks" the changes in the capacitance. Thus, a typical feature of two-capacitor generators is the optimum of the normalized power P/P_0 in the frequency range $f \sim 1/RC_0$, which is consistent with the results of the numerical analysis (see Fig.9). Under the optimum generation conditions, as the capacitance of the generating capacitor (in which energy conversion occurs in the time interval considered) decreases, a significant portion of the charge flows to the other capacitor, thus, recovering the state corresponding to the beginning of generation on this capacitor. As the capacitance modulation factor η increases, the generator power P/P_0 also increases, and its peak is shifted toward higher frequencies.

Let us consider the operation of the two-capacitor generator taking into account the charge losses due to leakage currents and its compensation from an external current source. In the equivalent circuit shown in Fig. 1e, the charge losses are shown as conductivities $\sigma_i(t)$ connected in parallel to the capacitors $C_i(t)$. A d. c. current source I_0 is used for compensation of these losses. The operation of such a generator is described by the system of differential equations (28) with initial conditions corresponding to the steady state of the system with the current I_0 flowing in the circuit.

Solving this system numerically, we determined the dimensionless values of the potentials $y(x)$ and $z(x)$ on the capacitors C_i and then the quantities characterizing the generator operation.

Let us first analyze the case with capacitance and conductivity modulation in accordance with an identical sine law, i.e., when the conditions (29) and (30) are satisfied.

Such modulation of conductivities is typical for real situation when leakages are proportional to the area of overlapping of the capacitors plates. Subtracting the corresponding components of the third equation of system (28) from the left and right sides of the second equation of the same system and taking into account equations (29,30,32) we obtain the expression of total charge:

$$\frac{dQ_\Sigma(t)}{dt} = -Q_\Sigma(t)\frac{\sigma_0}{C_0} + I_0 \qquad (41)$$

Using the initial condition that describes the charge of the capacitors from the current source I_0 in the steady state: $Q_\Sigma(0)=I_0C_0/\sigma_0$ we can easily show that equation (41) has only one unique solution (Fig.12, curve 1)

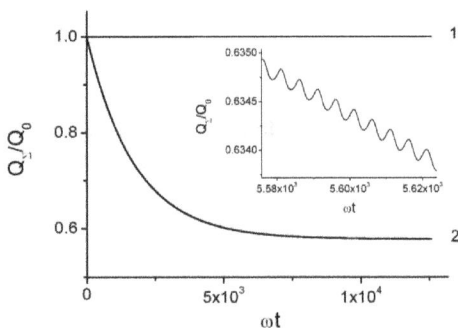

Figure 12. Time evolution of the total charge of the generator capacitances at $fRC_0 = 9.6$ and $\eta = 10$: modulation of leakages synchronized with modulation of the capacitances, equation (18)), - (1), and constant leakages (2) ($frC_0 = 9.6\ 10^2$). The inset shows a zoomed-in fragment of curve 2.

$$Q_\Sigma(t) = Q_\Sigma(0) = \frac{I_0 C_0}{\sigma_0} \qquad (42)$$

Then, all estimates of the maximum energy produced by the ideal generator during the period of energy conversion and the estimates of the generator power are valid at a certain effective value of the initial voltage

$$V_0^* = I_0 / \sigma_0 \qquad (43)$$

As an example, Fig. 10 shows the characteristic curve 2 of P_{max}/P_0 as a function of η, which completely coincides with curve 1 for the ideal generator. In the general case, the value of V_0^* is not equal to the real value of the initial voltage $V(0)$, because the value of $V(0)$ depends on the initial phase of oscillations, i.e., on particular values of the conductivities σ_1 and σ_2 at the time $t = 0$.

The second case also observed in practice is the case with constant leakages:

$$\sigma_1(t) = \sigma_2(t) = \sigma_0 / 2 \qquad (44)$$

A significant difference of this solution from the case of negligibly small leakage currents and also from the case of proportionality of the conductivity σ_i to the electrode overlapping area considered above is the initial decrease in the total charge Q_Σ in time (see Fig. 12, curve 2) and its low-amplitude oscillations (see the inset in Fig. 12) in accordance with the period of changes in the capacitances C_i: $T = 2\pi/\omega$. This effect is explained by the increase in the leakage currents in each period owing to the increase in the potentials V_i on the capacitances, which leads to a certain decrease in the charges Q_i (later on, the charges Q_i again increase when the potentials V_i pass through their minimum values owing to recharging from the source I_0). At the initial stage of the process, the leakage currents averaged over the cycle of generation are greater than the source current I_0, and the discharge of capacitors takes place. For this mode, the steady-state value of the charge $Q_\Sigma(\infty)$ cannot be estimated analytically; therefore, the decay of the charge in time was analyzed numerically: it grows with increasing of both η and the absolute value of the capacitance. In most realistic cases, however, the decrease in the total charge $k = Q_\Sigma(0)/Q_\Sigma(\infty)$ is not more than a factor of 2.

Assuming that the total charge decreases by a factor of k and taking into account the expression for the initial charge (33), we can easily obtain expressions for steady-state values of the minimum voltage on the capacitors $V_{min}(\infty)$ and the energy produced by the generator $W(\infty)$. These formulas are completely identical to the expressions for the ideal generator (34-36) with V_0 being replaced by $V(\infty) = I_0/(k\sigma_0)$.

Thus, a typical feature of the two-capacitor generator with constant charge leakages is an additional decrease in the generated voltage by a factor of k and, correspondingly an additional decrease in power by a factor of k^2, as compared with the above-described case where the leakages are proportional to the electrode overlapping area (30). The dependence of the power P/P_0 on frC_0 is qualitatively similar to the corresponding curves for the ideal generator with the only difference that it is additionally affected by the leakage currents (parameter frC_0): it increases linearly with decreasing leakages at a constant current I_0. In our scales, the value of P/P_0 is almost independent of frC_0 if the leakages are sufficiently small: $r \ll R$ (Fig. 13). As for the ideal generator, we see that P_{max}/P_0 increases with increasing capacitance modulation factor η and then reaches a "plateau" (see Fig. 10, curve 4). The frequency dependence of the power is qualitatively similar to the corresponding characteristic of the ideal generator.

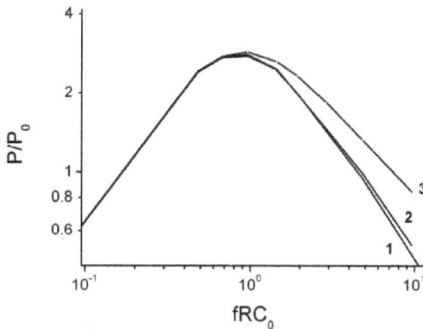

Figure 13. Characteristic curves of the generator power for different leakage resistances r: $frC_0 = 960$ (1), 96 (2), and 9.6 (3); $\eta = 10$.

The decrease in the total charge (see Fig. 12) and the generated power, which is determined by the charge redistribution after generator actuation at the beginning of modulation of the capacitances C_i, has an exponential character. The time needed for the generator to reach a steady-state mode is inversely proportional to the conductivity of the leakage currents. As energy generation is efficient only at $r \gg R$ [5], and the time constant of the decrease in power τ has the order of $rC_0/2$, this time can be sufficiently large (more than 10^2–10^3 s). Note that the quasi-steady mode of generator operation is not reached under real conditions (e.g., in the regime of harvesting the energy of microvibrations of environment) where the modulation frequencies change in a shorter time than τ; therefore, the maximum of the generated power lies between curves 1 and 4 in Fig. 10.

3.1.2. Two capacitor generator in vibration out-of-plane mode

If the two-capacitor generator works at the interelectrode gap modulation mode, when the electrode overlapping area remains constant, called "mode of vibrations", then the condition of the constant total capacitance of the capacitors in time (29) is not satisfied. In this case, the gaps of two capacitors are modulated in opposition in accordance with a sinusoidal law, while the capacitance of each capacitor is inversely proportional to the gap value (Fig. 14, curves 1 and 1'; the quantity C_0' has the meaning of a capacitance averaged over the period of vibrations). Therefore, the capacitance of each capacitor is close to the minimum value during the major part of the period of vibrations T. The greater the capacitance modulation factor η, the more pronounced this effect: when the plates of one capacitor become separated (curve 1, motion toward decreasing $C_1(t)$) and the voltage on this capacitor increases accordingly (curve 2), the charge from this capacitor flows to the second capacitor whose capacitance $C_2(t)$ is still low (curve 1'). Thus, in contrast to the lateral shift of the plates, the charge overflow is not matched with the motion of the plates of the second capacitor: the peak of $V_1(t)$ occurs earlier than the peak of $C_2(t)$, i.e., the charge from the first capacitor flows to the second capacitor mainly during the time when its capacitance is close to the minimum value.

The absence of "synchronization" of the charge exchange between the capacitors in the generator in the mode of vibrations reduces the generator power (see Fig. 10, curve 5, P_0 is determined by equations (36) and (39), in which C_0 is replaced by C_0'), which is manifested as a decrease in the ratio V_{max}/V_0 (see Fig. 14, curve 2). Because of the lack of synchronization, there appears a second peak (with a lower amplitude) on the curve $V_1(t)$ after the peak on the curve $C_2(t)$, which also decreases the efficiency of generation in this mode.

Speaking about the generator in the vibration mode, we should emphasize the ideal generator mode (Fig. 10, curves 7, 8). In contrast to the mode of the lateral shift of the electrodes, the generated energy here depends appreciably on the initial values of the capacitances C_i. In particular, if the plates of one capacitor are located at the minimum distance at the beginning of the vibration process, then the maximum of energy generated in one cycle normalized to its initial value (Fig.10, curve 7) is greater than the value typical for the generator whose operation principle is based on the shift of the plates (Fig.10, curve 1). In

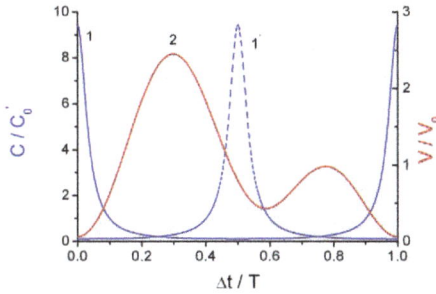

Figure 14. Time evolution of the capacitances of the first (1) and second (1′) capacitors and the voltage (2) for the first capacitor of the generator in the optimum power region. Mode of out-of-plane antiphase vibrations, $\eta = 10$, and $fRC_0 = 0.854$.

this case, the ultimate power for two capacitor generators is reached, because it is possible to ensure the minimum possible gaps between the electrodes of the generator capacitors and, therefore, the maximum possible values of the capacitance. However, if the beginning of the vibration process does not coincide with the instant when the maximum capacitance of one of the capacitors is reached, then the energy generation efficiency drastically decreases (Fig.10, curve 8). This behavior of the generated power is explained by the magnitude of the charge trapped at the beginning of the process: if one of the capacitances has the maximum value at the beginning of the process (Fig.10, curve 7), then the initial charge also reaches the maximum value. Under different initial conditions, the smaller charge is trapped first (intermediate curves between 7 and 8, Fig.10).

3.2. Compensation of charge losses with the use of a voltage source

The charge losses are compensated with the use of a voltage source by connecting the source for a short time to one capacitor only (see Fig. 1f). The charge losses on the second capacitor are compensated owing to the current flowing through the load resistance R in the process of generation with modulation of the capacitances. Breaux [37] considered another method: the charging of both capacitors. In this case, however, highly accurate synchronization of two switches is needed because even a small delay in commutation of switches leads to significant reduction of the generation efficiency.

The recharging voltage pulse should be applied at $V_1(t) < V_0$ and it should be finished at $C_1(t) = C_{max}$. An example of the optimum synchronization of the switch connecting the source V_0 in agreement with the capacitance modulation periods for the case of the lateral antiphase shift of the moving electrodes of the capacitors is shown in Fig. 15. Here the period of charging pulses T_{ch} was less than time constant of charge losses, no pronounced charge decay at $t < T_{ch}$ was observed, see Fig.15b.

When a series of recharging pulses is applied to compensate charge losses one can see the additional effect of charge growth by a factor of two compared to the case of ideal generator (Fig.15b) and corresponding growth of V_{min} value up to $V_{min} = V_0$ (Fig.15a).

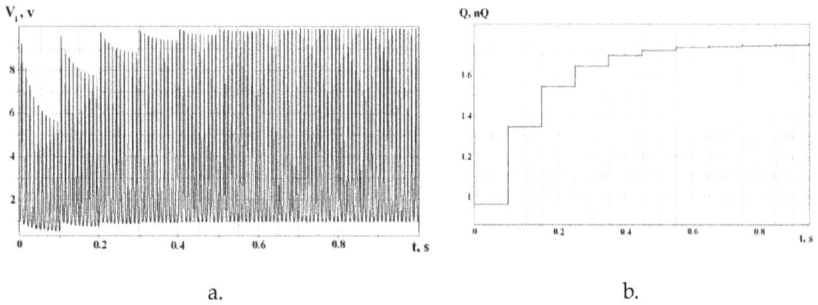

a.   b.

Figure 15. Example of solving the problem of compensation of charge losses by recharging the capacitor from a voltage source. Mode of the lateral shift of the capacitor plates: (a) voltage on the first capacitor; (b) total charge. Recharging pulse duration 10 μs, amplitude 1 V, supply period 100 ms, number of cycles per the period $N = 10$, frequency 100 Hz, $C_{min} = 87.6$ pF, $\eta = 10$, $R = 2 \cdot 10^8$ Ω, and $r = 10^{12}$ Ω.

Therefore, in this case the maximum energy transferred to the load resistance is greater than the energy of ideal generator by a factor of 4, and of single-capacitor generator (2) by a factor of 2; in the limit (at $\eta \gg 1$) it tends to the value $C_{max}V_0^2 \, \eta$ (see Fig. 10, curve 6).

Thus, the use of a switch performing synchronous recharging of the capacitor C_1, in addition to compensation of charge losses, increases the charge to the limiting value $2C_{max}V_0$, which involves an increase in the generated energy up to values exceeding the energy generated by the ideal generator by a factor of 4. Note that similar features are also observed for the two-capacitor generator in vibration out-of-plane mode, leading to an even more dramatic increase in the generator power in this case (cf. curves 5 and 6 in Fig. 10).

4. Experimental studies of two-capacitor rotational generator

To prove the possibility of electric energy generation under the action of mechanical forces with highly efficient utilization of the charge injected into the generator, we performed experimental studies using a macroscopic model of two-capacitor generator consisting of two stator plates and rotor plate located exactly between them. Each plate was metalized and divided into 12 sectors in such a way as to provide the central plate with the two series of the connected capacitors modulated in antiphase when central plate is rotating. The area of the electrodes was 25 cm². The moving electrode was fixed on the shaft of a d. c. motor rotated with a frequency of the order of 1–50 Hz, therefore, the capacitor modulation frequency was 10–600 Hz. All plates were insulated from the body and shaft of the motor with the use of insulators having a high resistance (above 100 GΩ). The gap between the plates was 100–200 μm, the capacitance C_0 had the order of 250–350 pF, and the modulation factor was $\eta = 1.8$–3.5; these parameters were determined by independent measurements. The current through the load resistance $I(t)$ and the voltage on it $V(t)$ were measured by a digital oscilloscope using the matching circuit.

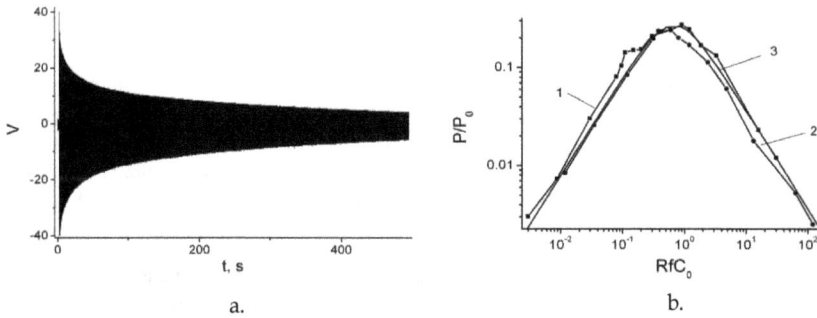

Figure 16. Experimental studies of two-capacitor generator: (a) - relaxation of voltage oscillations in the load resistance (C_0 = 300 pF, η = 1.8, R = 10 MΩ, and f = 400 Hz), (b) - characteristic loading curves of the ideal generator: f = 400 Hz (1), f = 100 Hz (C_0 = 300 pF and η = 1.8) (2), and calculation by the proposed model (3).

The oscillogram characterizing energy generation is shown in Fig. 16a. A charge of 2 10^{-8} C was initially injected at the time t = 0 into this structure by a short pulse of voltage equal to 80 V. After that, the flow of the current of up to 8 μA (acting value) through the load resistance R = 10 MΩ in the process of rotation of the moving electrode with the effective frequency of 400 Hz is determined by this charge. The initial time of voltage redistribution is of order RC_0 = 3·10^{-3} s, after that dynamically equilibrium mode of generation is established. Therefore, the measured initial amplitude of the voltage on the load resistance R (e.g., 40 V, based on the data in Fig. 16a) was used to calculate the power of the ideal generator in this mode on the basis of the proposed experimental model. Initial charging of the capacitors ensures energy generation for a long time (up to 1000 s). During this time, more than 4 10^5 cycles of energy conversion with the use of this charge take place, and the Joule energy released on the load resistance is much greater than the energy spent on initial charging of the generator capacitors. Based on the time constant of charge decay, we can easily estimate the leakage resistance; for the circuit considered, it is approximately 10^{12} Ω. The power developed by the generator with a 80-V starting voltage was 1 mW.

To confirm the main result of the developed generator model, i.e., the universal character of the dependences of the generated power on the load resistance R, we studied the specific features of energy generation at the initial stage of the process when the leakages could be neglected (in this period, the current amplitude is close to its value for the ideal generator). The loading curves plotted in the RfC_0–P/P_0 coordinates for different modulation frequencies of the generator capacitances were found to be almost coincident, i.e., to have a universal character and also to agree well with the model described above (Fig. 16b). Moderate disagreement is explained by a small difference in the modulation factors of two capacitors.

5. Peculiarities of microgenerators operation with vibrations of moving electrode in submicron gap above the surface of ferroelectric-metal structure

A technology of mutual shifting (vibration) of the surfaces of the microcircuit components in the submicron range was developed during the last 5–7 years in modern microelectronics, namely, in its most intensely developing direction: MEMS, e.g., gyroscopes, generators, frequency stabilizers, high-frequency filters. Recently we have demonstrated the possibility of use the nanogaps in electromechanical energy conversion for applications in micromotors and actuators [40-42], which is promising for use in inverse mechanic-to-electricity energy transformation also.

The characteristic feature of operation of electrostatic capacitance microgenerators, the capacitors of that having the structure consisting of substrate – metal - thin ferroelectric layer – moving electrode, is the possibility of creation of high electric field densities in the submicron gap between the surfaces of ferroelectric and moving electrode. Thus, high energy of electric field is stored, which is transformed then into the current. In this structure thin ferroelectric with high ε (more than 1000) plays a role of damping layer to supress a breakdown in the air gap, because the breakdown is controlled here by breakdown field strength of the ferroelectric (more than 10^7 V/m). Because the field strength distribution in the layers of the structure is inversly proportional to their ε ratio then the major portion of voltage is applied to the gap and the field in ferroelectric is much less than its breakdown value, even at high V. Therefore, the breakdown does not occure at high field strength in the gap reaching 10^{10} V/m at the gaps of order 10-100 nm and votages of 100 V, according to our experimental data.

The expressions for electric field strength in the gap E_1 and in the ferroelectric film E, and for the energy stored in the structure are the following:

$$E_1 = \frac{V}{d/\varepsilon + d_1}, \quad E = E_1/\varepsilon, \quad W = \frac{\varepsilon_0 V^2}{2(d/\varepsilon + d_1)} . \tag{45}$$

At $d_1 \gg d/\varepsilon$, the following expressions are always true in the field of parameters listed above:

$$E_1 \approx \frac{V}{d_1}, \quad E \approx \frac{V}{\varepsilon d_1}, \quad W \approx \frac{\varepsilon_0 V^2}{2d_1} = \frac{\varepsilon_0 V}{2} E_1 = \frac{\varepsilon_0 d_1 E_1^2}{2} . \tag{46}$$

Thus the maximum energy generated per one sycle could reach 1-4 J/m^2 during the operation of microgenerators at submicron gaps. At low frequencies of order 10-100 Hz the output power could reach up to 10-40 mW/cm^2. This estimate is true for all types of capacitance generators discribed above, in spite of a number of differencies in various types of their implementation.

6. Discussion and conclusions

1. The analysis of general laws of operation of microgenerators based on the use of multi-layer structure consisting of electrode – thin dielectric - air gap – moving electrode has

been performed taking into account the oscillatory motions with modulation of both the electrodes overlap area (including a rotational motion) and interelectrode gap.

2. It was shown that the use of the submicron gaps in these microgenerators gives rise to considerable increase of output power. To achieve high energy output of these microgenerators it is necessary to have high values of maximal capacitance of generating element C_{max} and electric field strength in the interelectrode gap. These generators can develop a power of 40 mW/cm² and more for the range of low-frequency vibrations characteristic of the vibrations of environment without the use of voltage sources.

    The closest manufactured analogs of such generators are electret in-plane devices (with lateral shift of moving electrode). They have rather large sizes (about 20*20 mm²) and small specific power (of order 100 μW/cm²) despite the fact that the value of power was considerably increased by means of multiple overlapping of strips in interdigitated comb structure (see [5]) of the generator with high displacements (of about 1 mm) in resonance mode of operation [36]. It should be noted that these devices have large interelectrode gaps (more than 20 μm), which prevent the essential decrease of the sizes of generators to reach the values needed for the microelectronics. Therefore, we believe that the only alternative to solve the problem is the development of out-of-plane (vibration) constructions of generators.

3. The mathematical model of the generators was developed, and the numerical solutions describing the process of generation were derived. The universal type of these solutions was confirmed, and the analytical description of the output maximum power in dependence of capacitance modulation depth was carried out.

    The main parameters controlling the efficiency of electret generator operation were determined to be the ratio of charge built in dielectric layer to its geometric capacitance $V_P = Q_P/C_F$ and the value of mean capacitance of air gap C_1. It was shown that for two-capacitor capacitance generators these parameters are the values of maximum capacitance C_{max} and capacitance modulation depth η.

4. Unlike the one-capacitor prototype it was shown that for these generators it is not necessary to recuperate the charge in each cycle of power generation. For the electret generator there is no need at all to turn on the source to compensate charge losses. However, unlike the electret generator working in the current mode, to enlarge the output power it is necessary to use the switch synchronized with the certain phase of oscillations to short-circuit the plates of capacitor to eliminate the parasitic induced charge.

    There is no need to renew the charge at each cycle of power generation for two-capacitor generator in which the charge serves as "working medium" for production the electric power. It is necessary to compensate only small charge losses arising due to leakage currents in the capacitors.

It was determined that for the charge recuperation there is no need to connect the charge sources to each capacitor, it is enough to connect the source only to one of the capacitors. It can be the current source (connected permanently) or the voltage source (connected for a short period of time in a certain phase of capacitance alteration and periodically after the high enough number of energy transformation cycles). In this case one can use the feedback circuit utilizing a small part of output energy for the recuperation of the initial charge.

Author details

Igor L. Baginsky and Edward G. Kostsov
Institute of Automation and Electrometry, Russian Academy of Sciences, Russia

7. References

[1] Roundy S, Wright P K, Rabaey J (2003) A study of low level vibrations as a power source for wireless sensor nodes. Computer Communs. 26: 1131-1144.

[2] Stephen N G (2006) On energy harvesting from ambient vibration. J. Sound Vib. 293: 409–425.

[3] Miyazaki M., Tanaka H, Ono G, Nagano T, Ohkubo N, Kawahara T (2004) Electric-energy generation using variable-capacitive resonator for power-free-LSI. IEICE Trans. Electron. E87: 549–555.

[4] Mitcheson P D, Miao P, Stark B H, Yeatman E M, Holmes A S, Green T C (2004) MEMS electrostatic micropower generator for low frequency operation. Sensors Actuators. A115:.523–529.

[5] Baginsky I L, Kostsov E G (2002) The possibility of creating a microelectronic electrostatic energy generator. Optoelectronics, Instrumentation and Data Processing. No.1: 89-102.

[6] El-Hami M, Glynne-Jones P, White N M, et al. (2001) Design and fabrication of a new vibration based electromechanical power generator. Sensors Actuators. A92: 335–342.

[7] Meninger S, Mur-Miranda J, Lang J, et al. (2001) Vibration to electric energy conversion. IEEE Trans. Very Large Scale Integration (VLSI) Syst.. 9: 64–76.

[8] Roundy S, Wright P K (2004) A piezoelectric vibration based generator for wireless electronics. Smart Mater. Struct.. 13: 1131–1142.

[9] Du Toit N E, Wardle B L, Kim S-G (2005) Design considerations for MEMS-scale piezoelectric mechanical vibration energy harvesters. Integrated Ferroelectrics. 71: 121–160.

[10] Chen C-T, Islam R A, Priya S (2006) Electric energy generator, ultrasonics, ferroelectrics and frequency control. IEEE Trans. Ultrasonics, Ferroelectrics Freq. Control, 53: 656–661.

[11] Dragunov V P, Kostsov E G (2009) Specific features of operation of electrostatic microgenerators of energy. Optoelectronics, Instrumentation and Data Processing. 45: 234–242.

[12] Beeby S P, Tudor M J, White N M (2006) Energy harvesting vibration sources for mycrosystems applications. Meas. Sci. Technol. 17: R175-R195.

[13] Lueke J, Moussa W A (2011) MEMS-based power generation techniques for implantable biosensing applications. Sensors. 11: 1433-1460.
[14] Mitcheson P D, Sterken T, He C, Kiziroglou M, Yeatman E M, Puers R. (2008) Electrostatic microgenerators. Meas. Cont. 41: 114-119.
[15] Moore A D, Ed. (1973) Electrostatics and its application. Wiley: N. Y. 481 p.
[16] Chang J, Kelly A J, Crowley J M, Ed. (1995) Handbook on electrostatic processes. Marcel Dekker Inc.: N.Y.
[17] Sessler G M, West J E (1987) Electrets, 2nd. ed. Topics in Applied Physics 33. Springer-Verlag.
[18] Chiu Y, Kuo C T, Chu Y S (2007) Design and fabrication of a micro electrostatic vibration-to-electricity energy conventer. Microsys. techn.-micro-nanosys.-inf. storage and processing systems. 13: 1663-1669.
[19] Basset P, Galayko D, Paracha A M, Marty F, Dudka A, and Bouroina T (2009) A bath fabricated and electret-free silicon electrostatic vibration energy harvester. J. Mictomech. Microeng. 19: 115025-115037.
[20] Hoffmann D, Folkmer B, Manoli Y (2009) Fabrication, characterization and modelling of electrostatic micro-generators. J. Micromech. Microeng. 19: 094001-094012.
[21] Kiziroglou M E, He C, Yeatman E M (2009) Rolling rod electrostatic microgenerator. IEEE Trans. Industrial Electron. 56: 1101-1108.
[22] Roundy S, Wright P K, Pister K S J (2002) Micro-electrostatic vibration-to-electricity converters. Proc. IMECE, November 17-22: New Orleans, Louisiana, 1-10.
[23] Baginsky I L, Kostsov E G, Sokolov A A (2010) Electrostatic microgenerators of energy with a high specific power. Optoelectronics, Instrumentation and Data Processing. 46: 580–592.
[24] Potter M D (2004) Electrostatic based power source and method thereof. US patent 6750590 B2.
[25] Naruse Y, Matsubara N, Mabuchi K, Izumi M, Suzuki S (2009) Electrostatic micro power generation from low-frequency vibration such as human motion. J. Micromech.Microeng. 19: 094002-094006.
[26] Grachevski S M, Funkenbush P D, Jia Z, Ross D S, Potter M D (2006) Design and modeling of a micro-energy harvester using embedded charge layer. J. Micromech. Microeng. 16: 235-241.
[27] Mizuno M, Chetwynd P G (2003) Investigation of resonance microgenerator. J.Micromech.Microeng. 13: 209-216.
[28] Okamoto H, Suzuki T, Mori K, Cao Z, Onuki T, Kuwano H (2009) The advantages and potential of electret-based vibration-driven micro energy harvesters. Int.J.Energy Res. 33: 1180-1190.
[29] Boland J, Chao Y H, Suzuki Y, Tai Y C (Jan. 19–23, 2003) Micro electret power generator. Proc. MEMS'03. Kyoto, Japan. 538–541.
[30] Masaki T, Sakurai K, Yokoyama T, Ikuta M, Sameshima H, Doi M, Seki T, Oba M. (2011) Power output enhancement of a vibration-driven electret generator for wireless sensor applications. J. Micromech. Microeng. 21: 104004- 104009.

[31] Ma W, Zhu R, Rufer L, Zohar Yi, Wong M (2007) An integrated floating-electrode electric microgenerator. J. Microelectromech. Sys. 16: 29-37.

[32] Mahmoud M A, El-Saadany E F, Mansour R R (Nov. 29 - Dec. 1, 2006) Planar electret based electrostatic micro-generator. The Sixth International Workshop on Micro and Nanotechnology for Power Generation and Energy Conversion Applications. Berkeley, U.S.A.: 223-226.

[33] Lo H, Tai Y-Ch (2008) Parylene-based electret power generators. J. Micromech. Microeng. 18 104006 – 104014.

[34] Suzuki Y (2011) Recent progress in MEMS electret generator for energy harvesting. IEEJ Trans. 6: 101–111.

[35] Lo H-W, Tai Y-Ch (2009) Electret power generator. US patent 0174281 A1.

[36] Sakane Y, Suzuki Y, Kasagi N (2008) The development of a high-performance perfluorinated polymer electret and its application to micro power generation. J. Micromech. Microeng. 18: 104011-104017.

[37] Breaux O P (1978) Electrostatic energy conversion system. US patent 4127804.

[38] Baginsky I L, Kamyshlov V F, Kostsov E G (2011) Specific features of operation of a two-capacitor electrostatic generator. Optoelectronics, Instrumentation and Data Processing. 47: 100 – 120.

[39] Lines M, Glass A (1977) Principles and Applications of Ferroelectrics and Related Materials. Oxford: Oxford University Press.

[40] Baginsky I L, Kostsov E G (2003) High-energy capacitive electrostatic micromotors. J. Micromech. Microeng. 13: 190–200.

[41] Baginsky I L, Kostsov E G (2004) Electrostatic micromotor based on ferroelectric ceramics. J. Micromech. Microeng. 14: 1569–1575.

[42] Baginsky I L, Kostsov E G (2007) High energy output MEMS based on thin layers of ferroelectric materials. Ferroelectrics. 351: 66–78.

Chapter 5

Electrostatic Conversion for Vibration Energy Harvesting

S. Boisseau, G. Despesse and B. Ahmed Seddik

Additional information is available at the end of the chapter

1. Introduction

"Everything will become a sensor"; this is a global trend to increase the amount of information collected from equipment, buildings, environments… enabling us to interact with our surroundings, to forecast failures or to better understand some phenomena. Many sectors are involved: automotive, aerospace, industry, housing. Few examples of sensors and fields are overviewed in Figure 1.

Transportations	Industry	Houses & buildings
Accelerometer (CEA-Leti)	Pressure sensor (CEA-Leti)	Force sensor (CEA-Leti)

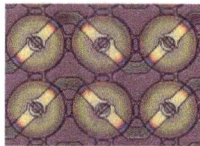

Figure 1. Millions sensors in our surroundings

Unfortunately, it is difficult to deploy many more sensors with today's solutions, for two main reasons:

1. Cables are becoming difficult and costly to be drawn (inside walls, on rotating parts)

2. Battery replacements in wireless sensor networks (WSN) are a burden that may cost a lot in large factories (hundreds or thousands sensor nodes).

As a consequence, industrialists, engineers and researchers are looking for developing autonomous WSN able to work for years <u>without any human intervention</u>. One way to proceed consists in using a green and theoretically unlimited source: ambient energy [1].

1.1. Ambiant energy & applications

Four main ambient energy sources are present in our environment: mechanical energy (vibrations, deformations), thermal energy (temperature gradients or variations), radiant energy (sun, infrared, RF) and chemical energy (chemistry, biochemistry).

These sources are characterized by different power densities (Figure 2). Energy Harvesting (EH) from outside sun is clearly the most powerful (even if values given in Figure 2 have to be weighted by conversion efficiencies of photovoltaic cells that rarely exceed 20%). Unfortunately, solar energy harvesting is not possible in dark areas (near or inside machines, in warehouses). And similarly, it is not possible to harvest energy from thermal gradients where there is no thermal gradient or to harvest vibrations where there is no vibration.

As a consequence, the source of ambient energy must be chosen according to the local environment of the WSN's node: <u>no universal ambient energy source exists</u>.

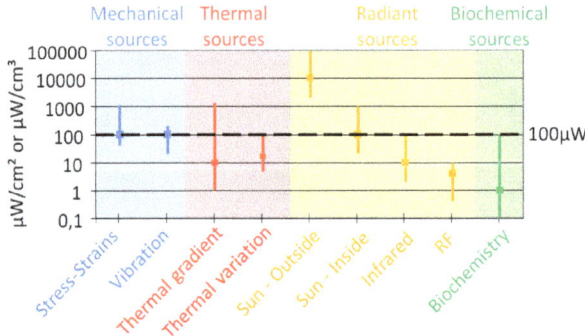

Figure 2. Ambient sources power densities before conversion

Figure 2 also shows that 10-100µW of available power is a good order of magnitude for a 1cm² or a 1cm³ energy harvester. Obviously, 10-100µW is not a great amount of power; yet it can be enough for many applications and especially WSN.

1.2. Autonomous wireless sensor networks & needs

A simple vision of autonomous WSN' nodes is presented in Figure 3(a). Actually, autonomous WSN' nodes can be represented as 4 boxes devices: (i) "sensors" box, (ii)

"microcontroller (μC)" box, (iii) "radio" box and (iv) "power" box. To power this device by EH, it is necessary to adopt a "global system vision" aimed at reducing power consumption of sensors, μC and radio.

Actually, significant progress has already been accomplished by microcontrollers & RF chips manufacturers (Atmel, Microchip, Texas Instruments…) both for working and standby modes. An example of a typical sensor node's power consumption is given in Figure 3(b). 4 typical values can be highlighted:

- 1-5μW: μC standby mode's power consumption
- 500μW-1mW: μC active mode's power consumption
- 50mW: transmission power peak
- 50-500μJ: the total amount of energy needed to perform a complete measurement and its wireless transmission, depending on the sensor and the RF protocol.

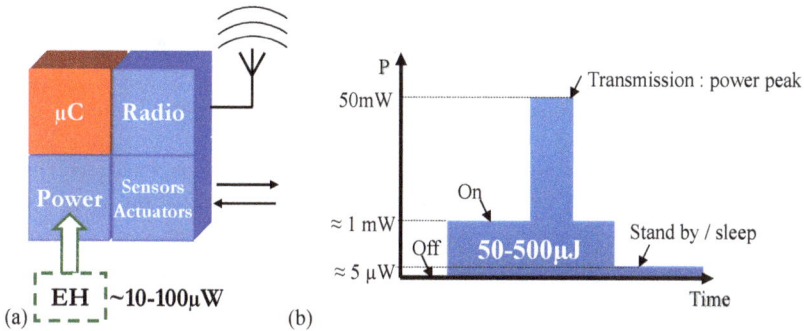

Figure 3. (a) Autonomous WSN node and (b) sensor node's power consumption

Then, the energy harvester has to scavenge at least 5μW to compensate the standby mode's power consumption, and a bit more to accumulate energy (50-500μJ) in a storage that is used to supply the following measurement cycle.

Today's small scale EH devices (except PV cells in some cases) cannot supply autonomous WSN in a continuous active mode (500μW-1mW power consumption vs 10-100μW for EH output power). Fortunately, thanks to an ultra-low power consumption in standby mode, EH-powered autonomous WSN can be developed by adopting an intermittent operation mode as presented in Figure 4. Energy is stored in a buffer (a) (capacitor, battery) and used to perform a measurement cycle as soon as enough energy is stored in the buffer (b & c). System then goes back to standby mode (d) waiting for a new measurement cycle.

Therefore, it is possible to power any application thanks to EH, even the most consumptive one; the main challenge is to adapt the measurement cycle frequency to the continuously harvested power.

As a consequence, Energy Harvesting can become a viable supply source for Wireless Sensor Networks of the future.

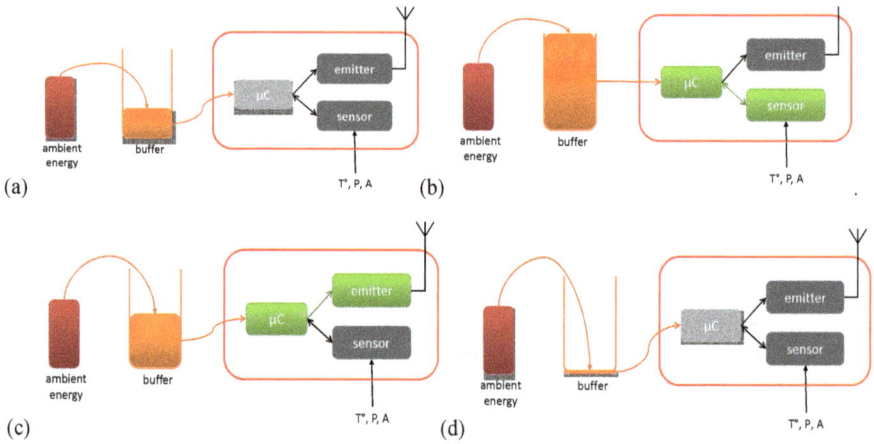

Figure 4. WSN measurement cycle

This chapter focuses on Vibration Energy Harvesting that can become an interesting power source for WSN in industrial environments with low light or no light at all. We will specifically concentrate on electrostatic devices, based on capacitive architectures, that are not as well-known as piezoelectric or electromagnetic devices, but that can present many advantages compared to them.

The next paragraph introduces the general concept of Vibration Energy Harvesters (VEH) and of electrostatic devices.

2. Vibration energy harvesting & electrostatic devices

Vibration Energy Harvesting is a concept that began to take off in the 2000's with the growth of MEMS devices. Since then, this concept has spread and conquered macroscopic devices as well.

2.1. Vibration energy harvesters – Overview

The concept of Vibration Energy Harvesting is to convert vibrations in an electrical power. Actually, turning ambient vibrations into electricity is a two steps conversion (Figure 5(a)). Vibrations are firstly converted in a relative motion between two elements, thanks to a mass-spring system, that is then converted into electricity thanks to a mechanical-to-electrical converter (piezoelectric material, magnet-coil, or variable capacitor). As ambient vibrations are generally low in amplitude, the use of a mass-spring system generates a phenomenon of resonance, amplifying the relative movement amplitude of the mobile mass compared to the vibrations amplitude, increasing the harvested power (Figure 5(b)).

Figure 5(c) represents the equivalent model of Vibration Energy Harvesters. A mass (m) is suspended in a frame by a spring (k) and damped by forces (f_{elec} and f_{mec}). When a vibration occurs $y(t) = Y\sin(\omega t)$, it induces a relative motion of the mobile mass $x(t) = X\sin(\omega t + \varphi)$ compared to the frame. A part of the kinetic energy of the moving mass is converted into electricity (modeled by an electromechanical force f_{elec}), while an other part is lost in friction forces (modeled by f_{mec}).

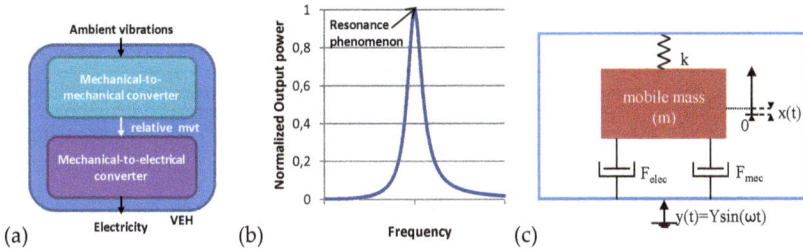

Figure 5. Vibration Energy Harvesters (a) concept (b) resonance phenomenon and (c) model

Newton's second law gives the differential equation that rules the moving mass's relative movement (equation 1). Generally, the mechanical friction force can be modeled as a viscous force $f_{mec} = b_m \dot{x}$. Then, the equation of movement can be simplified by using the natural angular frequency $\omega_0 = \sqrt{k/m}$ and the mechanical quality factor $Q_m = m\omega_0/b_m$.

$$m\ddot{x} + f_{meca} + kx + f_{elec} = -m\ddot{y} \Rightarrow \ddot{x} + \frac{\omega_0}{Q_m}\dot{x} + \omega_0^2 x + \frac{f_{elec}}{m} = -\ddot{y} \qquad (1)$$

Then, when the electromechanical and the friction forces can be modeled by viscous forces, $f_{elec} = b_e\dot{x}$ and $f_{mec} = b_m\dot{x}$, where b_e and b_m are respectively electrical and mechanical damping coefficients, William and Yates [2] have proven that the maximum output power of a resonant energy harvester submitted to an ambient vibration is reached when the natural angular frequency (ω_0) of the mass-spring system is equal to the angular frequency of ambient vibrations (ω) and when the damping rate $\xi_e = b_e/(2m\omega_0)$ of the electrostatic force f_{elec} is equal to the damping rate $\xi_m = b_m/(2m\omega_0)$ of the mechanical friction force f_{mec}. This maximum output power $P_{W\&Y}$ can be simply expressed with (2), when $\xi_e = \xi_m = \xi = 1/(2Q_m)$.

$$P_{W\&Y} = \frac{mY^2\omega_0^3 Q_m}{8} \qquad (2)$$

But obviously, to induce this electromechanical force, it is necessary to develop a mechanical-to-electrical converter to extract a part of mechanical energy from the mass and to turn it into electricity.

2.2. Converters & electrostatic devices – Overview

Three main converters enable to turn mechanical energy into electricity: piezoelectric devices, electromagnetic devices and electrostatic devices (Table 1).

- Piezoelectric devices: they use piezoelectric materials that present the ability to generate charges when they are under stress/strain.
- Electromagnetic devices: they are based on electromagnetic induction and ruled by Lenz's law. An electromotive force is generated from a relative motion between a coil and a magnet.
- Electrostatic devices: they use a variable capacitor structure to generate charges from a relative motion between two plates.

Piezoelectric converters	Electromagnetic converters	Electrostatic converters
Use of piezoelectric materials	Use of Lenz's law	Use of a variable capacitor structure

Table 1. Mechanical-to-electrical converters for small-scale devices

Obviously, each of these converters presents both advantages and drawbacks depending on the application (amplitudes of vibrations, frequencies…).

2.3. Advantages & Drawbacks of Electrostatic Devices

A summary of advantages and drawbacks of electrostatic devices is presented in Table 2. In most cases, piezoelectric and electrostatic devices are more appropriate for small scale energy harvesters (<1-10 cm³) while electromagnetic converters are better for larger devices.

This chapter is focused on electrostatic vibration energy harvesters. These VEH are well-adapted for size reduction, increasing electric fields, capacitances and therefore converters' power density capabilities. They also offer the possibility to decouple the mechanical structure and the converter (which is not possible with piezoelectric devices). Finally, they can be a solution to increase the market of EH-powered WSN by giving the possibility to develop "low-cost" devices as they do not need any magnet or any piezoelectric material that can be quite expensive.

The next paragraph is aimed at presenting the conversion principles of electrostatic devices. It covers both standard (electret-free) and electret-based electrostatic converters.

2.4. Conversion principles

Electrostatic converters are capacitive structures made of two plates separated by air, vacuum or any dielectric materials. A relative movement between the two plates generates a capacitance variation and then electric charges. These devices can be divided into two categories:

- Electret-free electrostatic converters that use conversion cycles made of charges and discharges of the capacitor (an active electronic circuit is then required to apply the charge cycle on the structure and must be synchronized with the capacitance variation).
- Electret-based electrostatic converters that use electrets, giving them the ability to directly convert mechanical power into electricity.

	Piezoelectric devices	Electromagnetic devices	Electrostatic devices
Advantages	-high output voltages -high capacitances -no need to control any gap	-high output currents -long lifetime proven -robustness	-high output voltages -possibility to build low-cost systems -coupling coefficient easy to adjust -high coupling coefficients reachable -size reduction increases capacitances
Drawbacks	-expensive (material) -coupling coefficient linked to material properties	-low output voltages -hard to develop MEMS devices -may be expensive (material) -low efficiency in low frequencies and small sizes	-low capacitances -high impact of parasitic capacitances -need to control μm dimensions -no direct mechanical-to-electrical conversion for electret-free converters

Table 2. Advantages and drawbacks of converters

2.4.1. Electret-free electrostatic converters

These first electrostatic devices are passive structures that require an energy cycle to convert mechanical energy into electricity. Many energy cycles enable such a conversion, but the most commonly-used are charge-constrained and voltage-constrained cycles (Figure 6). They both start when the converter's capacitance is maximal. At this point, a charge is injected into the capacitor thanks to an external source, to polarize it. Charge-constrained and voltage-constrained cycles are presented in the following sub-sections.

1. Charge-constrained Cycle

The charge-constrained cycle (Figure 7) is the easiest one to implement on electrostatic devices. The cycle starts when the structure reaches its maximum capacitance C_{max} (Q_1). In this position, the structure is charged thanks to an external polarization source: an electric charge Q_{cst} is stored in the capacitor under a given voltage U_{min}. The device is then let in open circuit (Q_2). The structure moves mechanically to a position where its capacitance is minimal (Q_3). As the charge Q_{cst} is kept constant while the capacitance C decreases, the voltage across the capacitor U increases. When the capacitance reaches its minimum (C_{min}) (or the voltage its maximum (U_{max})), electric charges are removed from the structure (Q_4).

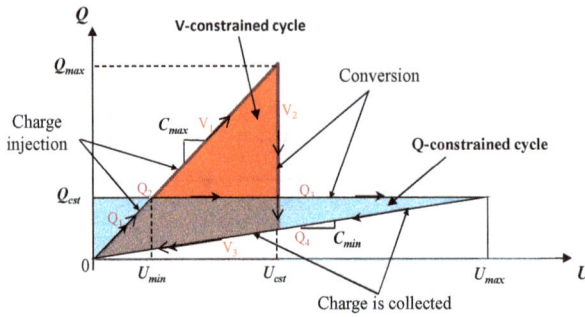

Figure 6. Standard energy conversion cycles for electret-free electrostatic devices

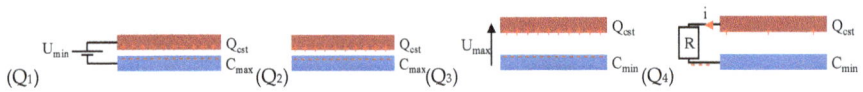

Figure 7. Charge-constrained cycle

The total amount of energy converted at each cycle is presented in equation (3).

$$E_{Q=cte} = \frac{1}{2}Q_{cst}^2 \left(\frac{1}{C_{min}} - \frac{1}{C_{max}}\right) \tag{3}$$

2.   Voltage-Constrained Cycle

The voltage-constrained cycle (Figure 8) also starts when the capacitance of the electrostatic converter is maximal. The capacitor is polarized at a voltage U_{cst} using an external supply source (battery, charged capacitor…) (V_1). This voltage will be maintained throughout the conversion cycle thanks to an electronic circuit. Since the voltage is constant and the capacitance decreases, the charge of the capacitor increases, generating a current that is scavenged and stored (V_2). When the capacitance reaches its minimum value, the charge Q still presents in the capacitor is completely collected and stored (V_3).

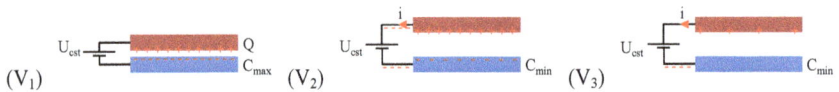

Figure 8. Voltage-constrained cycle

The total amount of energy converted at each cycle is presented in equation (4).

$$E_{U=cte} = U_{cst}^2 (C_{max} - C_{min}) \tag{4}$$

In order to maximize the electrostatics structures' efficiency, a high voltage polarization source (>100V) is required. Obviously, this is a major drawback of these devices as it implies that an external supply source (battery, charged capacitor) is required to polarize the capacitor at the beginning of the cycle or at least at the first cycle (as one part of the energy harvested at the end of a cycle can be reinjected into the capacitor to start the next cycle).

One solution to this issue consists in using electrets, electrically charged dielectrics, that are able to polarize electrostatic energy harvesters throughout their lives, avoiding energy cycles and enabling a direct mechanical-to-electrical conversion. Electrostatic energy harvesters developed today tend to use them increasingly.

2.4.2. Electret-based electrostatic converter

Electret-based electrostatic converters are quite similar to electret-free electrostatic converters. The main difference relies on the electret layers that are added on one (or two) plate(s) of the variable capacitor, polarizing it.

1.   Electrets

Electrets are dielectric materials that are in a quasi-permanent electric polarization state (electric charges or dipole polarization). They are electrostatic dipoles, equivalent to permanent magnets (but in electrostatic) that can keep charges for years. The word **electret** comes from "**electr**icity magn**et**" and was chosen by Oliver Heaviside in 1885.

2.   Definition and electret types

Electret's polarization can be obtained by dipole orientation (Figure 9(a)) or by charge injection (Figure 9(b)) leading to two different categories of electrets:

-   Oriented-dipole electrets (dipole orientation)
-   Real-charge electrets (excess of positive or negative charges on the electret's surface or on the electret's volume)

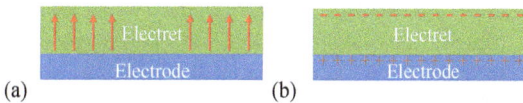

(a)   (b)

Figure 9. Standard electrets for electret-based electrostatic converters (a) dipole orientation and (b) charge injection

In the past, electrets were essentially obtained thanks to dipole orientation, from Carnauba wax for example [3]. Today, real-charge electrets are the most commonly used and especially in vibration energy harvesters because they are easy to manufacture with standard processes.

Indeed, oriented-dipole electrets and real-charge electrets are obtained from very different processes, leading to different behaviors.

a.   Fabrication Processes

The first step to make oriented-dipole electrets is a heating of a dielectric layer above its melting temperature. Then, an electric field is maintained throughout the dielectric layer when it is cooling down. This enables to orient dielectric layer's dipoles in the electric field's direction. Solidification enables to keep dipoles in their position. This manufacturing process is similar to the one of magnets.

As for real-charge electrets, they are obtained by injecting an excess of charges in a dielectric layer. Various processes can be used: electron beam, corona discharge, ion or electron guns.

Here, we focus on charge injection by a triode corona discharge, which is probably the quickest way to charge dielectrics. Corona discharge (Figure 10) consists in a point-grid-plane structure whose point is submitted to a strong electric field: this leads to the creation of a plasma, made of ions that are projected onto the surface of the sample to charge and whose charges are transferred to the dielectric layer's surface.

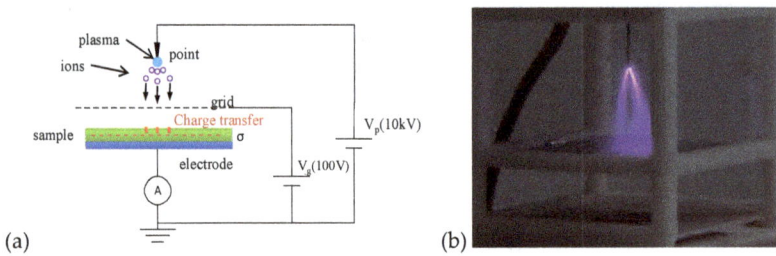

Figure 10. Corona discharge device (a) principle and (b) photo (CEA-LETI)

Corona discharge may be positive or negative according to the sign of the point's voltage. Positive and negative corona discharges have different behaviors as the plasma and the charges generated are different. Obviously positive corona discharges will lead to positively-charged electrets and negative corona discharges to negatively-charged electrets that have also different behaviors (stability, position of the charges in the dielectric). The grid is used to control the electret's surface voltage V_s that results from the charges injected. Actually, when the electret's surface voltage V_s reaches the grid voltage V_g, there is no potential difference between the grid and the sample any longer and therefore, no charge circulation anymore. So, at the end of the corona charging, the electret's surface voltage is equal to the grid voltage.

b.   Equivalent model of electrets

Charge injection or dipole orientation leads to a surface potential V_s on the electret (Gauss's law). It is generally assumed that charges are concentrated on the electret's surface and therefore, the surface potential can be simply expressed by: $V_s = \sigma d / \varepsilon \varepsilon_0$, with ε the electret's dielectric permittivity, σ its surface charge density and d its thickness.

The equivalent model of an electret layer (Figure 11(a) and (b)) is then a capacitor $C = \varepsilon\varepsilon_0 S / d$ in series with a voltage source whose value is equal to the surface voltage of the electret V_s (Figure 11(c)).

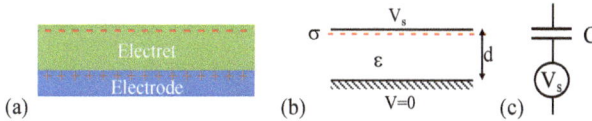

Figure 11. (a) electret layer (b) parameters and (c) equivalent model

c.   Charge stability and measurement

Nevertheless and unfortunately, dielectrics are not perfect insulators. As a consequence, some charge conduction phenomena may appear in electrets, and implanted charges can move inside the material or can be compensated by other charges or environmental conditions, and finally disappear. Charge stability is a key parameter for electrets as the electret-based converter's lifetime is directly linked to the one of the electret. Therefore, it is primordial to choose stable electrets to develop electret-based vibration energy harvesters.

Many measurement methods have been developed to determine the quantity of charges stored into electrets and their positions. These methods are really interesting to understand what happens in the material but are complicated to implement and to exploit. Yet, for vibration energy harvesters, the most important data is the surface potential decay (SPD), that is to say, the electret's surface voltage as a function of the time after charging.

In fact, the surface voltage can be easily measured thanks to an electrostatic voltmeter (Figure 12(a)). This method is really interesting as it enables to make the measurement of the surface voltage without any contact and therefore without interfering with the charges injected into the electret.

Figure 12(b) presents some examples of electrets' SPDs (good, fair and poor stability). Electret stability depends of course of the dielectric material used to make the electret: dielectrics that have high losses (high $\tan(\delta)$) are not good electrets and may lose their charges in some minutes, while materials such as Teflon or silicon dioxide (SiO_2) are known as stable electrets. But, other parameters, such as the initial surface voltage or the environmental conditions (temperature, humidity) have an important impact as well. Generally, for a given material, the higher the initial surface voltage is, the lower the stability becomes. For environmental conditions, high temperatures and high humidity tend to damage the electret stability.

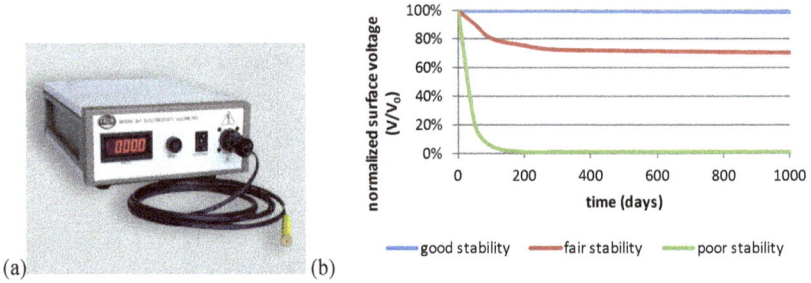

(a)   (b)

Figure 12. (a) Electrostatic voltmeter (Trek® 347) and (b) examples of Surface Potential Decays (SPDs)

The electret behaviors of many materials have been tested. The next sub-section gives some examples of well-known and stable electrets.

d.   Well-known electrets

Teflon [4-7], SiO$_2$ [8-9] and CYTOP [10-14] are clearly the most well-known and the most used electrets in electret-based electrostatic converters. Of course, many more electrets can be found in the state of the art. Table 3 presents some properties of these electrets. It is for example interesting to note that SiO$_2$-based electrets have the highest surface charge densities. As for the stability, it is quite complicated to provide a value. Actually, it greatly depends on the storage, the humidity, the temperature, the initial conditions, the thickness... Yet, the examples given below show a stability $V_{s90\%}$ (90% of the initial surface voltage) generally higher than 2-3 years.

Electret	Deposition method	Maximum thickness	Dielectric Strength (V/µm)	ε	Standard surface charge density (mC/m²)
Teflon (PTFE/FEP/PFA)	Films are glued	Some 100 µm	100-140	2.1	0.1-0.25
SiO$_2$-based electrets	Thermal oxidization of silicon wafers (+ LPCVD Si$_3$N$_4$)	Some µm (<3µm)	500	4	5-10
Parylene (C/HT)	PVD-like deposition method	Some 10µm	270	3	0.5-1
CYTOP	Spin-coating	20µm	110	2	1-2
Teflon AF	Spin-coating	20µm	200	1.9	0.1-0.25

Table 3. Well-known electrets from the state of the art

Added in capacitive structures, electrets enable a simple mechanical-to-electrical energy conversion.

3. Conversion principle

The conversion principle of electret-based electrostatic converters is quite similar to electret-free electrostatic converters and is tightly linked to variations of capacitance. But contrary to them, the electret-based conversion does not need any initial electrical energy to work; a structure deformation induces directly an output voltage, just like a piezoelectric material.

a. Principle

Electret-based converters are electrostatic converters, and are therefore based on a capacitive structure made of two plates (electrode and counter-electrode (Figure 13)). The electret induces charges on electrodes and counter-electrodes to respect Gauss's law. Therefore, Q_i, the charge on the electret is equal to the sum of Q_1 and Q_2, where Q_1 is the total amount of charges on the electrode and Q_2 the total amount of charges on the counter-electrode ($Q_i = Q_1 + Q_2$). A relative movement of the counter-electrode compared to the electret and the electrode induces a change in the capacitor geometry (e.g. the counter-electrode moves away from the electret, changing the air gap and then the electret's influence on the counter-electrode) and leads to a reorganization of charges between the electrode and the counter-electrode through load R (Figure 14). This results in a current circulation through R and one part of the mechanical energy (relative movement) is then turned into electricity.

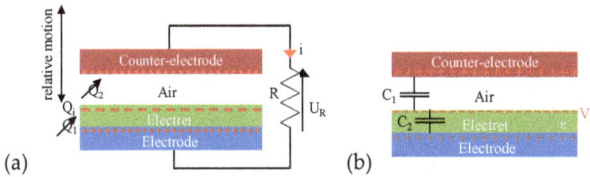

Figure 13. Electret-based electrostatic conversion – Concept

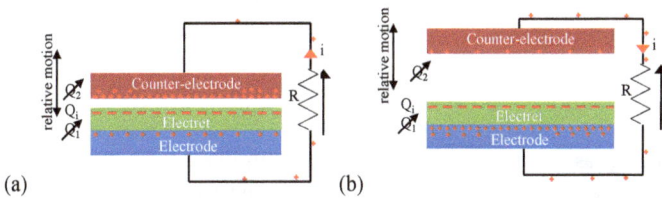

Figure 14. Electret-based electrostatic conversion – Charge circulation

The equivalent model of electret-based electrostatic converters is presented below.

b. Equivalent model and Equations

The equivalent model of electret-based electrostatic converters is quite simple as it consists in a voltage source in series with a variable capacitor. This model has been confronted to experimental data and corresponds perfectly to experimental results (see section 3.2.4).

Figure 15 presents an electret-based electrostatic converter connected to a resistive load R. As the capacitances of electret-based converters are quite low (often lower than 100pF), it is important to take parasitic capacitances into account. They can be modeled by a capacitor in parallel with the electret-based converter. And actually, only 10pF of parasitic capacitances may have a deep impact on the electret-based converter's output voltages and output powers.

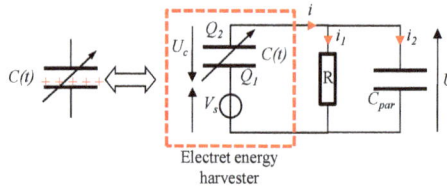

Electret energy
harvester

Figure 15. Electrical equivalent model of electret-based electrostatic converters

In the case of a simple resistive load placed at the terminals of the electret-based converter, the differential equation that rules the system is presented in equation (5).

$$\frac{dQ_2}{dt} = \frac{V_s}{R} - \frac{Q_2}{C(t)R} \qquad (5)$$

Taking parasitic capacitances into account, this model is modified into (6) [15].

$$\frac{dQ_2}{dt} = \frac{1}{\left(1 + \dfrac{C_{par}}{C(t)}\right)}\left(\frac{V_s}{R} - Q_2\left(\frac{1}{RC(t)} - \frac{C_{par}}{C(t)^2}\frac{dC(t)}{dt}\right)\right) \qquad (6)$$

Actually, electret-based converter's output powers (P) are directly linked to the electret's surface voltage V_s and the capacitance variation dC/dt when submitted to vibrations [16]. And, as a first approximation:

$$P \propto V_s^2 \frac{dC}{dt} \qquad (7)$$

As a consequence, as the electret-based converter's output powers is linked to the electret's surface voltage V_s and its lifetime to the electret's lifetime, we confirm that Surface Potential Decays (SPDs) are the most appropriate way to characterize electrets for an application in energy harvesting.

2.4.3. Capacitors and capacitances' models

Whether it is electret-free or electret-based conversion, electrostatic converters are based on a variable capacitive structure. This subsection is focused on the main capacitor shapes employed in electrostatic converters and on their models.

1.   Main capacitor shapes

Most of the electrostatic converters' shapes are derived from accelerometers. Actually, it is possible to count four main capacitor shapes for electrostatic converters (Figure 16).

(a) in-plane gap closing converter: interdigitated comb structure with a variable air gap between fingers and movement in the plane

(b) in-plane overlap converter: interdigitated comb structure with a variable overlap of the fingers and movement in the plane

(c) out-of-plane gap closing converter: planar structure with a variable air gap between plates and perpendicular movement to the plane

(d) in-plane converter with variable surface: planar structure with a variable overlap of the plates and movement in the plane. There is a great interest of developing patterned versions with bumps and trenches facing.

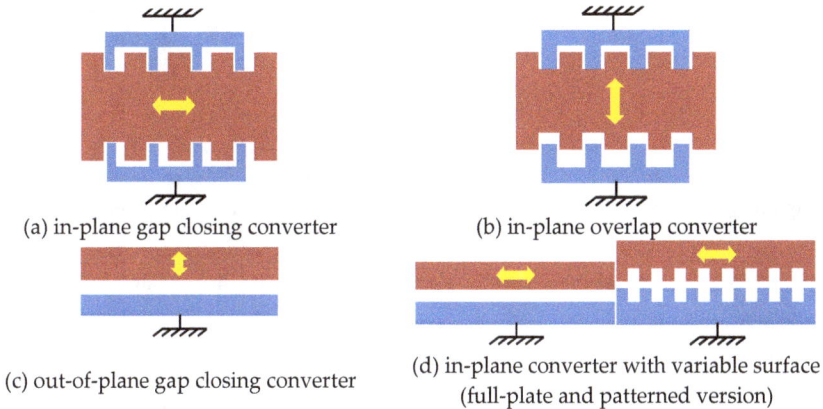

(a) in-plane gap closing converter

(b) in-plane overlap converter

(c) out-of-plane gap closing converter

(d) in-plane converter with variable surface (full-plate and patterned version)

Figure 16. Basic capacitor shapes for electrostatic converters

Obviously, these basic shapes can be adapted to electret-free and electret-based electrostatic converters. As capacitances and electrostatic forces are heavily dependent on the capacitor's shape and on its dimensions, it is interesting to know the capacitance and the electrostatic forces generated by each of these structures to design an electrostatic converter. Capacitances values and electrostatic forces for each shape are presented in the next subsection.

2.   Capacitances values and electrostatic forces

Capacitances values are all deduced from the simple plane capacitor model. In this subsection, the capacitance is computed with an electret layer. To get the capacitances for electret-free electrostatic converters, one has just to take d=0 (where d is the electret thickness).

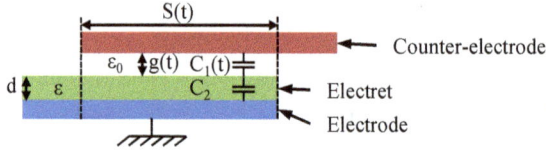

Figure 17. Capacitance of the simple plane capacitor

The total capacitance of the electrostatic converter presented in Figure 17 corresponds to two capacitances (C_1 and C_2) in series.

$$C(t) = \frac{C_1(t)C_2}{C_1(t)+C_2} = \frac{\varepsilon_0 S(t)}{g(t)+ d/\varepsilon} \tag{8}$$

The electrostatic force f_{elec} induced by this capacitor can be expressed by:

$$F_{elec} = \frac{d}{dx}(W_{elec}) = \frac{d}{dx}\left(\frac{1}{2}C(x)U_c(x)^2\right) = \frac{d}{dx}\left(\frac{1}{2}\frac{Q_c^2(x)}{C(x)}\right) \tag{9}$$

With W_{elec} the total amount of electrostatic energy stored in C, Q_c the charge on C, $U_c(x)$ the voltage across C and x the relative movement of the upper plate compared to the lower plate.

Capacitances and electrostatic forces of the four capacitor shapes are obtained by integrating equations (8) and (9).

a.   In-plane gap-closing converter

In-plane gap-closing converters are interdigitated comb devices with a variable air gap between fingers as presented in Figure 18.

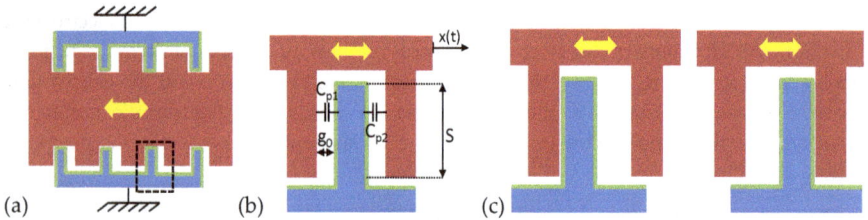

Figure 18. (a) In-plane gap-closing converters, (b) zoom on one finger with C_{min} position, (c) C_{max} positions

The capacitance of the converter corresponds to the two capacitors C_{p1} and C_{p2} in parallel and is expressed in equation (10).

$$C(x) = \frac{2N\varepsilon_0 S\left(g_0 + \frac{d}{\varepsilon}\right)}{\left(g_0 + \frac{d}{\varepsilon}\right)^2 - x^2}$$

(10)

Where N is the number of fingers of the whole electrostatic converter, and S the facing surface.

b.   In-plane overlap converter

In-plane overlap converters are interdigitated comb structure with a variable overlap of the fingers as presented in Figure 19. The whole structure must be separated into two variable capacitors C_{c1} and C_{c2} as the increase of C_{c1}'s capacitance leads to a decrease of C_{c2}'s capacitance and vice-versa.

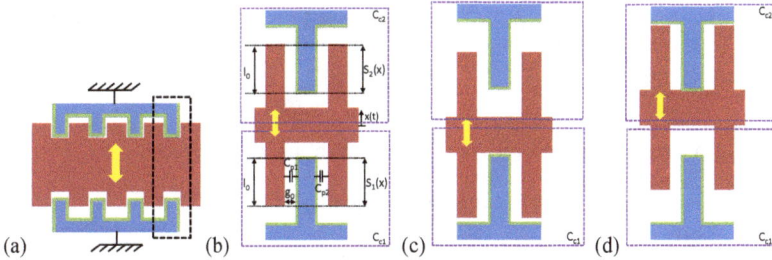

Figure 19. (a) In-plane overlap converters, (b) zoom on two fingers (c) C_{max} position for C_{c1} and C_{min} position for C_{c2} and (d) C_{min} position for C_{c1} and C_{max} position for C_{c2}.

The capacitance of C_{c1} and C_{c2} are expressed in equation (11).

$$C_{c1}(x) = \frac{\varepsilon_0 N w}{g_0 + \frac{d}{\varepsilon}}(l_0 - x) \quad \text{and} \quad C_{c2}(x) = \frac{\varepsilon_0 N w}{g_0 + \frac{d}{\varepsilon}}(l_0 + x)$$

(11)

Where N is the number of fingers, l_0 the facing length of C_{c1} and C_{c2} at the equilibrium position and w the thickness of the fingers (third dimension).

c.   Out-of-plane gap closing converter

In this configuration, the counter-electrode moves above the electrode, inducing a variation of the gap between the electret and the counter-electrode. The initial gap between the counter-electrode and the electret is g_0 and the surface is denoted by S (Figure 20(a)). C_{max} and C_{min} positions are presented in Figure 20(b, c).

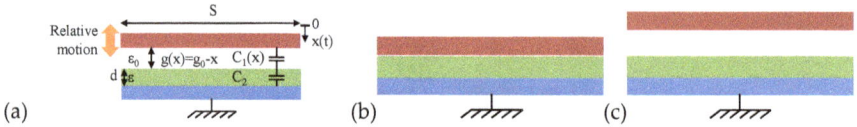

Figure 20. (a) Out-of-plane gap closing, (b) C_{max} position and (c) C_{min} position

The capacitance of the converter is expressed in equation (12).

$$C(x) = \frac{\varepsilon_0 S}{g_0 + d/\varepsilon - x} \tag{12}$$

d.   In-plane converter with variable surface

In in-plane converters with variable surface, it is a change in the capacitor's area that is exploited, as presented in Figure 21(a).

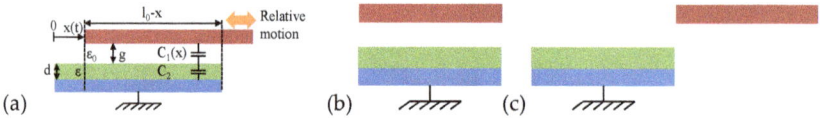

Figure 21. (a) In-plane with variable surface, (b) C_{max} position and (c) C_{min} position

The capacitance of the converter is expressed in equation (13).

$$C(x) = \frac{\varepsilon_0 w(l_0 - x)}{g + d/\varepsilon} \tag{13}$$

Where w is the converter's thickness (third dimension), l_0 the facing length between the plates at t=0.

So as to increase the capacitance variation for a given relative displacement x, it is interesting to pattern the capacitive structure, as presented in Figure 22.

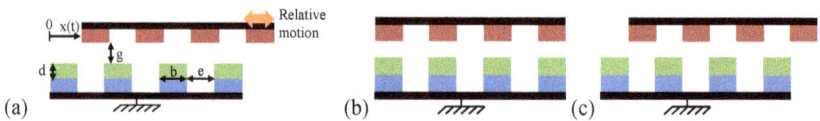

Figure 22. (a) In-plane converter with variable patterned surface, (b) C_{max} position and (c) C_{min} position

To develop efficient vibration energy harvesters able to work with low vibrations, it is necessary to use micro-patterned capacitive structure (small e and b). With such dimensions, fringe effects must be taken into account and it was proven [17] that the capacitance of the energy harvester can be simply modeled by a sine function presented in equation (14).

$$C(x) = \frac{C_{max} + C_{min}}{2} + \left(\frac{C_{max} - C_{min}}{2}\right) \times \cos\left(\frac{2\pi x}{e + b}\right) \qquad (14)$$

Where C_{max} and C_{min} are the maximal and the minimal capacitances of the energy harvester and computed by finite elements.

Electrostatic forces are deduced from the derivation of the electrostatic energy stored into the capacitor, as presented in equation (9). Concerning electret-based devices, the electrostatic force cannot be easily expressed as both capacitor's charge and voltage change when the geometry varies. Table 4 overviews the electrostatic forces for the various converters and their operation modes.

Converter	f_{elec} for charge-constrained cycle	f_{elec} for voltage-constrained cycle	f_{elec} for electret-based devices
In-plane gap closing	$\dfrac{Q_{cst}^2 x}{2\varepsilon_0 N\left(g_0 + \frac{d}{\varepsilon}\right)S}$	$\dfrac{2N\varepsilon_0\left(g_0 + \frac{d}{\varepsilon}\right)SxU_{cst}^2}{\left(\left(g_0 + \frac{d}{\varepsilon}\right)^2 - x^2\right)^2}$	$\dfrac{d}{dx}\left(W_{elec}\right)$
In-plane overlap (only C_{c2})	$\dfrac{Q_{cst}^2\left(g_0 + \frac{d}{\varepsilon}\right)}{2\varepsilon_0 Nw\left(l_0 + x\right)^2}$	$\dfrac{\varepsilon_0 NwU_{cst}^2}{2\left(g_0 + \frac{d}{\varepsilon}\right)}$	$\dfrac{d}{dx}\left(W_{elec}\right)$
Out-of-plane gap closing	$\dfrac{Q_{cst}^2}{2\varepsilon_0 S}$	$\dfrac{\varepsilon_0 SU_{cst}^2}{2\left(g_0 + \frac{d}{\varepsilon} - x\right)^2}$	$\dfrac{d}{dx}\left(W_{elec}\right)$
In-plane with variable surface (non patterned)	$\dfrac{Q_{cst}^2\left(g_0 + \frac{d}{\varepsilon}\right)}{2\varepsilon_0 w\left(l_0 - x\right)^2}$	$\dfrac{\varepsilon_0 wU_{cst}^2}{2\left(g_0 + \frac{d}{\varepsilon}\right)}$	$\dfrac{d}{dx}\left(W_{elec}\right)$

Table 4. Electrostatic forces according to the converter and its operation mode

These electrostatic converters are then coupled to mass-spring systems to become vibration energy harvesters.

3. Electrostatic Vibration Energy Harvesters (eVEH)

As presented in section 2.1, harvesting vibrations requires two conversion steps: a mechanical-to-mechanical converter made of a mass-spring resonator that turns ambient vibrations into a relative movement between two elements (presented in 2.1) and a mechanical-to-electrical converter using, in our case, a capacitive architecture (presented in 2.2) that converts this relative movement into electricity. Section 3 is aimed at presenting complete devices that gather these two converters. It is firstly focused on electret-free electrostatic devices before presenting electret-based devices.

3.1. Electret-free Electrostatic Vibration Energy Harvesters (eVEH)

3.1.1. Devices

The first MEMS electrostatic comb based VEH was developed at the MIT by Meninger et al. in 2001 [18]. This device used an in-plane overlap electrostatic converter. Operating cycles are described and it is proven that the voltage-constrained cycle enables to maximize output power (if the power management electronic is limited in voltage). Yet, for the prototype, a charge-constrained cycle was adopted to simplify the power management circuit even if it drives to a lower output power.

Electrostatic devices can be particularly suitable for Vibration energy harvesting at low frequencies (<100Hz). In 2002, Tashiro et al. [19] developed a pacemaker capable of harvesting power from heartbeats. The output power of this prototype installed on the heart of a goat was 58μW.

In 2003, Roundy [20] proved that the best structure for electrostatic devices was the in-plane gap closing and would be able to harvest up to 100μW/cm³ with ambient vibrations (2.25m/s²@120Hz). Roundy et al. then developed an in-plane gap closing structure able to harvest 1.4nJ/cycle.

In 2005, Despesse et al. developed a macroscopic device (Figure 23(a)) able to work on low vibration frequencies and able to harvest 1mW for a vibration of 0.2G@50Hz [21]. This prototype has the highest power density of eVEH ever reached. Some other MEMS devices were then developed by Basset et al [22] (Figure 23(b)) and Hoffmann et al. [23].

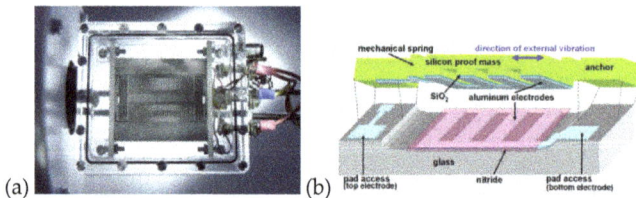

Figure 23. Electrostatic vibration energy harvesters from (a) Despesse et al. [21] and (b) Basset et al. [22].

3.1.2. State of the art – Overview

An overview of electret-free electrostatic vibration energy harvesters is presented in Table 5.

Many prototypes of electret-free electrostatic vibration energy harvesters have been developed and validated. Currently, the tendency is to couple these devices to electrets. The next subsection is focused on them.

3.2. Electret-Based Electrostatic VEH

Electret-based devices were developed to enable a direct vibration-to-electricity conversion (without cycles of charges and discharges) and to simplify the power management circuits.

Author	Ref	Output power	Surface	Volume	Polarization voltage	Vibrations
Tashiro	[19]	36 µW		15000 mm³	45V	1,2G@6Hz
Roundy	[24]	11 µW	100 mm²	100 mm³		0.23G@100Hz
Mitcheson	[25]	24 µW	784 mm²	1568 mm³	2300 V	0.4G@10Hz
Yen	[26]	1,8 µW	4356 mm²	21780 mm³	6 V	1560Hz
Despesse	[21]	1050 µW	1800 mm²	18000 mm³	3 V	0.3G@50Hz
Hoffmann	[23]	3.5 µW	30 mm²		50 V	13G@1300-1500Hz
Basset	[22]	61nW[1]	66 mm²	61.49mm³	8 V	0.25G@250Hz

Table 5. Electret-free electrostatic vibration energy harvesters from the state of the art

3.2.1. History

The idea of using electrets in electrostatic devices to make generators goes back to about 40 years ago. In fact, the first functional electret-based generator was developed in 1978 by Jefimenko and Walker [27]. From that time, several generators exploiting a mechanical energy of rotation were developed (Jefimenko [27], Tada [28], Genda [29] or Boland [16]). Figure 24 presents an example, developed by Boland in 2003 [16, 30] of an electret-based generator able to turn a relative rotation of the upper plate compared to the lower plate into electricity.

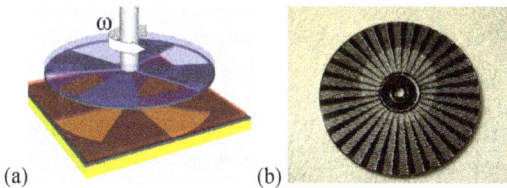

(a)   (b)

Figure 24. Boland's electret-based generator prototype [30] (a) perspective view and (b) stator

With the development of energy harvesting and the need to design autonomous sensors for industry, researchers and engineers have decided to exploit electrets in their electrostatic vibration energy harvesters as their everlasting polarization source.

3.2.2. Devices

Even if the four capacitor shapes presented in subsection 2.3.1 are suitable to develop electret-based vibration energy harvesters, only two architectures have been really exploited: out-of-plane gap closing and patterned in-plane with variable surface structures.

---

[1] latest results from ESIEE showed that higher output powers are reachable thanks to this device (up to 500nW).

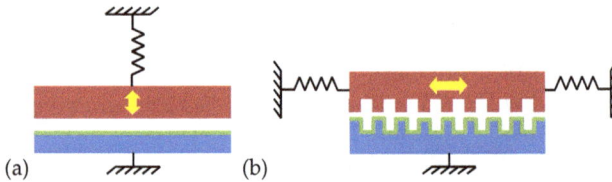

Figure 25. Standard architectures for electret-based Vibration Energy Harvesters

This section presents some examples of electret-based vibration energy harvesters from the state-of-the-art. We have decided to gather these prototypes in 2 categories: devices using full-sheet electrets (electret dimensions or patterning higher than 5mm) and devices using patterned electrets (electret dimensions or patterning smaller than 5mm). Indeed, it is noteworthy that texturing an electret is not an easy task as it generally leads to a weak stability (important charge decay) and requires MEMS fabrication facilities.

a.   Devices using full-sheet electrets

Full-sheet-electret devices can exploit a surface variation or a gap variation. In 2003, Mizuno [31] developed an out-of-plane gap closing structure using a clamped-free beam moving above an electret. This structure was also studied by Boisseau et al. [15] in 2011. This simple structure is sufficient to rapidly demonstrate the principle of vibration energy harvesting with electrets. Large amount of power can be harvested even with low vibration levels as soon as the resonant frequency of the harvester is tuned to the frequency of ambient vibrations.

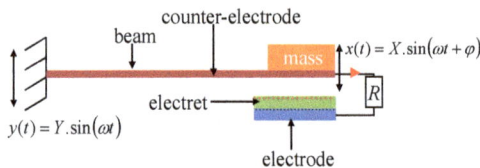

Figure 26. Cantilever-based electret energy harvesters [15]

The first integrated structure using full-sheet electrets was developed by Sterken et al. from IMEC [32] in 2007. A diagram is presented in Figure 27: a full-sheet is used as the polarization source. The electret layer polarizes the moving electrode of the variable capacitance (C_{var}). The main drawback of this prototype is to add a parasitic capacitance in series with the energy harvester, limiting the capacitance's variation and the converter's efficiency.

Today, most of the electret-based vibration energy harvesters use patterned electrets and exploit surface variation.

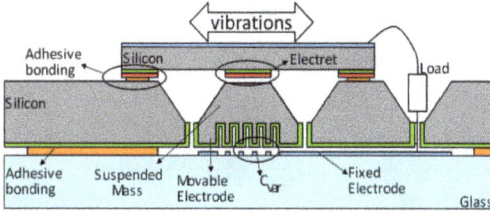

Figure 27. IMEC's first electret-based vibration energy harvester [32]

b.   Devices using patterned electrets

The first structure using patterned electrets was developed by the university of Tokyo in 2006 [33]. Many other devices followed, each of them, improving the first architecture [10, 34-38]. For example Miki et al. [39] improved these devices by developing a multiphase system and using non-linear effects. Multiphase devices enable to limit the peaks of the electrostatic force and thus to avoid to block the moving mass.

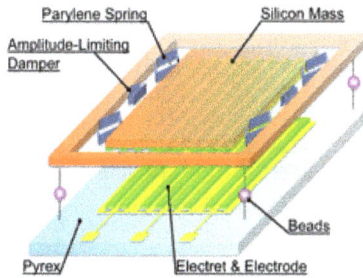

Figure 28. Multiphase electret energy harvester exploiting non-linear springs [38-39].

c.   Mechanical springs to harvest ambient vibrations

Developing low-resonant frequency energy harvesters is a big challenge for small-scale devices. In most cases, ambient vibrations' frequencies are below 100Hz. This leads to long and thin springs difficult to obtain by using silicon technologies (form factors are large and structures become brittle). Thus, to reduce the resonant frequency of vibration energy harvesters, keeping small dimensions, solutions such as parylene springs [40] were developed. Another way consists in using microballs that act like a slideway. Naruse has already shown that such a system could operate at very low frequencies (<2 Hz) and could produce up to 40 μW [37] (Figure 29).

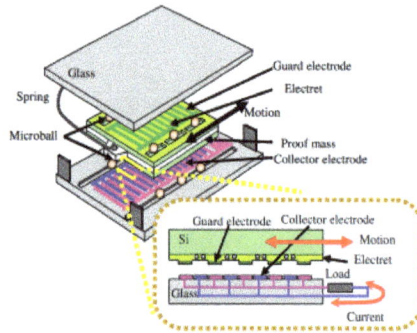

Figure 29. Device on microballs from [37]

Besides, a good review on MEMS electret energy harvesters can be found in [41]. The next subsection presents an overview of some electret-based prototypes from the state of the art.

3.2.3. State of the art – Overview

An overview of electret-based electrostatic vibration energy harvesters is presented in Table 6.

Author	Ref	Vibrations / Rotations	Active Surface	Electret Potential	Output Power
Jefimenko	[27]	6000 rpm	730 cm²	500V	25 mW
Tada	[28]	5000 rpm	90 cm²	363V	1.02 mW
Boland	[16]	4170 rpm	0.8 cm²	150V	25 µW
Genda	[29]	1'000'000 rpm	1.13 cm²	200V	30.4 W
Boland	[42]	7.1G@60Hz	0.12 cm²	850V	6 µW
Tsutsumino	[33]	1.58G@20Hz	4 cm²	1100V	38 µW
Lo	[43]	14.2G@60Hz	4.84 cm²	300V	2.26 µW
Sterken	[32]	1G@500Hz	0.09 cm²	10V	2nW
Lo	[34]	4.93G@50Hz	6 cm²	1500V	17.98 µW
Zhang	[35]	0.32G@9Hz	4 cm²	100V	0.13 pW
Yang	[44]	3G@560Hz	0.3 cm²	400V	46.14 pW
Suzuki	[40]	5.4G@37Hz	2.33 cm²	450V	0.28 µW
Sakane	[10]	0.94G@20Hz	4 cm²	640V	0.7 mW
Naruse	[37]	0.4G@2Hz	9 cm²		40µW
Halvorsen	[45]	3.92G@596Hz	0.48 cm²		1µW
Kloub	[46]	0.96G@1740Hz	0.42 cm²	25V	5µW
Edamoto	[36]	0.87G@21Hz	3 cm²	600 V	12µW
Miki	[39]	1.57G@63Hz	3 cm²	180V	1µW
Honzumi	[47]	9.2G@500Hz	0.01 cm²	52V	90 pW
Boisseau	[15]	0.1G@50Hz	4.16cm²	1400V	50µW

Table 6. Electret-based energy harvesters from the state of the art

Table 6 shows a significant increase of electret-based prototypes since 2003. It is also interesting to note that some companies such as Omron or Sanyo [48] started to study these devices and to manufacture some prototypes.

Thanks to simple cantilever-based devices developed for example by Mizuno [31] and Boisseau [15], the theoretical model of electret-based devices can be accurately validated.

3.2.4. Validation of theory with experimental data – Cantilever-based electret energy harvesters

The theoretical model of electret-based energy converters and vibration energy harvesters can be easily validated by experimental data with a simple cantilever-based electret vibration energy harvester [15].

a.   Device

The prototype presented in Figure 30 consists in a clamped-free beam moving with regards to an electret due to ambient vibrations. The mechanical-to-mechanical converter is the mass-beam system and the mechanical-to-electrical converter is made of the electrode-electret-airgap-moving counter-electrode architecture [15].

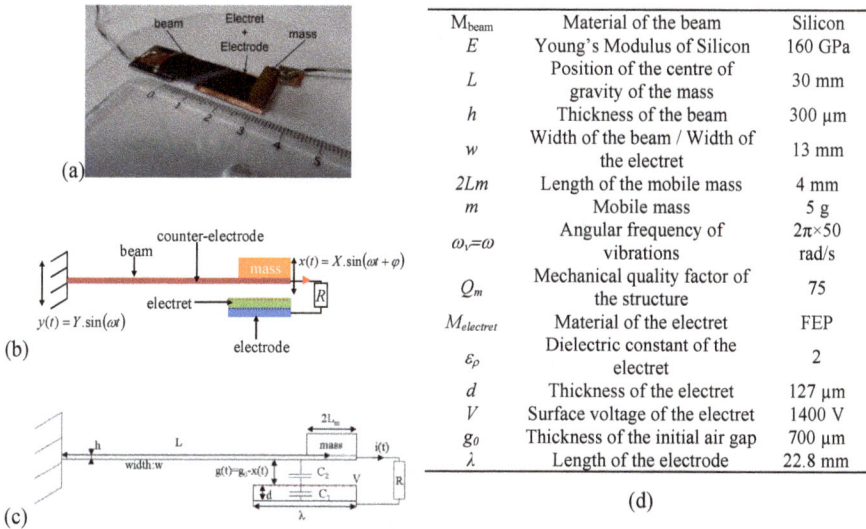

M_{beam}	Material of the beam	Silicon
E	Young's Modulus of Silicon	160 GPa
L	Position of the centre of gravity of the mass	30 mm
h	Thickness of the beam	300 µm
w	Width of the beam / Width of the electret	13 mm
$2Lm$	Length of the mobile mass	4 mm
m	Mobile mass	5 g
$\omega_y = \omega$	Angular frequency of vibrations	$2\pi \times 50$ rad/s
Q_m	Mechanical quality factor of the structure	75
$M_{electret}$	Material of the electret	FEP
ε_p	Dielectric constant of the electret	2
d	Thickness of the electret	127 µm
V	Surface voltage of the electret	1400 V
g_0	Thickness of the initial air gap	700 µm
λ	Length of the electrode	22.8 mm

Figure 30. Example of a simple out-of-plane electret-based VEH (cantilever) (a) prototype, (b) diagram, (c) parameters and (d) dimensions

This system can be modeled by equations developed in section 2.

b.   Model

From equations (1) and (6), one can prove that this device is ruled by the system of differential equations (15).

$$
\begin{cases}
m\ddot{x} + b_m \cdot \dot{x} + kx - \dfrac{d}{dx}\left(\dfrac{Q_2^2}{2C(t)}\right) - mg = -m\ddot{y} \\[4mm]
\dfrac{dQ_2}{dt} = \dfrac{1}{\left(1 + \dfrac{C_{par}}{C(t)}\right)}\left(\dfrac{V}{R} - Q_2\left(\dfrac{1}{RC(t)} - \dfrac{C_{par}}{C(t)^2}\dfrac{dC(t)}{dt}\right)\right)
\end{cases}
\tag{15}
$$

Obviously, this system cannot be solved by hand. Yet, by using a numerical solver (e.g. Matlab), this becomes possible. It is also imaginable to use Spice by turning this system of equations in its equivalent electrical circuit (Figure 31).

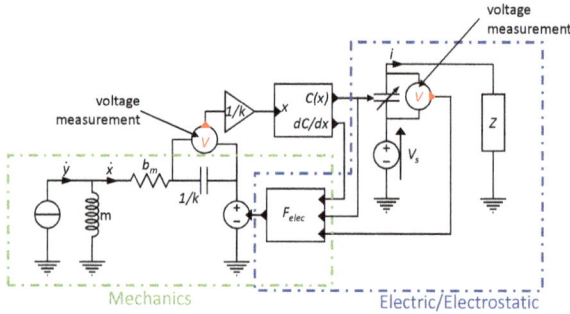

Figure 31. Equivalent electrical model of electret-based vibration energy harvesters

c.   Theory vs experimental data

The prototype presented in Figure 30 has been tested on a shaker at 0.1G@50Hz with two different loads (300MΩ and 2.2GΩ) and the corresponding theoretical results have been computed using a numerical solver. Theoretical and experimental output voltages are presented in Figure 32 showing an excellent match.

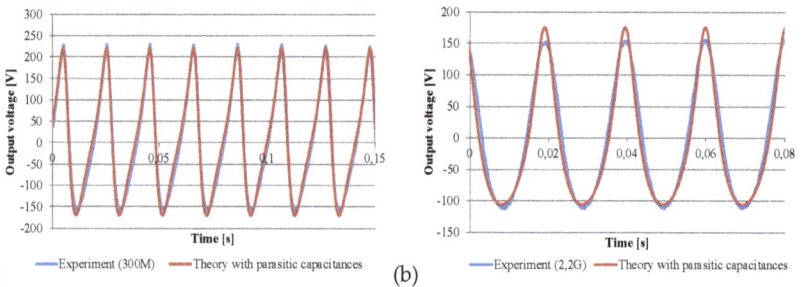

Figure 32. Validation of theory with a cantilever-based electret energy harvester (a) R=300MΩ and (b) R=2.2GΩ

This simple prototype enables to validate the model of electret-based vibration energy harvesters that was presented in section 2. It is also interesting to note that this simple prototype has an excellent output power that reaches 50μW with a low vibration acceleration of 0.1G@50Hz.

Section 3 is concluded by an overview of electret patterning methods. Actually, electret patterning can be a real challenge in electret-based devices because of weak stability problems.

3.2.5. Electret patterning

As presented in section 2, electret patterning is primordial to develop efficient and viable eVEH. Various methods from the state of the art to make stable patterned electrets in polymers and SiO$_2$-based layers are presented hereafter.

a.   Polymers

The problem of polymer electrets patterning has been solved for quite a long time [16, 49]. In fact, it has been proven that it is possible to develop stable patterned electrets in CYTOP by etching the electret layer before charging, as presented in Figure 33 [10]. The patterning size is in the order of 100μm.

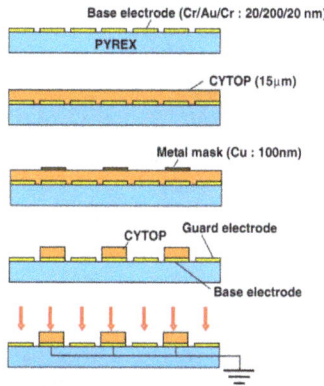

Figure 33. CYTOP electret patterning [10]

Equivalent results have been observed on Teflon AF [16].

However, making patterned SiO$_2$-based electrets is generally more complicated, leading to a strong charge decay and therefore an extremely weak stability.

b.   SiO$_2$-based electrets

In fact, an obvious patterning of electret layers would consist in taking full sheet SiO$_2$-based electrets (that have an excellent stability) and by etching them, like it is done on polymer electrets (Figure 34).

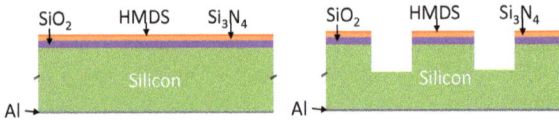

Figure 34. Obvious patterning that does not work

Unfortunately, this obvious patterning does not work because it makes the electret hard to charge (ions or electrons go directly in the silicon wafer) and the stability of these electrets is not good [37]. This is the reason why new and smart solutions have been developed to pattern SiO_2-based electrets.

- IMEC

The concept developed by IMEC to make SiO_2/Si_3N_4 patterned electrets is based on the observation that a single SiO_2 layer is less stable than a superposition of SiO_2 and Si_3N_4 layers. A drawing of the patterned electrets is provided in Figure 35. This method has been patented by IMEC [50].

These patterned electrets are obtained from a silicon wafer that receives a thermal oxidization to form a SiO_2 layer. A Si_3N_4 layer is deposited and etched with a patterning. This electret is then charged thanks to a corona discharge. Charges that are not on the SiO_2/Si_3N_4 areas are removed thanks to thermal treatments while charges that are on these areas stay trapped inside.

Figure 35. Patterned SiO_2/Si_3N_4 electrets from IMEC and used in a structure [45].

The stability of these electrets was proven down to a patterning size of 20µm.

- Sanyo and the University of Tokyo

Naruse et al. [37] developed SiO_2 patterned electrets thanks to a different concept. The electret manufacturing process starts with a SiO_2 layer on a silicon wafer. The SiO_2 layer is metalized with aluminum. The aluminum layer is then patterned and the SiO_2 layer is etched as presented in Figure 36. The sample is then charged.

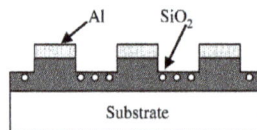

Figure 36. Patterned SiO_2 electrets from Sanyo and the University of Tokyo [37]

The guard electrodes of Aluminum form the low surface voltage and the charged areas of SiO_2 form the high surface voltage. This potential difference enables to turn mechanical energy into electricity. The hollow structure of SiO_2 prevents charge drifting to the guard electrode.

• CEA-LETI [51]

Contrary to the previous methods, the goal of this electret patterning method is to make continuous electret layers. Actually, instead of patterning the electret, it is the substrate of the electret (the silicon wafer) that is patterned thanks to a Deep Reactive Ion Etching (DRIE). The fabrication process of these patterned electrets is similar to the one of full sheet electrets in order to keep equivalent behaviors and above all equivalent stabilities.

The main difference between the two processes (full-sheet and patterned electrets) is the DRIE step that is used to geometrically pattern the electret. The main manufacturing steps are presented in Figure 37. The process starts with a standard p-doped silicon wafer (a). After a lithography step, the silicon wafer is etched by DRIE (b) and cleaned. Wafers are then oxidized to form a 1μm-thick SiO_2 layer (c). SiO_2 layer on the rear face is then removed by HF while front face is protected by a resin. A 100nm-thick LPCVD Si_3N_4 is deposited on the front face (d). Wafers receive a thermal treatment (450°C during 2 hours into N_2) and a surface treatment (vapour HMDS) (e). Dielectric layers are then charged by a standard corona discharge to turn them into electrets (f).

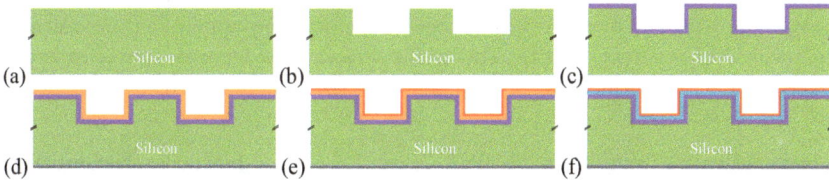

Figure 37. Fabrication process of CEA-LETI's DRIE-patterned electrets

Manufacturing results are presented in Figure 38 for a (e,b,h)=(100μm, 100μm, 100μm) electret (Figure 38(a)). SEM images in Figure 38(b, c, d) show the patterning of the samples and the different constitutive layers. It is interesting to note the continuity of the electret layer even on the right angle in Figure 38(d). The long-term stability of these patterned electrets has been proven thanks to various surface potential decays measurements.

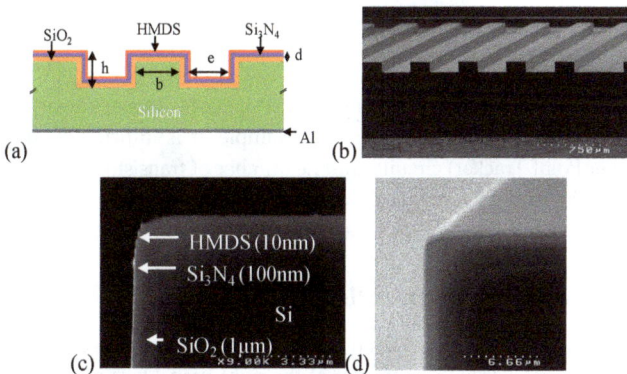

Figure 38. Patterned SiO_2/Si_3N_4 electrets from CEA-LETI [51]

We have presented in this section several prototypes of electrostatic VEH and their output powers that may reach some tens or even hundreds of microwatts. This is in agreement with WSN' power needs. Yet, the output voltages are not appropriate for supplying electronic devices as is. This is the reason why a power converter is required.

That power management unit is essential for Wireless Sensor Nodes; this is the topic of the next section.

4. Power Management Control Circuits (PMCC) dedicated to electrostatic VEH (eVEH)

The next section is aimed at presenting some examples of PMCC for electrostatic VEH.

4.1. Need for Power Management Control Circuit (PMCC)

As presented in section 3, electrostatic vibration energy harvesters are characterized by a high output voltage that may reach some hundreds of volts and a low output current (some 100nA). Obviously, it is impossible to power any application, any electronic device with such a supply source. This is the reason why a power converter and an energetic buffer are needed to develop autonomous sensors. Figure 39 presents the conversion chain.

Power Management Control Circuits (PMCC) can have many functions: changing eVEH resonant frequency, controlling measurement cycles... here, we focus on the power converter and on its control circuit.

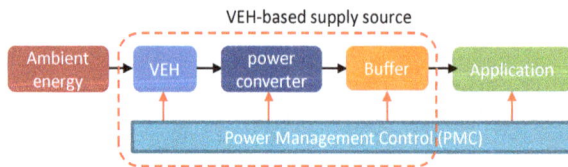

Figure 39. Power Management Control Circuit to develop viable VEH

As eVEH output powers are low (generally <100µW), Power Management Control Circuit must be simple and above all low power. For example, it is difficult to supply a MMPT (Maximum Power Point Tracker) circuits and the number of transistors and operations must be highly limited. We present in the next subsections some examples of Power Management Control Circuit for electret-free and electret-based eVEH.

4.2. PMCC for electret-free electrostatic VEH

As the mechanical-to-electrical conversion is not direct, electret-free eVEH need a PMCC able to charge and to discharge the capacitor at the right time. Once more, we will focus on voltage-constrained and charge-constrained cycles.

4.2.1. Voltage-constrained cycles

The voltage-constrained cycle is not often used, and no specific example is available. Yet, Figure 40 presents an example of a PMCC to implement voltage-constrained cycles on electret-free electrostatic converters.

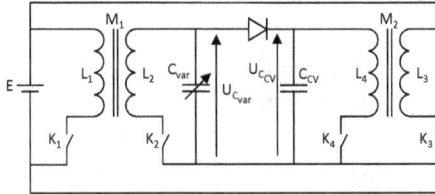

Figure 40. Example of a PMCC to implement voltage-constrained cycles

When the electrostatic converter's capacitance reaches its maximum, a quantity of energy is transferred from the electrical energetic buffer E and stored in the magnetic core M_1 by closing K_1 during few µs, negligible compared to the mechanical period. This energy is then transferred to the variable capacitor C_{var} by closing K_2 during few µs. The voltage U_{Cvar} through C_{var} reaches U_{cv}, the constant voltage. The mechanical movement induces a decrease of the electrostatic structure's capacitance C_{var} and a charge $\Delta Q = U_{CV}\Delta C_{var}$ is transferred to the constant voltage storage C_{cv}. To maintain U_{cv} approximately constant, a second electrical converter is used. When U_{cv} becomes higher than a threshold voltage, a quantity of energy is transferred from the constant voltage capacitor C_{cv} to the energetic buffer E by closing K_4 and then K_3. Finally, when the electrostatic structure's capacitance reaches its minimum, the remaining energy stored in the electrostatic structure C_{var} is sent to the energetic buffer by closing K_2 and then K_1.

Even though this PMCC works, it nevertheless requires two electrical converters that cost in price, space, losses and complexity. In order to have only one electrical converter, the MIT proposed in 2005 the following structure that applies a partial constant voltage cycle [26]:

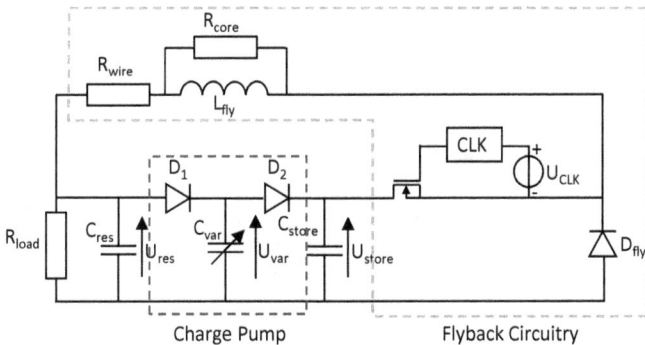

Figure 41. Example of a PMCC from MIT for voltage-constrained cycles [26]

This electronic circuit keeps the electrostatic structure's voltage between two values (V_{res} and V_{store}). When the electrostatic structure's capacitance C_{var} increases, its voltage decreases and finishes to reach the low voltage storage U_{res}. Then, diode D_1 becomes conductive and a current is transferred from C_{res} (storage capacitor) to the electrostatic device. When C_{var}'s capacitance decreases, its voltage increases and finally reaches the high voltage storage U_{store}. Then diode D_2 becomes conductive and a current is transferred from the electrostatic structure to C_{store}. This structure works as a charge pump from C_{res} to C_{store}. And, in order to close the cycle, one part of the energy transferred to C_{store} is transferred to C_{res} by using an inductive electrical converter. Although this structure uses only one inductive component, it requires a complex electronic circuit to drive the floating transistor connected to the high voltage.

Finally, the constant voltage cycle is not frequently used due to the complex electronic circuits associated.

4.2.2. Charge-constrained cycles

Charge-constrained cycles are easier to implement than voltage-constrained cycles as the conversion consists in charging the capacitor when the capacitance is maximal and to let it in open-circuit till it reaches its minimum. On the minimal capacitance, corresponding to the maximal voltage, charges are collected from the converter.

Usually, to reach a high conversion power density, the capacitor must be polarized at a high voltage ($V_1 > 100V$). Yet, in autonomous devices, only 3V supply sources are available: a first DC-to-DC converter (step-up) is therefore needed to polarize the capacitor at a high voltage (step 1). In the same way, the output voltage on the capacitor after the mechanical-to-electrical conversion (step 2) may reach several hundreds of volts ($V_2 > 200-300V$) and is therefore not directly usable to power an application: a second converter (step-down) is then necessary (step 3). Obviously, to limit the number of sources, it is interesting to use the same 3V-supply source to charge the electrostatic structure and to collect the charges at the end of the mechanical-to-electrical conversion. Figure 42 sums up the 3-steps conversion process with the two DC-to-DC conversions (DC-to-DC converters) and the mechanical-to-electrical conversion (energy harvester).

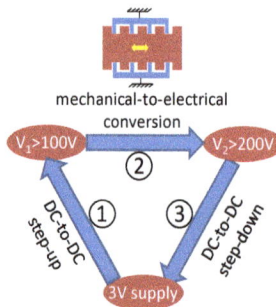

Figure 42. DC-to-DC conversions needed to develop an operational electret-free electrostatic converter and conversion steps

Furthermore, in order to limit the size and the cost of the power converters and the power management control circuit, it is worth combining the step-up and the step-down converters into a single DC-to-DC converter: a bidirectional converter is then used. The two most well-known bidirectional converters are the buck-boost and the flyback converters.

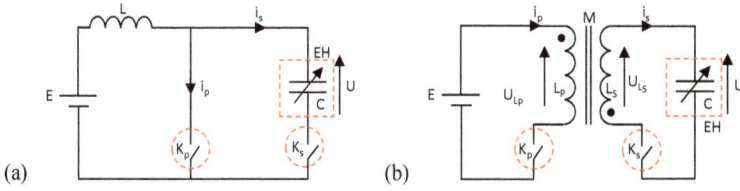

Figure 43. Bidirectional DC-to-DC converters (a) buck-boost and (b) flyback

c.   Bidirectional buck-boost converter

The operating principle of the bidirectional buck-boost converter (Figure 43(a)) is summed up below:

Step 1.   Capacitor charging

K_p is closed for a time t_1. The energy E_c, that has to be sent to the energy harvester to polarize it, is transferred from the supply source E to the inductance L.

K_p is open, and K_s is closed till current i_s becomes equal to 0, corresponding to the time needed to transfer the energy stored in inductance L to the capacitor of the energy harvester C.

Step 2.   Mechanical-to-electrical conversion step

K_p and K_s are open to let the electrostatic converter in open circuit so that the voltage across C may vary freely.

Step 3.   Capacitor discharging

K_s is closed for a time t_2, to transfer the energy stored in the capacitor C to inductance L and the storage element E.

K_s is open and K_p is closed till i_p becomes equal to 0 corresponding to the time needed to transfer the energy stored in L to the storage element E.

The waveforms of currents in buck–boost converters are presented in Figure 44.

This converter has a good conversion efficiency that can reach up to 80-90%. Yet, Flyback converters are generally more suitable for electrostatic energy harvesters where conversion ratios are higher than 30.

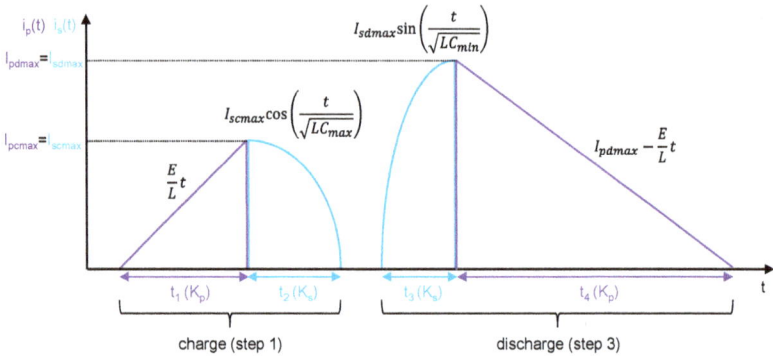

Figure 44. Waveforms of currents in buck–boost converters

d.   Bidirectional flyback converter

The operating principle of the bidirectional flyback converter (Figure 43(b)) is summed up below:

Step 1.   Capacitor charging

K_p is closed for a time t_1. The energy E_c, that has to be sent to the energy harvester to polarize it, is transferred from the supply source E to the inductance L_p that charges the magnetic core M.

K_p is open, and K_s is closed till current i_s becomes equal to 0, corresponding to the time needed to transfer the energy stored in the magnetic core M to the capacitor of the energy harvester C.

Step 2.   Mechanical-to-electrical conversion step

K_p and K_s are open to let the energy harvester in open circuit so that the voltage across C may vary freely.

Step 3.   Capacitor discharging

K_s is closed for a time t_2, to transfer the energy stored in the capacitor C to the magnetic core M through L_s.

K_s is open and K_p is closed till i_p becomes equal to 0 corresponding to the time needed to transfer the energy stored in the magnetic core M to the storage element E.

The waveforms of currents in flyback converters are presented in Figure 45.

Contrary to buck-boost converters, flyback converters do not need bidirectional transistors (K_s must be bidirectional in buck-boost converters) that complicate the power management circuit and increase losses. Moreover, flyback converters enable to optimize both the windings for the high voltages and the low voltages (while buck-boost converters have only one winding).

These two DC-to-DC conversions (step-up and step-down) can be simplified by using electret-based devices. The next sub-section is focused on the power converters and the power management control circuits for these energy harvesters.

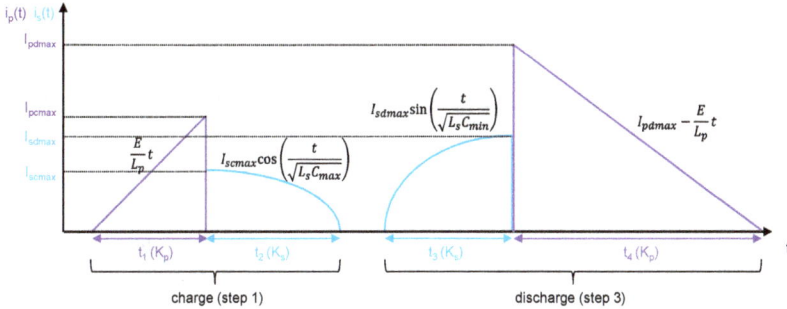

Figure 45. Waveforms of currents in flyback converters

4.3. PMCC for Electret-Based Electrostatic VEH

Electret-based eVEH enable to have a direct mechanical-to-electrical conversion without needing any cycles of charges and discharges. As a consequence, it is possible to imagine two kinds of power converters.

4.3.1. Passive power converters

Passive power converters are the easiest way to turn the AC high-voltage low-current eVEH output into a 3V DC supply source for WSN. An example of these circuits is presented in Figure 46(a). It consists in a diode bridge and a capacitor that stores the energy from the eVEH.

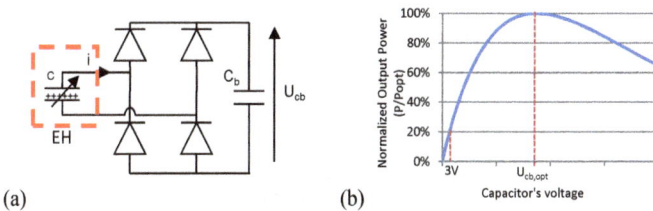

(a)   (b)

Figure 46. (a) Simple passive power converter – diode bridge-capacitor and (b) optimal output voltage on U_{cb}

Such a power converter does not need any PMCC as the energy from the energy harvester is directly transferred to the capacitor. This power conversion is quite simple, but the drawback is the poor efficiency.

Actually, to maximize power extraction from an electret-based electrostatic converter, the voltage across C_b must be close to the half of the eVEH's output voltage in open circuit. This optimal value ($U_{cb,opt}$) is generally equal to some tens or hundreds of volts. To power an electronic device, a 3V source is required: this voltage cannot be maintained directly on the capacitor as it greatly reduces the conversion efficiency of the energy harvester (Figure 46(b)).

The solution to increase the efficiency of the energy harvester consists in using active power converters.

4.3.2. Active power converters

As eVEH' optimal output voltages are 10 to 100 times higher than 3V, a step-down converter is needed to fill the buffer. The most common step-down converters are the buck, the buck-boost and the flyback converters. We focus here on the flyback converter that gives more design flexibilities (Figure 47).

Many operation modes can be developed to turn the eVEH high output voltages into a 3V supply source. Here, we focus on two examples: (i) energy transfer on maximum voltage detection and (ii) energy transfer with a pre-storage to keep an optimal voltage across the electrostatic converter.

a.   Energy transfer on a maximum voltage detection

The concept of this power conversion is to send the energy from the energy harvester to the 3V energy buffer when the eVEH output voltage reaches its maximum.

The power management control circuit is aimed at finding the maximum voltage across the energy harvester and to close K_p (Figure 48) to send the energy from the eVEH to the magnetic circuit. Then K_s is closed to send the energy from the magnetic circuit to the buffer C_b. The winding ratio m is determined from the voltage ratio between the primary and the secondary.

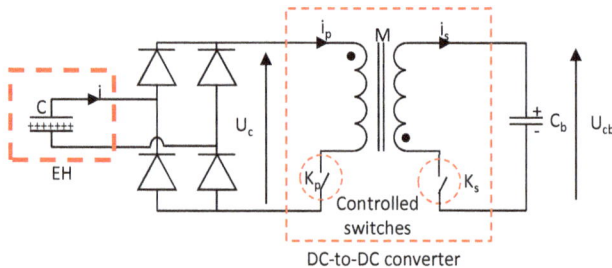

Figure 47. Energy transfer on maximum voltage detection

Figure 48 presents the voltages and the currents on the primary and on the secondary during the power transfer.

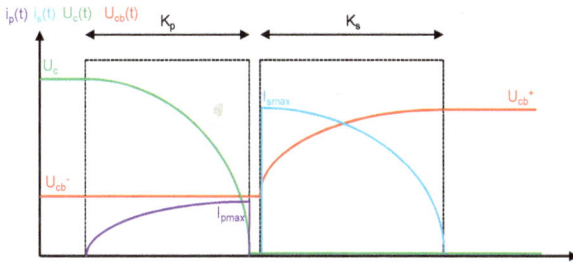

Figure 48. Voltages and currents during power conversion

As eVEH capacitances are quite small, parasitic capacitances of the primary winding may have a strong negative impact on the output powers, increasing conversion losses. An alternative consists in using a pre-storage capacitor.

b.   Energy transfer with a pre-storage capacitor

In this operation mode, a pre-storage capacitor C_p is used to store the energy from the eVEH and to maintain an optimal voltage across the diode bridge in order to optimize the energy extraction from the eVEH.

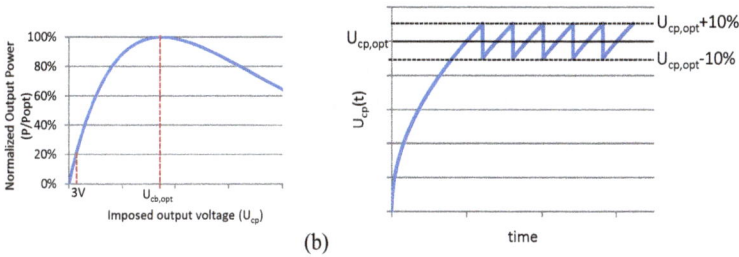

(a)   (b)

Figure 49. (a) eVEH output power vs imposed output voltage and (b) $U_{cp}(t)$

The goal of the PMCC is to maintain the voltage quite constant across the diode bridge (+/- 10% $U_{cp,opt}$). Then, when U_{cp} reaches $U_{cp,opt}+10\%$, one part of the energy stored in C_p is sent to C_b through the flyback converter.

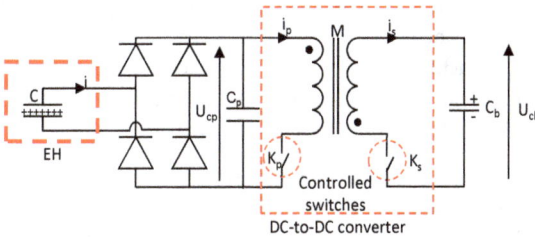

Figure 50. Energy transfer with pre-storage

Voltages and currents during the electrical power transfer are presented in Figure 51.

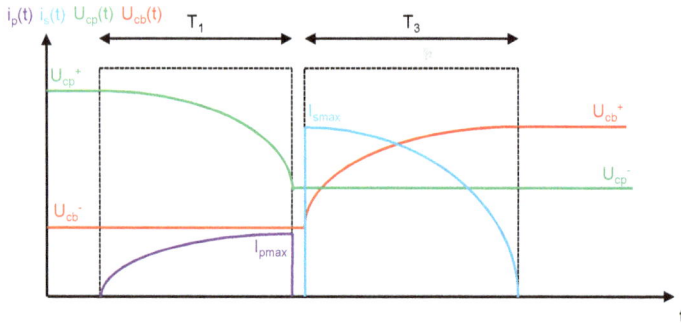

Figure 51. Voltages and currents during power conversion

As C_P can be in the order of some tens to hundreds of nanofarads, transformer's parasitic capacitances have smaller impacts on eVEH's output power.

This power conversion principle also enables to use multiple energy harvesters in parallel with only one transformer and above all only one PMCC (which is not the case with the maximum voltage detection).

We have presented some examples of power converters able to turn the raw output powers of the energy harvesters into supply sources able to power electronic devices. Thanks to this, and low power consumptions of WSN' nodes, it is possible to develop autonomous wireless sensors using the energy from vibrations from now on. The last section gives an assessment of this study.

5. Assessments and perspectives

In this last section, we present our vision of eVEH and their perspectives for the future.

5.1. Assessments

Electrostatic VEH are doubtless the less known vibration energy harvesters, and especially compared to piezoelectric devices. Yet, these devices have undeniable advantages: the possibility to develop structures with high mechanical-to-electrical couplings, to decouple the mechanical-to-mechanical converter and the mechanical-to-electrical converter, to develop low-cost devices able to withstand high temperatures...

Moreover, even if these devices have incontestable drawbacks as well, such as low capacitances, high output voltages and low output currents, it has been proven that they can be compatible with WSN needs as soon as a power converter is inserted between the VEH and the device to supply.

5.2. Limits

Obviously, eVEH have drawbacks and limitations. We present in this subsection the four most important limits of these devices.

i.   Integration of devices. The question of size reduction is common to all VEH. Actually, as the output power is proportional to the mobile mass, it is not necessarily useful to reduce VEH' dimensions at any cost. Moreover, it becomes particularly difficult to design devices with a resonant frequency lower than 50Hz when working with small-scale devices. As a consequence, to have a decent output power (>10µW) and a robust device, it is hard to imagine devices smaller than 1cm².

ii. Working frequency and frequency bandwidth. Ambient vibrations are characterized by a low frequency, generally lower than 100Hz. Moreover, when looking at the vibrations spectra, it appears that they are spread over a wide frequency range. This implies that we need to develop low-frequency broadband devices; this may rise to many problems in the design and the manufacturing of the springs. Indeed, to build low-frequency devices, especially with small-scale devices, thin and long guide beams are needed. They are particularly fragile and are moreover submitted to high strains and stresses.

iii. Gap control. eVEH output powers are greatly linked to the capacitance variations, that must be maximized. Therefore, the air gap must be controlled precisely and minimized to reach high capacitances. Yet, it is also important to take care of pull-in and electrical breakdown problems.

iv. Electret stability. Electret stability may also be critical. Actually, electret stability is strongly linked to environmental conditions, for example to humidity and temperature. Moreover, contacts between electrets and electrodes must be avoided as they generally lead to breakdown and discharge of electrets.

5.3. Perspectives

Like all VEH (piezoelectric, electromagnetic or electrostatic), the most critical point to improve is the frequency bandwidth that must be largely increased to develop viable and adaptable devices.

Indeed, a wide frequency bandwidth is firstly necessary to develop robust devices. VEH are submitted to a large amount of cycles (16 billion cycles for a device that works at 50Hz during 10 years), that may change the resonant frequency of the energy harvester due to fatigue. Then, the energy harvester's resonant frequency is not tuned to the ambient vibrations' frequency anymore. Therefore, it is absolutely primordial to develop devices able to maintain their resonant frequency equal to the vibration frequency.

Wideband energy harvesters are also interesting to develop adaptable devices, able to work in many environments and simple to set up and to use. There is a real need for Plug and Play devices.

Figure 52 presents our vision of VEH today: VEH market as a function of the time and the two technological bottlenecks linked to working frequency bandwidths. In our opinion, today's VEH are yet suitable for industry; increasing working frequency bandwidths and developing plug and play devices are the only way to conquer new markets.

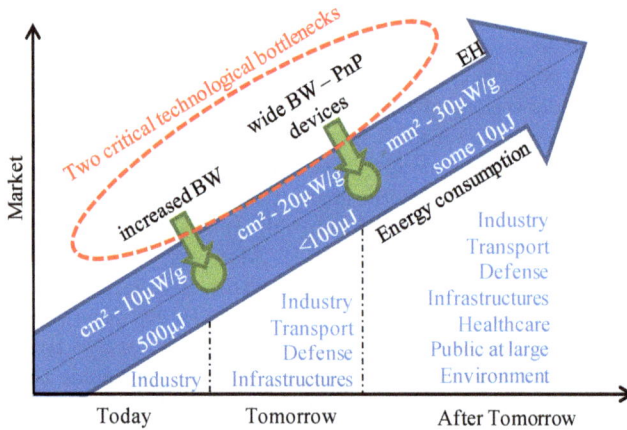

Figure 52. Vibration Energy Harvesters – Perspectives [52]

6. Conclusions

We have presented in this chapter the basic concepts and theories of electrostatic converters and electrostatic vibration energy harvesters together with some prototypes from the state of the art, adopting a "global system" vision.

Electrostatic VEH are increasingly studied from the early 2000s. Unfortunately, no commercial solution is on the market today, dedicating these devices to research.

We believe that this is a pity because they have undeniable advantages compared to piezoelectric or electromagnetic devices. The first in importance is probably the possibility to manufacture low cost devices (low cost and standard materials). Obviously, the limited frequency bandwidth of vibration energy harvesters does not help the deployment of these devices, even if some solutions are currently under investigation. Yet, with this increasing need to get more information from our surroundings, we can expect that these systems will match industrial needs and find industrial applications.

Anyway, electrostatic converters and electrostatic vibration energy harvesters remain an interesting research topic that gathers material research (electrets), power conversion, low consumption electronics, mechanics and so on...

Author details

S. Boisseau, G. Despesse and B. Ahmed Seddik

LETI, CEA, Minatec Campus, Grenoble, France

Acknowledgement

The authors would like to thank their VEH coworkers: J.J. Chaillout, A.B. Duret, P. Gasnier, J.M. Léger, S. Soubeyrat, S. Riché and S. Dauvé for their contributions to this chapter.

7. References

[1]   Boisseau S, Despesse G. Energy harvesting, wireless sensor networks & opportunities for industrial applications. EE Times 2012. http://www.eetimes.com/design/smart-energy-design/4237022/Energy-harvesting--wireless-sensor-networks---opportunities-for-industrial-applications

[2]   Williams C B, Yates R B. Analysis Of A Micro-electric Generator For Microsystems. Proc. Eurosensors 1995;1: 369-72.

[3]   Eguchi M. On the Permanent Electret. Philosophical Magazine 1925; 49:178.

[4]   Kotrappa P. Long term stability of electrets used in electret ion chambers. Journal of Electrostatics 2008;66: 407-9.

[5]   Kressmann R, Sessler G, Gunther P. Space-charge electrets. Transactions on Dielectrics and Electrical Insulation 1996;3: 607-23.

[6]   Gunther P, Ding H, Gerhard-Multhaupt R. Electret properties of spin-coated Teflon-AF films. Proc. Electrical Insulation and Dielectric Phenomena 1993: 197-202.

[7]   Sessler G. Electrets [3rd Edition] [in Two Volumes]. Laplacian Press, Morgan Hill, 1999.

[8]   Gunther P. Charging, long-term stability and TSD measurements of SiO_2 electrets. Transactions on Electrical Insulation 1989;24(3): 439-42.

[9]   Leonov V, Fiorini P, Van Hoof C. Stabilization of positive charge in SiO_2/Si_3N_4 electrets. IEEE transactions on dielectrics and electrical insulation. 2006;13(5): 1049-56.

[10]   Sakane Y, Suzuki Y, Kasagi N. The development of a high-performance perfluorinated polymer electret and its application to micro power generation. IOP Journal of Micromechanics and Microengineering 2008;18(104011). http://dx.doi.org/10.1088/0960-1317/18/10/104011

[11]   Rychkov D, Gerhard R. Stabilization of positive charge on polytetrafluoroethylene electret films treated with titanium-tetrachloride vapor. Appl. Phys. Lett. 2011; 98(122901).

[12]   Schwödiauer R, Neugschwandtner G, Bauer-Gogonea S, Bauer S, Rosenmayer T. Dielectric and electret properties of nanoemulsion spin-on polytetrafluoroethylene films. Appl. Phys. Lett. 2000;76(2612).

[13]   Kashiwagi K, Okano K, Morizawa Y, Suzuki Y. Nano-cluster-enhanced High-performance Perfluoro-polymer Electrets for Micro Power Generation. Proc. PowerMEMS 2010:169-72.

[14]   Kashiwagi K, Okano K, Miyajima T, Sera Y, Tanabe N, Morizawa Y, Suzuki Y. Nano-cluster-enhanced High-performance Perfluoro-polymer Electrets for Micro Power Generation. IOP J. Micromech. Microeng.2011;21(125016). http://dx.doi.org/10.1088/0960-1317/21/12/125016

[15] Boisseau S, Despesse G, Ricart T, Defay E, Sylvestre A. Cantilever-based electret energy harvesters. IOP Smart Materials and Structures 2011; 20(105013). http://dx.doi.org/ 10.1088/0964-1726/20/10/105013

[16] Boland J, Chao Y, Suzuki Y, Tai Y. Micro electret power generator. Proc. MEMS 2003:538-41.

[17] Boisseau S, Despesse G, Sylvestre A. Optimization of an electret-based energy harvester. Smart Materials and Structures 2010;19(075015). http://dx.doi.org/ 10.1088/0964-1726/19/7/075015

[18] Meninger S, Mur-Miranda J O, Amirtharajah R, Chandrakasan A, Lang J. Vibration-to-electric energy conversion. IEEE transactions on very large scale integration (VLSI) 2011;9(1): 64-75.

[19] Tashiro R, Kabei N, Katayama K, Tsuboi E, Tsuchiya K. Development of an electrostatic generator for a cardiac pacemaker that harnesses the ventricular wall motion. Journal of Artificial Organs 2002;5:239-45.

[20] Roundy S. Energy Scavenging for Wireless Sensor Nodes with a Focus on Vibration to Electricity Conversion. PhD Thesis. University of California, Berkeley, 2003.

[21] Despesse G, Chaillout J J, Jager T, Léger J M, Vassilev A, Basrour S, Charlot B. High damping electrostatic system for vibration energy scavenging. Proc. sOc-EUSAI 2005:283-6.

[22] Basset P, Galayko D, Paracha A, Marty F, Dudka A, Bourouina T. A batch-fabricated and electret-free silicon electrostatic vibration energy harvester. IOP Journal of Micromechanics and Microengineering 2009;19(115025). http://dx.doi.org/10.1088/0960-1317/19/11/115025

[23] Hoffmann D, Folkmer B, Manoli Y. Fabrication and characterization of electrostatic micro-generators. Proc. PowerMEMS 2008: 15.

[24] Roundy S. Energy Scavenging for Wireless Sensor Networks with Special Focus on Vibrations. Hardcover, Springer, 2003.

[25] Mitcheson P, Green T C, Yeatmann E M, Holmes A S. Architectures for vibration-driven micropower generators. J. of Microelect. Systems 2004;13: 429-40.

[26] Chih-Hsun Yen B, Lang J. A variable capacitance vibration-to-electric energy harvester. IEEE Trans. Circuits Syst. 2006;53: 288-95.

[27] Jefimenko O, Walker D K. Electrostatic Current Generator Having a Disk Electret as an Active Element. Transactions on Industry Applications 1978;IA-14: 537-40.

[28] Tada Y. Experimental Characteristics of Electret Generator, Using Polymer Film Electrets. Japanese Journal of Applied Physics 1992;31: 846-51.

[29] Genda T, Tanaka S, Esashi M. High power electrostatic motor and generator using electrets. Proc. Transducers 2003;1: 492-5.

[30] Boland J. Micro electret power generators. PhD thesis. California Institute of Technology. 2005.

[31] Mizuno M, Chetwynd D. Investigation of a resonance microgenerator. IOP Journal of micromechanics and Microengineering 2003;13: 209-16. http://dx.doi.org/10.1088/0960-1317/13/2/307

[32] Sterken T, Fiorini P, Altena G, Van Hoof C, Puers R. Harvesting Energy from Vibrations by a Micromachined Electret Generator. Proc. Transducers 2007: 129-32.

[33] Tsutsumino T, Suzuki Y, Kasagi N, Sakane Y. Seismic Power Generator Using High-Performance Polymer Electret. Proc. MEMS 2006: 98-101.
[34] Lo H W, Tai Y C, Parylene-HT-based electret rotor generator. Proc. MEMS 2008:984-7.
[35] Zhang X, Sessler G M. Charge dynamics in silicon nitride/silicon oxide double layers. Applied Physics Letters 2001;78: 2757-9.
[36] Edamoto M, Suzuki Y, Kasagi N, Kashiwagi K, Morizawa Y, YokoyamaT, Seki T, Oba M. Low-resonant-frequency micro electret generator for energy harvesting application. Proc. MEMS 2009: 1059–62.
[37] Naruse Y, Matsubara N, Mabuchi K, Izumi M, Suzuki S. Electrostatic micro power generation from low-frequency vibration such as human motion. IOP Journal of Micromechanics and Microengineering 2009;19(094002). http://dx.doi.org/10.1088/0960-1317/19/9/094002
[38] Suzuki Y, Miki D, Edamoto M, Honzumi M. A MEMS Electret Generator with Electrostatic Levitation for Vibration-Driven Energy Harvesting Applications. IOP J. Micromech. Microeng. 2010;20(104002). http://dx.doi.org/10.1088/0960-1317/20/10/104002
[39] Miki D, Honzumi M, Suzuki S, Kasagi N. Large-amplitude MEMS electret generator with nonlinear spring. Proc. MEMS 2010: 176-9.
[40] Suzuki Y, Edamoto M, Kasagi N, Kashwagi K, Morizawa Y. Micro electret energy harvesting device with analogue impedance conversion circuit. Proc. PowerMEMS 2008: 7-10.
[41] Suzuki Y. Recent Progress in MEMS Electret Generator for Energy Harvesting. IEEJ Trans. Electr. Electr. Eng. 2011;6(2): 101-11.
[42] Boland J, Messenger J, Lo K, Tai Y. Arrayed liquid rotor electret power generator systems. Proc. MEMS 2005: 618-21.
[43] Lo H W, Whang R, Tai Y C. A simple micro electret power generator. Proc. MEMS 2007: 859-62.
[44] Yang Z, Wang J, Zhang J. A micro power generator using PECVD SiO2/Si3N4 double layer as electret. Proc. PowerMEMS 2008:317-20.
[45] Halvorsen E, Westby E R, Husa S, Vogl A, Østbø N P, Leonov V, Sterken T, Kvisterøy T. An electrostatic Energy harvester with electret bias. Proc. Transducers 2009:1381–4.
[46] Kloub H, Hoffmann D, Folkmer B and Manoli Y. A micro capacitive vibration Energy harvester for low power electronics. Proc. PowerMEMS 2009:165–8.
[47] Honzumi M, Ueno A, Hagiwara K, Suzuki Y, Tajima T, Kasagi N. Soft-X-Ray-charged vertical electrets and its application to electrostatic transducers. Proc. MEMS 2010:635–8.
[48] Shimizu N. Omron, Sanyo Prototype Mini Vibration-Powered Generators, Nikkei Electronics Asia, Feb 16, 2009. [Online] 2009. http://techon.nikkeibp.co.jp/article/HONSHI/ 20090119/164257/.
[49] Suzuki Y, Miki D, Edamoto M, Honzumi M. A MEMS Electret Generator With Electrostatic Levitation For Vibration-Driven Energy Harvesting Applications. IOP J. Micromech. Microeng. 2010;20(104002). http://dx.doi.org/10.1088/0960-1317/20/10/ 104002
[50] Leonov V. Patterned Electret Structures and Methods for Manufacturing Patterned Electret Structures. Patent 2011/0163615. 2011.

[51] Boisseau S, Duret A B, Chaillout J J, Despesse G. New DRIE-Patterned Electrets for Vibration Energy Harvesting. Proc. European Energy Conference 2012.

[52] Boisseau S, Despesse G. Vibration energy harvesting for wireless sensor networks: Assessments and perspectives. EE Times 2012. http://www.eetimes.com/design/smart-energy-design/4370888/Vibration-energy-harvesting-for-wireless-sensor-networks--Assessments-and-perspectives

Chapter 6

Electrostrictive Polymers for Vibration Energy Harvesting

Mickaël Lallart, Pierre-Jean Cottinet, Jean-Fabien Capsal,
Laurent Lebrun and Daniel Guyomar

Additional information is available at the end of the chapter

1. Introduction

Recent progresses in microelectronics that enabled the design ultra-low consumption, fully operative electronic systems, have permitted the disposal of autonomous wireless devices ([1]). However, primary batteries, that initially promoted the development of such systems, have nowadays become a break to the spreading of long-lifetime autonomous apparatus, mainly because of their limited lifespan (typically one year under classical working conditions) as well as their complex recycling process that raises environmental issues ([2]).

In order to counteract this drawback, many researches have been carried out on ambient vibration energy harvesting over the last decade ([3]). However, although such investigations have been promoted by a growing demand from industries in terms of left-behind, self-powered wireless sensors and sensor networks, there is still a significant need of improving the conversion and harvesting abilities of microgenerators to dispose of truly working, reliable self-powered wireless systems.

In particular, when dealing with vibrations that are available in many environments for scavenging mechanical energy, many studies considered the use of piezoelectric elements for small-scale energy harvesting, as such materials present relatively high energy density and high intrinsic electromechanical coupling ([4]). Nevertheless, the high stiffness of such materials prevent them to be directly used as most of the available vibrating sources feature low frequencies (*e.g.*, human motions), high strain characteristics, and therefore the use of intermediate mechanical structures is mandatory to ensure a frequency matching for maximizing the input energy in the electroactive device (Figure 1). However, when adding such an additional conversion stage, the global coupling coefficient is dramatically reduced, leading to decreased harvesting abilities, and the compactness is compromised.

From Figure 1, it can be shown that, when the source presents high strain, low frequency behavior, electrostrictive materials are of premium choice to ensure a good mechanical

(a) Frequency contents

(b) Stress-strain curves

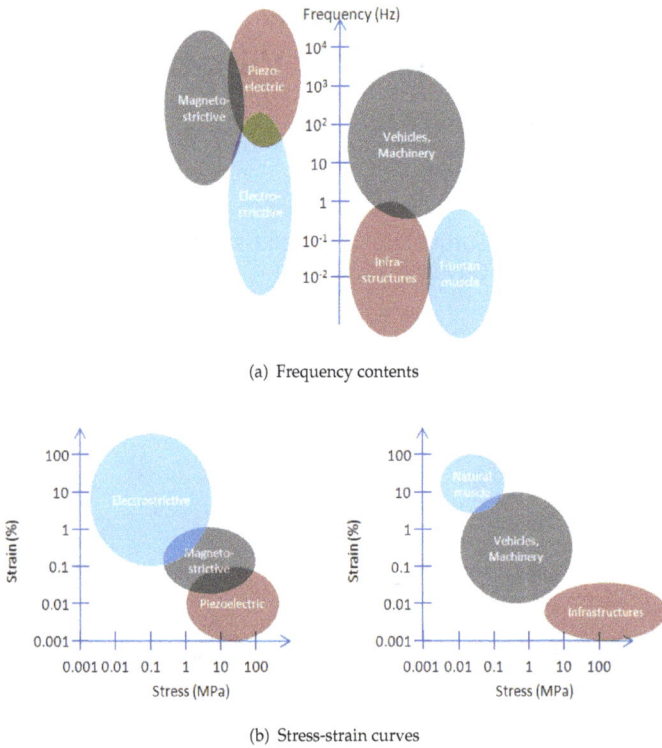

Figure 1. Comparison of (a) frequency contents and (b) stress-strain curves of electromechanical systems and typical applications

matching, thanks to their flexibility (Young's modulus in the range of a few MPa to hundreds of MPa ([5]) - which is much less than piezoelectric polymers). In addition to this high strain capabilities, electrostrictive polymers are cheap and also present high conformability, simple processing, and can be obtained in various shapes over large surfaces.

Hence, some recent studies have considered the use of such materials for harvesting energy from ambient vibrating sources. Then purpose of this chapter is therefore to give an overview of energy harvesting principles using electrostrictive polymers as well as enhancement possibilities both in terms of materials and techniques. The chapter is organized as follows. Section 2 aims at exposing the basic mechanisms of electrostriction allowing the derivation of the constitutive equations. Then material elaboration and enhancement will be exposed in Section 3, together with a figure of merit dedicated to energy harvesting ability assessment allowing a fair comparison of intrinsic material characteristics. Interfaces for efficiently harvesting the converted energy and optimization principles will be exposed in Section 4, as well as realistic implementation issues. Finally, a short conclusion highlighting the main topics and results exposed in this chapter will be summarized in Section 5.

2. Electrostrictive polymers

2.1. Phenomenological approach

Electrostriction effect is defined as a second-order relationship between strain and electrical polarization ([6]). Formulation of the constitutive relationships in terms of polarization is popular within the materials science community, but application-oriented engineers tend to prefer writing constitutive equations in terms of electric field ([7]). The constitutive relationships may include hyperbolic tangents or algebraic powers ([8]); each of these forms being merely variations of the thermodynamic potential. This part starts by the investigation using thermodynamic formalism considering the symmetries inherent to electrostrictive materials. The constitutive relationships are parameterized in terms of electric field. Higher-order algebraic terms are then simplified to quadratic functions.

2.1.1. Thermodynamics formalism

It is possible to describe an electrothermomechanical system by three independent variables chosen from the pairs (stress, T, and strain, S), (electric field E, and electric displacement, D) and (temperature, θ, and entropy, s) ([7]). The other three variables become the dependent variables of the system, which can be found through thermodynamic considerations.

The first law of the thermodynamics describes the conservation of energy in a unit volume. The change of the internal energy, dU, is given by:

$$dU = dQ + dW, \tag{1}$$

where dQ is the infinitesimal quantity of heat and dW is the total work done on the unit volume.

Assuming reversibility, the second law of the thermodynamics relates the increment of heat to the absolute temperature, θ, and the system's entropy, s, by:

$$dQ = \theta ds. \tag{2}$$

The infinitesimal work done by the system is the sum of the mechanical and electrical contributions:

$$dW = T_{ij}dS_{ij} + E_m dD_m \tag{3}$$

Substituting Eqs. (2) and (3) into (1), the change of the internal energy can therefore be expressed as:

$$dU = T_{ij}dS_{ij} + E_m dD_m + \theta ds. \tag{4}$$

Clearly, if S, D and s are chosen as the independent variables, then the dependent variables are:

$$T_{ij} = \left(\frac{\partial U}{\partial S_{ij}}\right)^{D,s} ; \quad E_m = \left(\frac{\partial U}{\partial D_m}\right)^{S,s} ; \quad \theta = \left(\frac{\partial U}{\partial s}\right)^{S,D} , \tag{5}$$

where the superscript indicates that the designated variables are held constant.

2.1.2. Gibbs theory

The first question that must be addressed when writing the constitutive equations is what are the preferred independent variables. For material characterization, it is easier if the independent parameters are the temperature, stress and electrical displacement ([7]). In fact strain is more easily measured than stress and electric field is more easily specified than electrical displacement. This is why the Gibbs free energy function (dG) is typically used:

$$dG = -sd\theta - S_{ij}dT_{ij} - D_m dE_m \tag{6}$$

The direct electrical and mechanical effects are clearly expressed in Eq. (6) but the form of the electromechanical coupling is yet unknown. The electrostrictive term for the direct electrostriction effect is defined by ([9]):

$$M_{ijmn} = \frac{1}{2}\frac{\partial^2 S_{ij}}{\partial E_m \partial E_n}, \tag{7}$$

and converse electrostriction effect is characterized by:

$$M_{mnij} = \frac{1}{2}\frac{\partial^2 D_m}{\partial T_{ij}\partial E_n}. \tag{8}$$

The other electromechanical coupling are defined in the same way.

Assuming a polynomial expansion for all of the internal energies and by neglecting temperature effect, the change in the full Gibbs free energy function becomes:

$$\begin{aligned}
\Delta G = &-\tfrac{1}{2}\epsilon_{mn}E_m E_n - \tfrac{1}{3}\epsilon_{mno}E_m E_n E_o - \tfrac{1}{4}\epsilon_{mnop}E_m E_n E_o E_p - \ldots \\
&-\tfrac{1}{2}s_{ijkl}T_{ij}T_{kl} - \tfrac{1}{3}s_{ijklmn}T_{ij}T_{kl}T_{mn} - \ldots \\
&-u_{mijkl}E_m T_{ij}T_{kl} - r_{mnijkl}E_m E_n T_{ij}T_{kl} - n_{mnoijkl}E_m E_n E_o T_{ij}T_{kl} - \ldots \\
&-d_{mij}E_m T_{ij} - M_{mnij}E_m E_n T_{ij} - g_{mnoij}E_m E_n E_o T_{ij} - h_{mnopij}E_m E_n E_o E_p T_{ij} - \ldots
\end{aligned} \tag{9}$$

where constants have been added to the first two lines for simplicity in later developments. The first line of the Gibbs energy represents the electrical energy terms and the mechanical energy is represented in the second line. The last two lines of Eq. (9) show the coupling between mechanical and electrical energies.

The expressions of the electrical displacement and mechanical strain are then obtained from the partial derivatives of Eq. (6):

$$\left(\frac{\partial G}{\partial E_m}\right)^T = -D_m \text{ and } \left(\frac{\partial G}{\partial T_{ij}}\right)^E = -S_{ij}. \tag{10}$$

Hence, it is then possible to express the constitutive relationships as:

$$\begin{aligned}
D_m = &\epsilon_{mn}E_n + \epsilon_{mno}E_n E_o + \epsilon_{mnop}E_n E_o E_p + \ldots \\
&+u_{mijkl}T_{ij}T_{kl} + 2r_{mnijkl}E_n T_{ij}T_{kl} + 3n_{mnoijkl}E_n E_o T_{ij}T_{kl} + \ldots \\
&+d_{mij}T_{ij} + 2M_{mnij}E_n T_{ij} + 3g_{mnoij}E_n E_o T_{ij} + 4h_{mnopij}E_n E_o E_p T_{ij} + \ldots
\end{aligned} \tag{11}$$

$$\begin{aligned}
S_{ij} = &s_{ijkl}T_{kl} + s_{ijklmn}T_{kl}T_{mn} + \ldots \\
&+u_{mijkl}E_m T_{kl} + 2r_{mnijkl}E_m E_n T_{kl} + 3n_{mnoijkl}E_m E_n E_o T_{kl} + \ldots \\
&+d_{mij}E_m + M_{mnij}E_m E_n + g_{mnoij}E_m E_n E_o + h_{mnopij}E_m E_n E_o E_p + \ldots
\end{aligned}$$

The form of the constitutive relationships in Eq. (11) is very general and, consequently, are not very useful for describing electrostrictive material behavior when used as actuators or microgenerators. The knowledge of the material behavior thus needs to be introduced. The energy formulation for a purely electrostrictive material is simplified by the material symmetry in the perovskite structure, where all odd-rank permittivity terms in the Gibbs energy are necessarily zero ([6, 7]); additionally, $M_{ijmn} = M_{mnij}$. As a result, the piezoelectric terms, d and g, the elastostriction terms, u and n and many of the electrical energy terms are equal to zero ([7]). Neglecting these, the constitutive relationships of an electrostrictive material become:

$$D_m = \epsilon_{mn} E_n + \epsilon_{mnop} E_n E_o E_p + 2r_{mnijkl} E_n T_{ij} T_{kl} + \dots$$
$$+2M_{mnij} E_n T_{ij} + 4h_{mnopij} E_n E_o E_p T_{ij} + \dots$$

$$S_{ij} = s_{ijkl} T_{kl} + s_{ijklmn} T_{kl} T_{mn} + 2r_{mnijkl} E_m E_n T_{kl} + \dots$$
$$+M_{mnij} E_m E_n + h_{mnopij} E_m E_n E_o E_p + \dots$$

$$(12)$$

In the literature ([6]), higher-order terms are typically suppressed from the electrostrictive equation as the associated effect may be neglected, and it then possible to express the constitutive equations as:

$$D_m = \epsilon_{mn}^T E_n + 2M_{mnij} E_n T_{ij}$$

$$S_{ij} = s_{ijkl}^E T_{kl} + M_{mnij} E_m E_n$$

$$(13)$$

The dielectric permittivity, ϵ_{mn}^T, indicates the charge stored in the capacitive element of the electrostrictive material at constant stress. The electrostrictive coefficient, M_{mnij}, is the electromechanical coupling term. The compliance, s_{ijkl}^E, relates stress and strain relationship under constant electric field.

The quadratic model is the form most often quoted in the electrostrictive literature ([5, 10]), since it is very easily measured experimentally. For example, the electrostrictive coefficient M, is found by applying and electric field on an unconstrained (i.e., zero stress) material and measuring the induced strain, or by measuring the short-circuit current delivered by a material submitted to a given strain level.

2.2. Electrostriction using Debye/Langevin formalism

Recently, Capsal et al. also proposed a physical model based on dipolar orientation using a Debye/Langevin formalism for evaluating the actuation abilities of an electrostrictive polymer film ([11]). Using such an approach, it has been demonstrated that the expression of the polarization $\mathcal{P}$ as a function of the electric field E is no longer linear and is given by:

$$\mathcal{P} = N\mu \left[\coth\left(\frac{\mu E}{k_b \theta}\right) - \frac{k_b \theta}{\mu E} \right].$$

$$(14)$$

with N the dipole density, μ the mean dipolar moment of the molecules or particle, θ the temperature and k_b the Boltzmann's constant. Hence, such an approach allows relating the polarization saturation effect that limits electrostriction for high electric fields. Eq. (14) may also be re-written using the low-field susceptibility χ and equivalent saturation electric field

E_{sat} as:

$$\mathcal{P} = 3\chi\epsilon_0 E_{sat} \left[\coth\left(\frac{E}{E_{sat}}\right) - \frac{E_{sat}}{E}\right] \text{ with } \chi = \frac{N\mu^2}{3k_b\theta} \text{ and } E_{sat} = \frac{k_b\theta}{\mu}. \tag{15}$$

Considering that the electrostrictive strain is generated through Maxwell's forces on the material, the electric field-induced strain is thus given by:

$$S = \frac{\epsilon_0}{Y}\left\{1 + 3\chi\left[\left(\frac{E_{sat}}{E}\right)^2 - \mathrm{csch}\left(\frac{E}{E_{sat}}\right)^2\right]\right\}E^2. \tag{16}$$

where csch is the hyperbolic cosecant function and Y the Young's modulus, yielding the equivalent electric-field induced electrostrictive coefficient M_{33}:

$$M_{33} = \frac{\epsilon_0}{Y}\left\{1 + 3\chi\left[\left(\frac{E_{sat}}{E}\right)^2 - \mathrm{csch}\left(\frac{E}{E_{sat}}\right)^2\right]\right\}. \tag{17}$$

whose low-field value for $E \ll E_{sat}$ may be approximated by:

$$M_{33} \approx \frac{(1+\chi)\epsilon_0}{Y}. \tag{18}$$

However, the polarization saturation leads to a decrease of the apparent electrostrictive as the electric field is getting closer to the saturation electric field and which tends to zeros for high electric field values (Figure 2).

3. Material aspect and comparison

This Section aims at exposing the elaboration and enhancement of electrostrictive polymers for energy harvesting purposes. In addition, a figure of merit relating the harvesting abilities of the considered materials from their intrinsic properties ([12]) will be presented and discussed.

3.1. Material properties and enhancement

Electrostrictive polymers are a novel class of electroactive polymers (EAP) that recently became the subject of interest thanks to their high actuation properties and harvesting capabilities ([5, 13–15]). Their lightweight, flexibility, and low mechanical impedance make

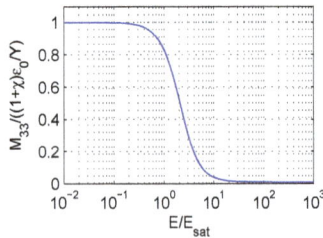

Figure 2. Evolution of the electrostrictive coefficient as a function of the electric field from Debye/Langevin analysis.

them suitable for the development of low-power sensors and actuators. Thereby, this new class of EAP can potentially replace piezoelectric ceramics commonly used as active materials of energy harvesting systems when high flexibility is required, such as smart textiles ([16]).

The main drawback concerns the need of applying high electrical fields to induce polarization when such materials are used as active materials for energy harvesting ([17, 18]). It is thus clear that the intrinsic dielectric properties of the polymer are of prior importance, and a trade-off must be found between stretchability and dielectric properties of the polymer.

Several studies have analyzed and enhanced the energy conversion performance of electrostrictive polymers, both in terms of actuation and energy harvesting ([12, 19, 20]). An ideal approach in order to obtain polymers with specific improved dielectric properties is represented by a challenging synthesis of new molecular architectures. There exist various approaches for obtaining polymer-like blends of known polymers, or copolymerization, and so on. Lehmann *et al.* ([21]) developed a process for synthetically modifying the dielectric properties of liquid-crystalline elastomers; in this type of material, the polarization phenomena can be enhanced by the rearrangement of the lateral group chains and the creation of crystalline regions.

For instance, it has been demonstrated that the easiest way to enhance the dielectric properties of a polymer is the use of inorganic nano-fillers dispersed in a polymer matrix. It significantly increases the harvested energy by increasing the dielectric permittivity ([10, 12]). Two kinds of inorganic fillers are commonly used. In one hand highly dielectric particles allows an increase of the dielectric permittivity without significant modification of the dielectric losses ([22, 23]). Figure 3 presents the volume fraction influence of the ceramic nano-fillers on the relative dielectric permittivity of Barium Titanate/polyamide 11 composite ([22]). The polyamide matrix have a low dielectric permittivity with $\epsilon = 2.5\epsilon_0$ at a frequency of $f = 1$ kHz. Introducing Barium Titanate leads to a four times increase of the dielectric permittivity of the composite. However, because of the significant difference of the dielectric permittivity between the inorganic and organic phase, high content of particles is usually required.

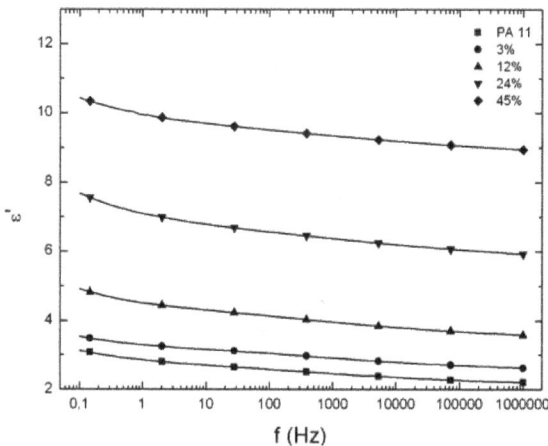

Figure 3. Room temperature dielectric permittivity (ϵ') versus frequency for BaTiO$_3$/Polyamide 11 composites with volume fraction ranging from 0%, to 45%.

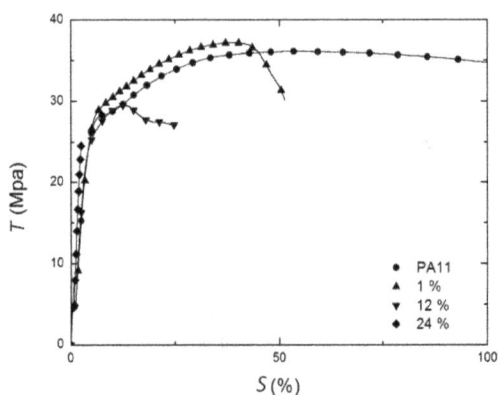

Figure 4. Room temperature stress versus strain measurements for BaTiO$_3$/Polyamide 11 composites with volume fraction ranging from 0%, to 24%.

Incorporating high volume fraction of ceramic fillers in the polymer matrix highly influences the mechanical properties of the polymer. In Figure 4 is depicted the stress versus strain measurements of the ceramic/polymer composites for various volume fraction of fillers ([24]). It can be easily deduced that dispersing high content of fillers not only increases the elastic modulus of the polymer but also highly reduces the breakdown strain. The elastic modulus of the polymer matrix is $E = 400$ MPa and increases to $E = 1.5$ GPa for 24% vol. of inorganic particles. Meanwhile, the strain at break is reduced from 175% for PA11 to 2.5% at 24% vol. of inorganic particles. These composites are therefore not suitable for stretchable energy harvesting systems.

On the other hand, conductive fillers can be used to increase the macroscopic dielectric permittivity. In that case, free charges not only contribute to conduction, but also possibly give rise to Maxwell-Wagner-Sillars (MWS) polarization. MWS polarization is characterized by a huge increase of the low frequency (below 10 Hz) dielectric permittivity at temperature above the glass transition of the polymer, because of charge trapping at heterogeneities ([25]). Conductive particles/polymer composites are prone to show losses with a percolative behavior above a critical weight fraction of conductive particles that depends on the aspect ratio of the fillers. At the percolation threshold, hopping conductive paths are formed between close particles within the matrix ([26, 27]). Unfortunately, the maximum increase in composite permittivity is achieved close to the percolation threshold. According to these results, reducing the stiffening introduced by inorganic fillers and simultaneously exploiting the dielectric enhancement when conductive fillers are introduced to a polymer matrix is very interesting. Many studies have demonstrated that, by carefully controlling the aspect ratio of the particles, the percolation threshold can be lowered down to 5 wt% ([28]) which is an evident advantage in terms of mechanical properties. The filling of the polymer must be done without reaching the percolation threshold and without decreasing the breakdown voltage too much. These two parameters not only depend on the fillers morphology and size ([29]) and on the polymer matrix but also on the dispersion of the fillers in the matrix. Some results obtained by filling highly electrostrictive matrices with conductive nano-fillers are summarized in Table 1. Depending upon the types of fillers (including organic and inorganic conductive fillers), a huge increase of the dielectric permittivity is reported at low filler content. Blending of different polymers with a conductive polymer can result in novel materials with potentially attractive properties.

The different methods available for enhancing the dielectric permittivity of polymers are listed in Table 2 which also gives the advantages and drawbacks of each technique. Random composites represent readily applicable approaches suitable for increasing the dielectric permittivity of polymers. In the long run, the challenge consists in synthesizing a new highly polarizable polymer.

Finally, another approach for greatly reducing the applied voltage consists in using a stack of multilayers of a few microns in thickness. Such a multilayer device has been developed by Choi et al. in [30]. This system was driven at a voltage level of $V = 40$ V, corresponding to

Polymer[a]	Content Fillers	(vol. %)	Relative dielectric permittivity (ϵ/ϵ_0)	M_{33} (10^{-15} m^2.V^{-2})	Ref
PU	None	-	6.8^b	-1^b	[20]
PU	SiC	0.5	10.9^b	-2.5^b	[15]
PU	CB	1	15.4^b	-4^b	[20]
P(VDF-TrFE-CFE)	None	-	65^b	-1.1^b	[12]
P(VDF-TrFE-CFE)	CB	1	95^b	-2.4^b	[12]
P(VDF-TrFE-CFE)	PANI	23	2000^c	-0.15^d	[19]

[a] SiC: silicon carbide; CB: carbon black; PANI: polyaniline
[b] measurements done at 0.1 Hz
[c] measurements done at 1000 Hz
[d] measurements done at 1 Hz

Table 1. Effect of nano-fillers on material properties

	Type of Filler	Advantages	Drawbacks
Random Composites	Inorganic/Dielectric	• High dielectric permittivity	• High filler content • Increase of the elastic modulus
	Inorganic/Dielectric	• High dielectric permittivity for low fillers content	• Increase of the conductivity • Decrease of the voltage breakdown
Polymer Blend	Organic	• No mechanical reinforcement • Very high dielectric permittivity	• Complex process of realization

Table 2. Comparison between the different methods for enhancing the dielectric permittivity

an electric field of $E = 50$ V.μm^{-1}, allowing to overcome all the problems inherent with the use of high voltage power supplies. Such an approach also permits increasing the breakdown electric field according to Paschen's law.

3.2. Material comparison

When comparing the harvesting performance of several energy harvesting devices featuring electrostrictive polymers (Table 3), significant difference can be observed between performance in terms of energy harvesting abilities of electrostrictive polymer-based system, even though the used materials may be very similar. However, as electrostriction requires a mean of activation through the application of an electric field and because the electrical activity is dependent on the mechanical solicitation, external parameters such as maximum electric field and strain applied to the system significantly affect the output power of the device. Hence, in order to have a fair comparison in terms of material aspects, it is mandatory to develop a figure of merit taking into account the intrinsic parameters of the material only, independently from external environmental parameters.

In order to assess the energy harvesting abilities of a given electrostrictive element independently from external applied parameters, it is considered that the material is connected to a constant voltage generator trough a load that is used to mimic the connected electrical system[1] (Figure 5). Considering such a scheme and from the linear constitutive equations of electrostriction (Eq. (13)) as a function of the strain, it is possible to express the current I delivered by the polymer as ([12]):

	Ren et al. ([17])	Cottinet et al. ([18])	Lallart et al. ([12])
Material	Irradiated copolymer $(PVDF - TrFE)$	Polyurethane	Terpolymer $(PVDF - TrFE - CFE)$ + 1% CB
Strain level (%)	3	8×10^{-3}	0.7
Maximum electric field (V.μm^{-1})	67	5	10
Energy density (J.cm^{-3})	40×10^{-3}	20×10^{-12}	170×10^{-6}

Table 3. Energy harvesting performance of electrostrictive polymer-based systems

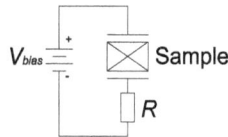

Figure 5. Energy harvesting circuit

---

[1] Although energy harvesting systems usually requires DC output voltage for realistic applications, the use of a single load is employed here as an approximation. Furthermore, some DC harvesting systems may use AC to DC converters that are are seen as a purely resistive load by the active element ([31]).

$$I = \frac{2\Lambda M_{31} Y E_{dc}}{1 + 2j\pi f \frac{\Lambda \epsilon_{33}^T}{l} R} 2j\pi f S_1, \tag{19}$$

where Λ, l, f, S_1 and E_{DC} respectively refer to the sample surface area, sample thickness, frequency, longitudinal strain and bias electric field, and assuming small-signal behavior (low current and electric field AC components). Hence, it is possible to derive the harvested power P on the load, yielding:

$$P = \frac{2R \left(\Lambda 2\pi f M_{31} Y E_{DC} \right)^2}{1 + \left(\frac{\Lambda \epsilon_{33}^T}{l} 2R\pi f \right)^2} S_M^2, \tag{20}$$

with S_M the strain magnitude. Hence, the maximum power at the optimal load is given by ([12]):

$$P_{max} = \frac{2\pi}{\epsilon_{33}^T} (M_{31} Y)^2 \Lambda l f E_{DC}^2 S_M^2. \tag{21}$$

Figure 6 presents the comparison of experimental maximum harvested power for well-known electrostrictive materials as well as the comparison with the predicted harvesting abilities (obtained from experimentally measured electrostrictive coefficient, permittivity and Young's modulus[2]), showing a very good agreement between measured and theoretically estimated data.

From the previous expression, it can be seen that the right part of the right side member refers to external parameters (dimensions, frequency, bias electric field and strain magnitude), while the left part allows defining a material figure of merit F from its intrinsic parameters as:

$$F = \frac{2\pi}{\epsilon_{33}^T} (M_{31} Y)^2, \tag{22}$$

which depends on the inverse permittivity, squared electrostrictive coefficient and squared Young's modulus, and whose dimensions are $J.m^{-3}.(m/m)^{-2}.(V/m)^{-2}.cycle^{-1}$ (energy density per squared strain level per squared electric field magnitude per cycle), or $J.m^{-1}.V^{-2}.cycle^{-1}$ in contracted form.

It is also possible to represent such a figure of merit in a graphical way, by plotting the squared product of the electrostrictive coefficient by the Young's modulus as a function of the inverse permittivity, leading to the chart depicted in Figure 7. From this Figure, it can be seen that the terpolymer outperforms the other considered samples, although the high Young's modulus of such a material limits the maximum strain that can be applied to the device. As well, the enhancement offered by the previously exposed permittivity increase approach using nano-filler incorporation can be demonstrated through the proposed criterion, both for polyurethane and terpolymer.

In order to assess the correctness of the proposed figure of merit, Table 4 shows the comparison of several other criteria with the proposed one (normalized with results for pure polyurethane), demonstrating the ability of the exposed figure of merit for accurately predicting the harvesting abilities of a given electrostrictive material compared to a reference one, while other factors do not relate quite well the actual performance, as they are not based on the direct evaluation of energy harvesting capabilities. It can also be noted that,

---

[2] The value of the electrostrictive coefficient has been obtained from short-circuit current measurement, while permittivity and Young's modulus were evaluated using a LCR meter and force-displacement monitoring.

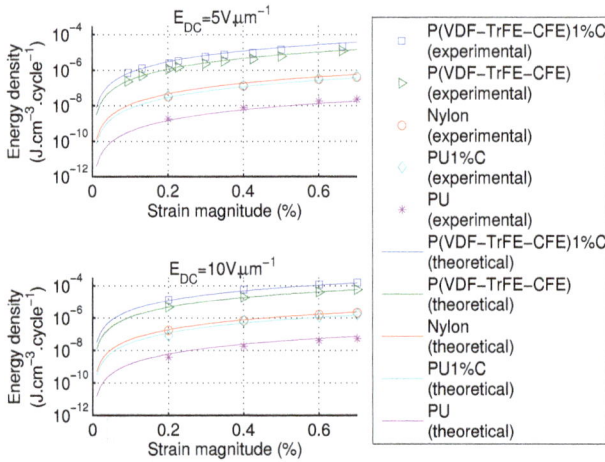

(a) Constant bias electric field

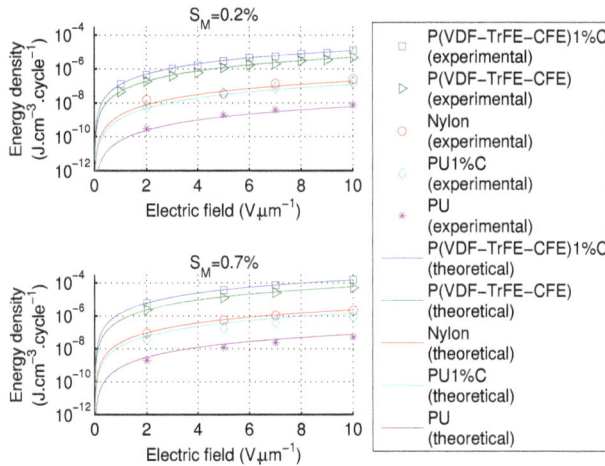

(b) Constant strain level

Figure 6. Experimental and predicted maximal harvested power using several electrostrictive polymers considering different bias electric fields and strains (frequency: 100 Hz).

as considered polymers belongs to different classes, the empirical law exposed by Eury *et al.* in ([5]) stating that the product of the electrostrictive coefficient by the Young's modulus $M_{31}Y$ is proportional to the squared product of the difference between material permittivity and vacuum permittivity (ϵ_0) divided by the material permittivity $\left(\epsilon_{33}^T - \epsilon_0\right)^2 / \epsilon_{33}^T$ leads here to inaccurate results.

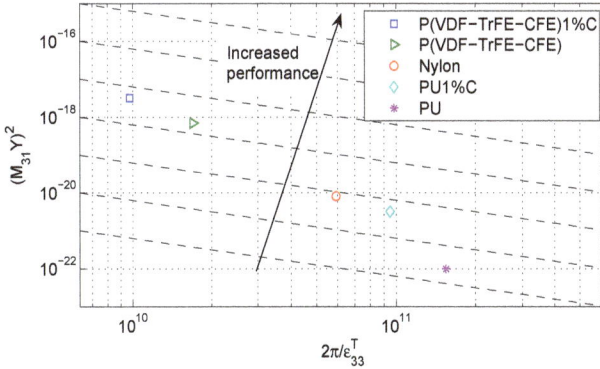

Figure 7. Comparison of several electrostrictive polymers using the energy harvesting figure of merit.

Material	Type of figure of merit				Experimental harvested power
	ϵ_{33}^T (energy conversion)	$(M_{31}Y)^2$ (power at constant load - [32])	$\frac{2\pi}{\epsilon_{33}^T}(M_{31}Y)^2$ (harvested energy - [12])	$\frac{\left(\epsilon_{33}^T-\epsilon_0\right)^4}{\epsilon_{33}^{T\,3}}$ (harvested energy considering Eury's law - [5])	
Polyurethane	1	1	1	1	1
Polyurethane +1%C	1.63	32	20	2.45	21.5
Nylon	2.61	83	31.7	4.91	32.9
Terpolymer	9.13	7056	773	22.1	731
Terpolymer +1%C	15.9	32400	2040	40	2060

Table 4. Comparison of several figures of merit for the evaluation of normalized energy harvesting performance.

Hence, applying this figure of merit to Table 3 by dividing the energy density by the squared electric field and squared strain level leads to the new table presented in Table 5, which reflects in a much better way the intrinsic material abilities for harvesting energy. Finally, as the previous development assumed linear behavior, it can also be noted that other parameters such as maximum admissible electric field, maximum strain level or saturation electric field may additionally be taken into account to precisely evaluate the performance in terms of energy scavenging from material aspect.

4. Energy harvesting techniques

The goal of this Section is to expose energy harvesting interfaces for efficiently extracting the converted energy to the storage stage. Basically, two global approaches can be adopted for such a purpose: either the electroactive material can be submitted to charge and discharge

	Ren *et al.* ([17])	Cottinet *et al.* ([18])	Lallart *et al.* ([12])
Material	Irradiated copolymer $(PVDF - TrFE)$	Polyurethane	Terpolymer $(PVDF - TrFE - CFE)$ + 1% CB
Figure of merit $(J.m^{-1}.V^{-2}.cycle^{-1})$	10×10^{-9}	125×10^{-12}	34×10^{-9}

Table 5. Energy harvesting performance of electrostrictive polymer-based systems

cycles (in a similar fashion that electrostatic devices - [33]), or a bias electric field can be applied, which allows an equivalent piezoelectric behavior in dynamic mode. In the following development, it will be considered that the system is submitted to a constant strain level, and backward coupling that limits the strain value under a given stress magnitude will be neglected, as the coupling in electrostrictive polymers is usually low for moderate electric fields. In addition, it will be considered that the strain levels are quite low ($< 10\%$), so that the thickness and surface changes are limited, and thus the modifications in the electric field and electric displacement due to changes in sample dimensions may be neglected.

4.1. Charge/discharge cycles

Because of the capacitive behavior of electrostrictive dielectric polymers, classical electrostatic cycles as exposed in [33] can be applied or adapted, which consist in electric field application and release cycles. The basic operations of such an energy harvesting approach can either consider constant electric field (Ericsson cycle) or constant charge (Stirling cycle), as depicted in Figure 8. In both cases however, the electrical charge has to be applied when the capacitance is highest and released when it is the lowest. Considering the electrical constitutive equation in Eq. (13) when the material is submitted to longitudinal strain, with T and D used as independent variables:

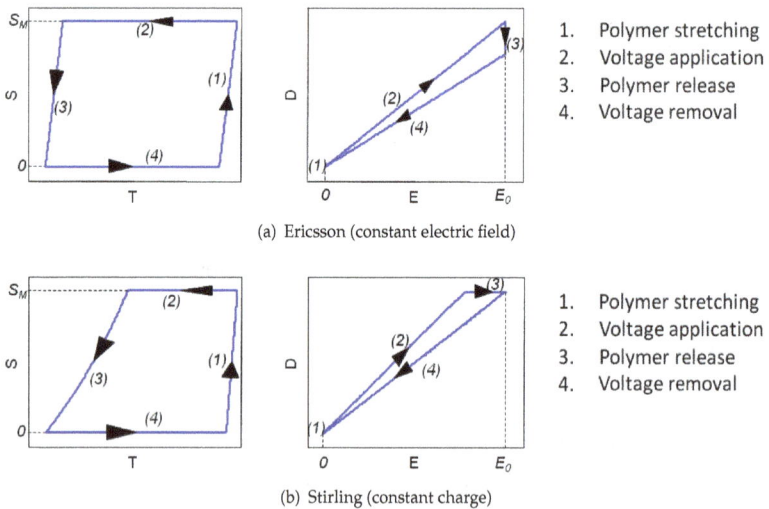

1. Polymer stretching
2. Voltage application
3. Polymer release
4. Voltage removal

(a) Ericsson (constant electric field)

1. Polymer stretching
2. Voltage application
3. Polymer release
4. Voltage removal

(b) Stirling (constant charge)

Figure 8. Electrostatic energy harvesting cycles and mechanical cycles for electrostrictive polymers.

$$D_3 = \epsilon_{33}^S E_3 + 2M_{31}YE_3S_1 \text{ with } M_{31} > 0, \tag{23}$$

the charge-voltage relationship is therefore given as:

$$Q = \left(\frac{\epsilon_{33}^S \Lambda}{l} + 2\frac{M_{31}Y\Lambda}{l}S_1 \right) V \text{ with } \frac{M_{31}Y\Lambda}{l} > 0, \tag{24}$$

with Q and V denoting the electrical charge and voltage, respectively and Λ and l the sample surface area and sample thickness.

Hence, the charge should be done when the strain is maximum (maximum capacitance) and the discharge should occur when the polymer is released (minimum capacitance). When doing so, it can be demonstrated that the harvested energy density per cycle is given by ([34, 35]):

$$W_V = M_{31}YS_M E_0^2$$
$$W_Q = \left(1 + 2\frac{M_{31}Y}{\epsilon_{33}^S}S_M\right)M_{31}YS_M E_0^2, \tag{25}$$

where W_V and W_Q refer to the harvested energy densities using Ericsson and Stirling cycles, respectively, and E_0 denotes the applied electric field.

However, such cycles may also be adapted specifically to electrostrictive material considering a non-zero initial electric field. In this case, E_0^2 is replaced by $(E_0^2 - E_{init}^2)$ in Eq. (25), with E_{init} denoting the initial electric field when the longitudinal strain is zero. Obviously, this would lead to reduced energy harvesting abilities. However, the application of an initial electric field permits a cycle combining Ericsson and Stirling approaches using constant voltage stretching and constant charge release (Figure 9), yielding a harvested energy density W_{QV}^3 ([34, 35]):

$$W_{QV} = 2\frac{(M_{31}Y)^2}{\epsilon_{33}^S}S_M^2 E_{init}^2. \tag{26}$$

However, the main drawback of these approaches is the need of continuously controlling a voltage source or the polymer voltage, which may compromise the operation of the system as the energy requirements for driving the voltage source may be greater than the harvested energy, yielding a negative energy balance and hence unrealistic operations. In order to counteract this drawback, it has been proposed in ([34–36]) a purely passive cycle consisting of two voltage sources a two diodes as depicted in Figure 10.

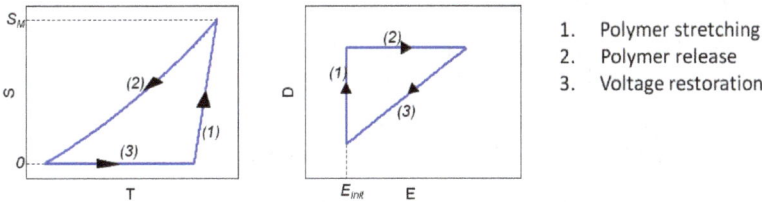

1. Polymer stretching
2. Polymer release
3. Voltage restoration

Figure 9. Energy harvesting cycle using hybrid Stirling/Ericsson combination.

---

[3] It may be interesting to note that such the expression of W_{QV} explicitly makes the figure of merit exposed in the previous section appearing.

(a) Schematic

1-2. Polymer stretching
3-4. Polymer release

(b) Associated cycle

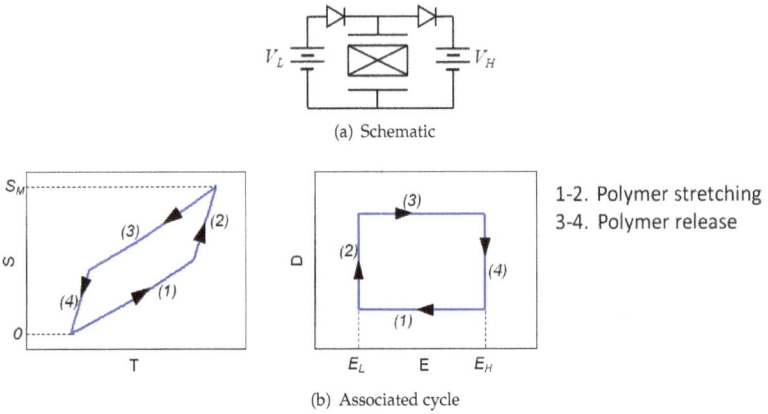

Figure 10. Passive energy harvesting cycle.

With such an approach, the voltage on the electrostrictive polymer is decreasing as it is stretched (as the system is operating at constant charge), until it reaches the low voltage value V_L. As the strain is further increased the polymer is charged by V_L until it is totally stretched. Then, as the longitudinal strain is decreased, the material voltage increases as well until it reaches the high voltage V_H ($V_H > V_L$), where a charge flow appears from the electroactive device to V_H, yielding an energy extraction process. Considering that E_L and E_H are the electric fields respectively associated to V_L and V_H, the harvested energy density is given by:

$$W_{passive} = \frac{(M_{31}Y)^2}{\epsilon_{33}^S + 2M_{31}YS_M}S_M{}^2 E_H{}^2. \tag{27}$$

However, in order to effectively reach E_H and therefore allowing the energy harvesting process, the following inequality between maximum strain and constant voltage source values has to be fulfilled:

$$V_H < 2\frac{M_{31}Y}{\epsilon_{33}^S}S_M V_L \tag{28}$$

4.2. Pseudo-piezoelectric mode

In the charge/discharge energy harvesting cycles, the use of voltage sources that need to be tuned may compromise the realistic implementation of the harvester[4] . In order to avoid such an issue, it is also possible to keep the bias electric field applied on the sample and consider dynamic operations. When doing so, the constitutive equations of electrostriction in such a dynamic mode with D and T as independent variables turn to:

$$dD = \epsilon_{33}^S d\,(E_{DC} + E_{AC}) + 2M_{31}Yd\,[(E_{DC} + E_{AC})\,S]$$
$$dT = YdS - M_{31}Yd\,(E_{DC} + E_{AC})^2, \tag{29}$$

---

[4] This statement is not true for the passive circuit which however features modest energy harvesting abilities as it will be shown in Section 4.3.

where the electric field is decomposed into its bias and time-dependent components ($E = E_{DC} + E_{AC}$). Assuming that the DC component is much higher than the AC one, these expressions may be approximated by:

$$dD \approx \left(\epsilon_{33}^S + 2M_{31}YS\right) dE_{AC} + 2M_{31}YE_{DC}dS$$

$$dT \approx YdS - 2M_{31}YE_{DC}dE_{AC},$$

(30)

which is very close to constitutive equations of piezoelectricity with an equivalent piezoelectric coefficient $e = 2M_{31}YE_{DC}$. Hence, because of this similarity, it is possible to apply any existing technique available for piezoelectric energy harvesting to electrostrictive materials undergoing a bias electric field and considering dynamic operations.

4.2.1. AC mode

The simplest way for harvesting energy is to directly connect a purely resistive load R to the material (Figure 11). Assuming sine excitation, the harvested power on the load yields[5] ([12, 35]):

$$P_{AC} \approx \frac{2R\left(2\pi f \Lambda M_{31}Y\right)^2}{1+\left(2\frac{\Lambda \epsilon_{33}^S}{l}R\pi f\right)^2}E_{DC}{}^2 S_M{}^2,$$

(31)

with S_M the strain magnitude. Cancelling the derivative of this expression with respect to the load gives the optimal load $R_{AC}|_{opt}$:

$$R_{AC}|_{opt} = \frac{1}{2\pi \frac{\Lambda \epsilon_{33}^S}{l} f}$$

(32)

that leads to the maximum power ([12, 35]):

$$P_{AC}|_{max} \approx \frac{2\pi}{\epsilon_{33}^S}\left(M_{31}Y\right)^2 \Lambda l f E_{DC}{}^2 S_M{}^2,$$

(33)

and thus the maximum harvested energy density per cycle is given by:

$$W_{AC}|_{max} \approx \frac{\left(M_{31}Y\right)^2}{\epsilon_{33}^S}E_{DC}{}^2 S_M{}^2,$$

(34)

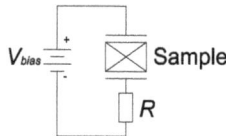

Figure 11. AC Energy harvesting circuit

---

[5] see Section 3.2 for the full development

The corresponding energy cycles are given in Figure 12, where the mean value of the electric field is approximately E_{DC}. Hence, unlike the previous cycles that consisted in changing the electrical boundaries at constant mechanical excitation and conversely, the use of the pseudo-piezoelectric mode leads to a continuous change in the electrical and mechanical quantities and therefore no curve breaking appears in the mechanical and electrical cycles.

4.2.2. DC mode

However, for the realistic application of energy harvesting devices, a DC output is often desirable. Although some AC/DC converters that are seen as resistive loads by the material have been proposed in the literature ([31]), most of the used architectures rely on a simple rectifier bridge with a smoothing capacitor, as depicted in Figure 13(a). The load may also be replaced by DC/DC converters operating in discontinuous mode for impedance matching ([37–39]). The principles consist in filtering the DC component introduced by the bias voltage source used for polarization purpose (through capacitance C_d) and then rectifying the voltage and filtering it. Instead of using a full diode voltage rectifier, the use of a voltage doubler in Figure 13(a) allows limiting the losses introduced by the voltage gaps of discrete components. In addition, in order to avoid a dynamic short circuit, a high value series resistance R_S is added between the polymer and the bias voltage source. Such operations therefore lead to the energy cycles shown in Figure 13(b).

When using such an approach, it can be shown that the harvested power may be approximated by ([35, 40]):

$$P_{DC} \approx \frac{(8f\Lambda M_{31}Y)^2 R}{\left(1 + 4\frac{4\Lambda \epsilon_{33}^S}{l}Rf\right)^2} E_{DC}{}^2 S_M{}^2, \tag{35}$$

and the maximal energy density per cycle value is given by:

$$W_{DC}|_{max} \approx 4\frac{(M_{31}Y)^2}{\epsilon_{33}^S} E_{DC}{}^2 S_M{}^2 \tag{36}$$

obtained for the optimal load $R_{DC}|_{opt}$:

$$R_{DC}|_{opt} = \frac{1}{4\frac{\Lambda \epsilon_{33}^S}{l}f} \tag{37}$$

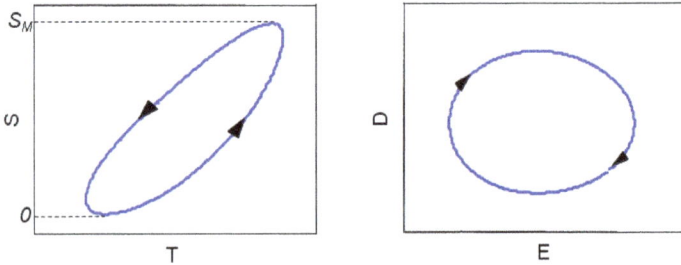

Figure 12. Energy harvesting cycle using AC pseudo-piezoelectric mode.

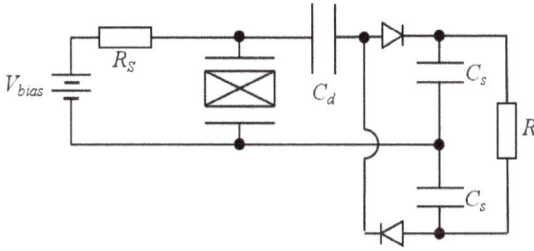

(a) Schematic

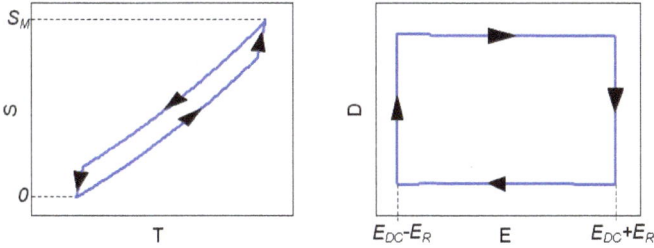

(b) Energy cycles[a]

Figure 13. Pseudo-piezoelectric DC energy harvesting: (a) schematic; (b) energy cycles.

[a]E_R is the equivalent DC electric field accross the load

4.2.3. Nonlinear conversion enhancement in pseudo-piezoelectric mode

Because of the similarities between electrostrictive polymers operating in dynamic mode and piezoelectric element, it is also possible to apply nonlinear processing to artificially enhance the conversion abilities of the material ([41–47]). The principles of this treatment consist in (imperfectly) inverting the active element voltage (with reference to the bias voltage) each time a maximum or a minimum strain value is reached (Figure 14), by briefly connecting the material to an inductance (hence shaping a resonant electrical network). When using such an approach, it can be shown that the harvested power is given by ([35, 48]):

$$P_{AC\_sw} \approx \frac{R(2\Lambda M_{31}Y)^2}{1+\left(2\frac{\Lambda e_{33}^S}{l}R\pi f\right)^2}$$

$$\times \left[\frac{\left(2\frac{\Lambda e_{33}^S}{l}R\pi f\right)^3}{1+\left(2\frac{\Lambda e_{33}^S}{l}R\pi f\right)^2}\frac{(1+\gamma)}{\left(e^{2\frac{\pi}{\frac{\Lambda e_{33}^S}{l}}R\pi f}-\gamma\right)^2}\frac{\left(e^{\frac{\pi}{\frac{\Lambda e_{33}^S}{l}}R\pi f}-1\right)^2}{\pi}+1\right]E_{DC}^2 S_M^2, \quad (38)$$

with γ the inversion coefficient giving the absolute ratio of the voltage after the inversion process over the voltage before the inversion (referenced to V_{bias}) and denoting the losses during the switch ($0 \leq \gamma \leq 1$).

(a) Schematic

(b) Energy cycles

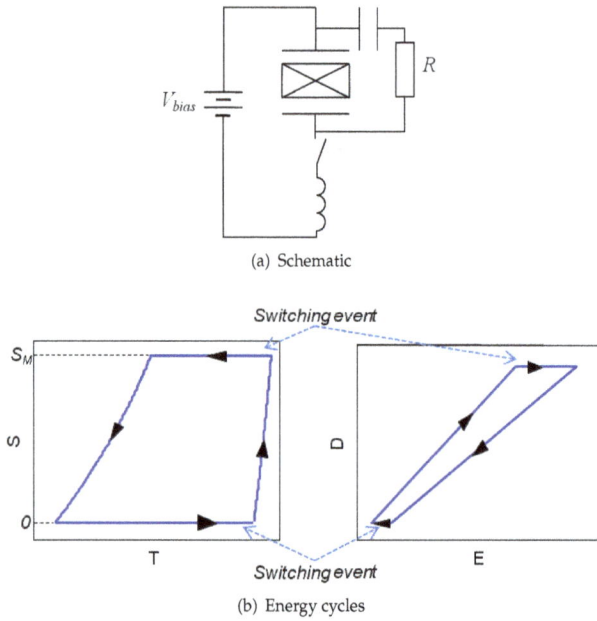

Figure 14. Pseudo-piezoelectric AC energy harvesting using nonlinear treatment: (a) schematic; (b) energy cycles.

Obviously, the combination of the DC approach with the nonlinear treatment is possible (Figure 15), leading to the harvested power expression ([35, 40]):

$$P_{DC\_sw} \approx \frac{(8f\Lambda M_{31}Y)^2 R}{\left(1 + 2(1-\gamma)\frac{\Lambda\epsilon_{33}^S}{l}Rf\right)^2} E_{DC}{}^2 S_M{}^2, \tag{39}$$

yielding the maximal harvested energy density per cycle:

$$W_{DC}|_{max} \approx \frac{8}{(1-\gamma)} \frac{(M_{31}Y)^2}{\epsilon_{33}^S} E_{DC}{}^2 S_M{}^2 \tag{40}$$

which is $2/(1-\gamma)$ times higher and obtained for an optimal load $R_{DC\_sw}|_{opt}$:

$$R_{DC\_sw}|_{opt} = \frac{1}{2(1-\gamma)\frac{\Lambda\epsilon_{33}^S}{l}f} \tag{41}$$

In the previous analysis, it was considered that the switching circuit is placed in parallel with the harvesting circuit, leading to an inversion process occurring after the harvesting process. However, this element can also be connected in series between the active material and the AC/DC conversion stage, yielding a harvesting process that happens at the same time than the switching event, and thus to a pulsed energy extraction system ([43]). Although the

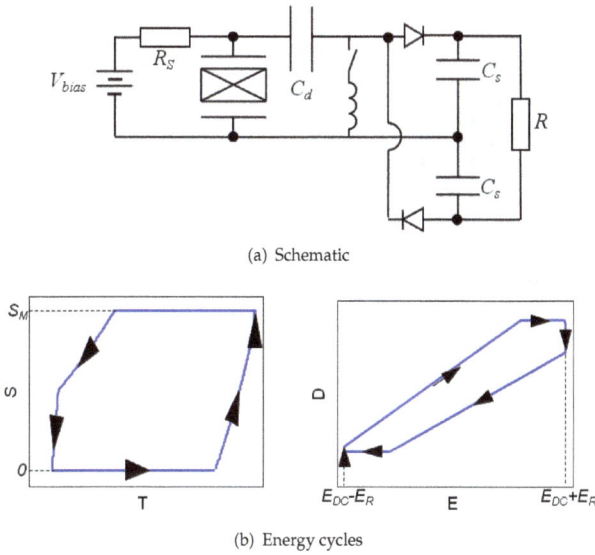

(a) Schematic

(b) Energy cycles

Figure 15. Pseudo-piezoelectric DC energy harvesting using nonlinear treatment: (a) schematic; (b) energy cycles.

maximum energy harvested with the series configuration is slightly less than in the parallel case (the gain compared to the standard case being $(1+\gamma)/(1-\gamma)$ instead of $2/(1-\gamma)$), the optimal load is much less which could be advantageous for limiting the losses and ensuring a better load adaptation.

4.3. Comparison, discussion & implementation issues

Figure 16 presents the theoretical performance comparison between the previously exposed harvesting techniques. Obviously, the electrostatic-derived cycles perform best, followed by the pseudo-piezo DC interface using the nonlinear treatment. Although very simple, the passive cycle using diodes features the lowest energy harvesting abilities.

Nevertheless, this comparison is obtained by neglecting the losses within the system. In particular, the cyclic voltage application and release in electrostatic cycles would lead to significant losses that may compromise the realistic implementation of the techniques. Hence, assuming that the energy transfer from the source to the electrostrictive polymer is done with an efficiency η_{prov} and that the energy extraction has an efficiency of η_{extr}, it can be shown that the harvested energy density is then given by, in the case of the Ericsson cycle[6] :

$$W_{harvested}|_{Ericsson} = \frac{\eta_{extr}}{2\eta_{prov}\eta_{extr}}\left[(\eta_{extr}\eta_{prov}-1)\,\epsilon_{33}^S + 2\,(2\eta_{extr}\eta_{prov}-1)\,M_{31}YS_M\right]E_0{}^2 \quad (42)$$

---

[6] Although being a little bit less efficient than the Stirling cycle, the Ericsson cycle is often preferred as it permits controlling the maximum electric field applied on the sample.

156 Energy Harvesting

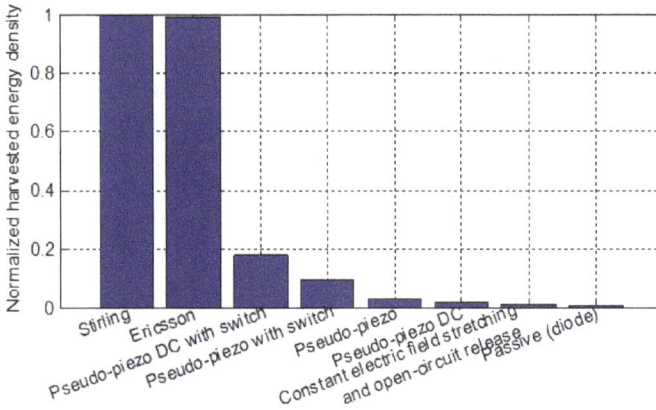

Figure 16. Maximum energy density (normalized with respect to the maximum one) of the energy harvesting techniques ($\gamma = 0.8$ for the nonlinear processing techniques).

which becomes negative as soon as:

$$\eta_{prov}\eta_{harv} < \frac{\epsilon_{33}^S + 2M_{31}YS_M}{\epsilon_{33}^S + 4M_{31}YS_M},\tag{43}$$

Figure 17 depicts the minimum value of the product of these efficiencies as a function of the strain level in order to have a positive energy balance in the case of a polyurethane PS 2000 polymer ([49]). For low strain magnitude values, this product should be close to 1, meaning that the energy transfer from the source and to the storage stage should be perfect. In addition, for high strain levels, the minimum efficiency product has to be greater than 0.5 (which can also be shown by Eq. (43) as $S_M \rightarrow \infty$), placing a significant constraint on the system design. Especially, directly applying a step voltage on the polymer yields an efficiency of 50%, and thus no energy can be harvested using such an approach, and consequently a careful attention has to be placed on the way to apply the electric field when charging the polymer.

On the other hand, when using pseudo-piezoelectric approaches and assuming no significant voltage gap of the discrete components, the origin of losses lies in the static application of the electric field, yielding a current flow because of the intrinsic losses in the polymer. The energy lost per cycle in this case is therefore a function of the equivalent parallel resistance R_p of the sample and is given by:

$$W_{lost} = \frac{V_{bias}^2}{fR_p},\tag{44}$$

which has to be less than the harvested energy (see Section 4.2) to have a positive energy balance. In particular, this energy loss tends to zero as the frequency increases, meaning that pseudo-piezoelectric mode is very well adapted to relatively high frequency operations. As an example, it has been estimated in ([18]) that, in the case of a polyurethane material with a bias electric field of 5 V.μm^{-1} for polarization purposes operating at a frequency of 20 Hz, the losses represent less than 0.5% of the harvested energy.

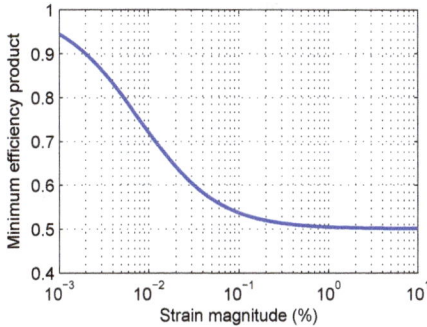

Figure 17. Minimum product of the injection and extraction efficiencies as a function of the strain for polyurethane PS 2000 material ($M_{31} = 5 \times 10^{-18}$ m^2.V^{-2}; $\epsilon_{33}^S = 6.1\epsilon_0$; $Y = 33.8$ MPa - [49]).

Therefore, although electrostatic-based harvesting schemes seem to be the most appealing ones, the losses when using such charge/discharge approaches may compromise the realistic operations of the system, yielding a negative energy balance. On the other hand, pseudo-piezoelectric operations feature reduced losses, especially at relatively high frequency, making them more suitable under some circumstances.

Finally, another concerns about the use of electrostrictive polymers for energy harvesting is the necessity of applying relatively high voltage to activate the electromechanical behavior of the material. Although very few research has been conducted on the subject, the use of efficient integrated DC/DC converters ([50]) or hybridation with piezoelectric materials ([35]) has been proposed.

5. Conclusion

This chapter exposed the use of electrostrictive polymers for mechanical energy harvesting. Thanks to their lightweight, flexibility and easy fabrication process, such materials are of premium choice for harvesting energy from high strain, low frequency systems.

First, the constitutive equations of electrostriction have been presented from a phenomenological approach using either Gibbs approach or Debye/Langevin formalism, giving a physical meaning to electrostriction.

Then, material aspect has been discussed. It has been shown that the simplest way for enhancing the electrostrictive activity lies in the incorporation of nano-fillers which allows increasing interfacial effects and thus electromechanical conversion abilities, although decreasing mechanical and electrical strengths. Another approach would consists in the synthesis of new polymer architectures, which is however more complex. A figure of merit allowing the comparison of electrostrictive materials in terms of energy harvesting abilities independent from external parameters has also been developed, emphasizing the parameters to optimize for the elaboration of efficient materials for energy harvesting purposes.

Finally, techniques for efficiently extracting and harvesting the converted energy have been exposed. In particular, two kinds of techniques have been considered, whether the system is subjected to charge and discharge cycles, or operating in pseudo-piezoelectric mode in a dynamical fashion. It has therefore been shown that, although electrostatic-based cycles feature the highest conversion abilities, losses within the system may compromise the realistic operation of the device because of a negative energy balance, while pseudo-piezoelectric operations present limited losses that make them particularly suitable for relatively high frequency operations (> 1 Hz).

In summary, electrostrictive polymers are particularly interesting materials for harvesting energy for large stroke systems (such as human motions), but their real application still requires significant advances, both in terms of materials or electrical interfaces, especially for the application of the bias electric field (using multilayer structures or efficient electronic interfaces for example).

Author details

Mickaël Lallart, Pierre-Jean Cottinet, Jean-Fabien Capsal,
Laurent Lebrun and Daniel Guyomar
Université de Lyon, INSA-Lyon, LGEF EA 682, F-69621, Villeurbanne, France

6. References

[1] Lallart M, Guyomar D, Jayet Y, Petit L, Lefeuvre E, Monnier T, Guy P, Richard C. Synchronized Switch Harvesting applied to Selfpowered Smart Systems: Piezoactive Microgenerators for Autonomous Wireless Receiver. Sens. Act. A: Phys. 2008; 147(1): 263-272. doi: 10.1016/j.sna.2008. 04.006

[2] Roundy S, Wright PK, Rabaey J. A study of low level vibrations as a power source for wireless sensor nodes. Comp. Comm. 2003; 26: 1131-1144.

[3] Beeby SP, Tudor MJ, White NM. Energy harvesting vibration sources for microsystems applications. Meas. Sci. Technol. 2006; 17: R175-R195.

[4] Anton SR, Sodano HA. A review of power harvesting using piezoelectric materials (2003Ű2006). Smart Mater. Struct. 2007; 16(3): R1-R21.

[5] Eury S, Yimniriun R, Sundar V, Moses PJ. Converse Electrostriction in Polymers and Composites. Mat. Chem. Phys. 1999; 61(1): 18-23.

[6] Devonshire AF. Theory of Ferroelectrics. Adv. Phys. 1954; 3: 85-130.

[7] Damjanovic D. Ferroelectric, dielectric and piezoelectric properties of ferroelectric thin films and ceramics. Rep. Prog. Phys. 1998; 61: 1267. doi:10.1088/0034-4885/61/9/002

[8] Sterkenburg SWPv, Kwaaitaal T, van den Eijnden WMMM. A double Michelson interferometer for accurate measurements of electrostrictive constants. Rev. Sci. Instrum. 1990; 61(9): 2318-2322. http://dx.doi.org/10.1063/1.1141357

[9] Blackwood G, Ealey MA. Electrostrictive behavior in lead magnesium niobate (PMN) actuators. Part I: materials perspective. Smart Mater. Struct. 1993; 2: 123-133.

[10] Guiffard B, Guyomar D., Seveyrat L, Chowanek Y, Bechelany M, Cornu D & Miele P. Enhanced Electroactive Properties of Polyurethane Films Loaded With Carbon-Coated SiC Nanowires. J. Phys. D.: Appl. Phys. 2009; 42(5): 055503.1-055503.6.

[11] Capsal JF, Lallart M, Cottinet PJ, Galineau J, Sébald G, Guyomar D. Evaluation of macroscopic polarization and actuation abilities of electrostrictive dipolar polymers using microscopic Debye/Langevin formalism. J. Phys. D.: Appl. Phys. 2012; 45(20): 205401.

[12] Lallart M, Cottinet PJ, Lebrun L, Guiffard B, Guyomar D. Evaluation of Energy Harvesting Performance of Electrostrictive Polymers and Carbon-filled Terpolymer Composites. J. Appl. Pol. Sci. 2010; 108(3): 034901.

[13] Liu R, Zhang Q, Cross LE. Experimental Investigation of Electrostrictive Polarization Biased Direct Apparent Piezoelectric Properties in Polyurethane Elastomer under Quasistatic Conditions. J. Appl. Pol. Sci. 1999; 73: 2603-2609.

[14] Klein RJ, Runt J, Zhang QM. Influence of Crystallization Conditions on the Microstructure and Electromechanical Properties of Poly(vinylidene fluoride-trifluoroethylene-chlorofluoroethylene) Terpolymers. Macromol. 2003; 36(19): 7220-7226.

[15] Guiffard B, Severat L, Sebald G, Guyomar D. Enhanced Electric Field Induced Strain in Non Percolative Carbon Nanopowder/Polymer Composites. J. Phys. D: Appl. Phys. 2006; 39: 3053-3056.

[16] De Rossi D, Carpi F, Galantini F. Functional Materials for Wearable Sensing, Actuating and Energy Harvesting. Adv. Sci. Tech. 2008; 57: 247-256.

[17] Ren K, Liu Y, Hofmann HF Zhang QM, Blottman J. An Active Energy Harvesting Scheme with an Electroactive Polymer. Appl. Phys. Lett. 2007; 91(13): 132910.

[18] Cottinet PJ, Guyomar D, Guiffard B, Putson C, Lebrun L. Modelling and Experimentations on an Electrostrictive Polymer Composite for Energy Harvesting. IEEE Trans. UFFC 2010; 57(4): 774-784.

[19] Huang C, Zhang QM, Su J. High Dielectric Constant All Polymer Percolative Composite. Appl. Phys. Lett. 2003; 82: 3502.

[20] Wongtimnoi K, Guiffard B, Bogner Van de Moortèle A, Seyverat L, Gauthier C, Cavaillé JY. Improvement of Electrostrictive Properties of a Polyether-based Polyurethane Elastomer Filled with Conductive Carbon Black. Comp. Sci. Tech. 2011; 71(6): 885-891.

[21] Lehmann W, Skupin H, Tolksdorf C, Gebharde E, Zentel R, Krüger P, Lôsche M, Kremer F. Giant Lateral Electrostriction in Ferroelectric Liquid-Crystalline Elastomers. Nature 2001; 410(6827): 447-450.

[22] Capsal JF, Dantras E, Lacabanne C. Molecular Mobility in Piezoelectric Hybrid Nanocomposites with 0-3 Connectivity: Volume Fraction Influence. J. Non Cryst. Sol. 2011; 357(19): 3410-3415.

[23] Capsal JF, Dantras E, Lacabanne C. Physical Structure of P(VDF-TrFE)/Barium Titanate Submicrn Composites. J. Non Cryst. Sol. 2012 ; 358(4): 794-798.

[24] Capsal JF, Pousserot C, Dantras E, Lacabanne C. Dynamic Mechanical Behaviour of Polyamide 11/Barium Titanate Composites. Polymer 2010; 51(22): 5207-5211.

[25] Kremer F, Schonals A. Broadband Dielectric Spectroscopy, Berlin: Springer; 2003.

[26] Barrau S, Demont P, Lacabanne C. Macromol. 2003; 36(14): 5187-5194.

[27] Tishkova V, Raynal PI, Puech P. Electrical Conductivity and Raman Imaging of Double Wall Carbon Nanotubes in a Polymer Matrix. Comp. Sci. Tech. 2011; 71(10): 1326-1330.

[28] Bauhofer W, Kovacs JZ. A Review and Analysis of Electrical Percolation in Carbon Nanotube Composites. Comp. Sci. Tech. 2009; 69(10): 1486-1498.

[29] Lonjon A, Laffont L, Demont P., Lacabanne C. New Highly Conductive Nickel Nanowire-Filled P(VDF-TrFE) Copolymer Nanocomposites: Elaboration and Structural Study. J. Phys. Chem. C 2009; 113(28): 12002-12006.

[30] Choi ST, Lee JY, Kwon JO, Seungwan L, Woonbae K. Liquid-Filled Varifocal Lens on a Chip. Proceedings of SPIE, the International Society for Optical Engineering, 27-28 January 2009, San Jose, USA, 7208: 1-9.

[31] Kong N, Ha DS, Erturk E, Inmand DJ. Resistive impedance matching circuit for piezoelectric energy harvesting. J. Intell. Mat. Syst. Struct. 2010; 21(13): 1293-1302.

[32] Lebrun L, Guyomar D, Guiffard B, Cottinet PJ, Putson C. The Characterisation of the harvesting capabilities of an electrostrictive polymer composite. Sens. Act. A: Phys. 2009; 153: 251-257.

[33] Meninger S, Mur-Miranda JO, Amirtharajah R, Chandrakasan AP, Lang LH. Vibration-to-Electric Energy Conversion. IEEE Trans. VLSI 2001; 9(1): 64-76.

[34] Liu Y, Ren KL, Hofmann HF, Zhang Q. Investigation of Electrostrictive Polymers for Energy Harvesting. IEEE Trans. UFFC 2005; 52(12): 2411-2417.

[35] Lallart M, Cottinet PJ, Guyomar D, Lebrun L. Electrostrictive polymers for mechanical energy harvesting. Pol. Phys. 2012; 50(8): 523-535.

[36] Ashley S. Artificial muscles. Sc. Am. 2003; 289(4): 34-41.

[37] Ottman GK, Hofmann HF, Bhatt AC, Lesieutre GA. Adaptive Piezoelectric Energy Harvesting Circuit for Wireless Remote Power Supply. IEEE Trans. Power Elec. 2002; 17(5): 669-676.

[38] Lefeuvre E, Audigier D, Richard C, Guyomar D. Buck-boost converter for sensorless power optimization of piezoelectric energy harvester. IEEE Trans. Power Elec. 2007; 22(5): 2018-2025.

[39] Lallart M, Inman DJ. Low-cost integrable self-tuned converter for piezoelectric energy harvesting optimization. IEEE Trans. Power Elec. 2010; 25(7): 1811-1819.

[40] Cottinet PJ, Guyomar D, Lallart M. Electrostrictive conversion enhancement of polymer composites using a nonlinear approach. Sens. Act. A: Phys. 2011; 172: 497-503.

[41] Guyomar D, Badel A, Lefeuvre E, Richard C. Towards energy harvesting using active materials and conversion improvement by nonlinear processing. IEEE Trans. UFFC 2005; 52: 584-595.

[42] Lefeuvre E, Badel A, Richard C, Guyomar D. Piezoelectric energy harvesting device optimization by synchronous electric charge extraction. J. Intell. Mat. Syst. Struct. 2005; 16(10): 865-876.

[43] Lefeuvre E, Badel A, Richard C, Petit L, Guyomar D. A comparison between several vibration-powered piezoelectric generators for standalone systems, Sens. Act. A: Phys. 2006; 126: 405-416.

[44] Lallart, M.; Garbuio, L.; Petit, L.; Richard, C. & Guyomar, D. (2008a) Double Synchronized Switch Harvesting (DSSH) : A New Energy Harvesting Scheme for Efficient Energy Extraction, IEEE Trans. UFFC, Vol. 55,(10), 2119-2130.

[45] Guyomar D, Sébald G, Pruvost S, Lallart M, Khodayari A, Richard C. Energy Harvesting From Ambient Vibrations and Heat.J. Intell. Mater. Syst. Struct. 2009; 20(5): 609-624.

[46] Lallart M, Guyomar D, Richard C, Petit L. Nonlinear optimization of acoustic energy harvesting using piezoelectric devices. J. Acoust. Soc. Am. 2010; 128(5): 2739-2748.

[47] Lallart M, Guyomar D. Piezoelectric conversion and energy harvesting enhancement by initial energy injection. Appl. Phys. Lett. 2010; 97: 014104.

[48] Guyomar D, Lallart M, Cottinet PJ. Electrostrictive conversion enhancement of polymer composites using a nonlinear approach. Phys. Lett. A 2011; 375: 260-264.

[49] Guillot FM, Balizer E. Electrostrictive Effect in Polyurethanes. J. Appl. Pol. Sci. 2003; 89: 399-404.

[50] Emco miniature power converter A and Q series.
http://www.emcohighvoltage.com/pdfs/aseries.pdf;
http://www.emcohighvoltage.com/pdfs/qseries.pdf.

Chapter 7

Piezoelectric MEMS Power Generators for Vibration Energy Harvesting

Wen Jong Wu and Bor Shiun Lee

Additional information is available at the end of the chapter

1. Introduction

1.1. Background

Over the past few years, the development of wireless sensor network application has generated much interest. Research on the various ways to power wireless sensor devices has gradually become important [1-3]. Unlike portable devices such as cell phones and PDAs where the batteries can be recharged or replaced regularly, most micro sensors are powered by embedded batteries. Therefore, the life of a battery is a major constraint when trying to extend the convenience of micro sensors. With the advent of low-power electronic designs and improvements in fabrication, technology has progressed towards the possibility of self-powered sensor nodes and micro sensors [4].

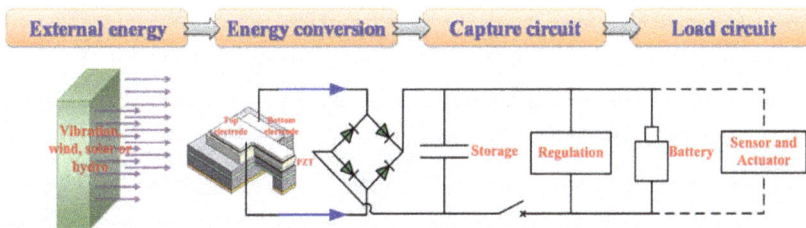

Figure 1. Schematic diagram of a typical power harvesting system

Figure 1 shows a typical power harvesting system for self-powered sensor nodes and micro sensors. It includes an external energy source, a transducer to convert energy from external energy to electric power, a harvesting circuit to optimize the harvesting efficiency and a storage battery or a load circuit. Much research has been focused on harvesting electric power from various ambient energy sources, including solar power, thermal gradients and

Energy Source	Power Density	Energy Density
Batteries (znic-air)		1050-1560mWh/cm³
Batteries (rechargeable lithium)		300 mWh/cm³ (3-4V)
Solar (outdoors)	15mW/cm² (direct sun) 0.15mW/cm² (cloudy day)	
Solar (indoors)	0.006 mW/cm² (standard office desk) 0.057 mW/cm²(<60W desk lamp)	
Vibrations	0.01-0.1 mW/cm³	
Acoustic Noise	3E-6 mW/cm² at 75 dB 9.6E-4 mW/cm² at 100 dB	
Passive Human-Powered Systems	1.8mW (shoe inserts)	
Nuclear Reaction	80 mW/cm³	1E6mWh/cm³

Table 1. A comparison of energy sources [2] [1]

vibrations [5]. When comparing all possible energy sources, mechanical vibration is a potential power source that can be easily accessed through adopting micro-electromechanical systems (MEMS) technology [6, 7]. Table 1 shows a comparison of various energy sources [2]. Mechanical vibration energy can be converted into usable electrical energy through piezoelectric [3, 8, 9], electromagnetic [4, 10, 11] and electrostatic [12-14] transducers. The piezoelectric transducer is considered a potential choice when compared with electromagnetic and electrostatic transducers due to its high energy density [15]. Such comparison is given in table 2. [2]

Type	Energy Density (mJ cm⁻³)	Equation	Assumptions
Piezoelectric	35.4	$(1/2)\sigma_y^2\,k^2/2c$	PZT 5H
Electromagnetic	24.8	$(1/2)B^2/\mu_0$	0.25 T
Electrostatic	4	$(1/2)\varepsilon_0E^2$	3 x 10⁷ V m⁻¹

Table 2. Summary of maximum energy densities of three kinds of transducers [15]

1.2. Literature review

Several researches have been focused on the piezoelectric power generators for vibration power harvesting. T. Starner [16] et. al have concluded that power generation through walking can easily generate power when needed, and 5–8W of power may be recovered

---

[1] Values are estimates from literatures, analyses and few experiments; Values are highly dependent on amplitude and frequency of the driving vibrations

[2] There were already many successful vibration harvesting devices reported of different structures and interface circuits [7, 16, 17]. Piezoelectric material that has been found to have the ability to convert vibration energy to electric power has sparked much attention as it was attractive for use in MEMS applications [16, 18, 19, 20, 21, 22].

while walking at a brisk pace. N. S. Shenck and J. A. Paradiso [8] at the MIT Media Lab then demonstrated a shoe-mounted device to scavenge electricity from the forces exerted on a shoe during walking. Further researches on improvement in the structures and circuits for the shoe-mounted devices were published at [17-19].

Figure 2. Piezoelectric-powered RFID shoes with mounted electronics.

To realize the power supplement of wireless sensor net work, S. Roundy and P. K. Wright [15] demonstrated a vibration based piezoelectric generator. The device is a piezoelectric bimorph cantilever beam type with proof mass to adjust the resonance frequency. An optimized design demonstrated a power output of 375μW from a vibration source of 2.5m/s² at 120Hz. It could be used to power a custom designed 1.9 GHz radio transmitter from the same vibration source. [3]

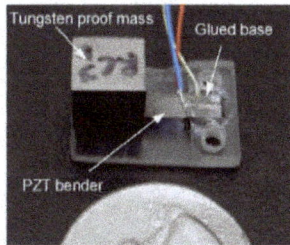

Figure 3. An optimized piezoelectric generator with a 1.5 cm length constraint

Since the mechanical vibration of a piezoelectric element generates an alternating voltage across its electrodes, most of the proposed electrical circuits include an AC–DC converter to provide the electrical energy to its storage device. Guyomar *et al.* [24], Lefeuvre *et al.* [25-27] and Badel *et al.* [28] have developed a new power flow optimization principle based on the extraction of the electric charge produced by a piezoelectric element, synchronized with the mechanical vibration operated at the steady state. They have claimed that the harvested

---

[3]Similar works based on cantilever beam devices using piezoelectric materials to scavenge vibration energy can be found at [17, 20-23].

electrical power may be increased by as much as 900% over the standard technique. Then, Sue et al. [29] detailed the analysis for the performance of a piezoelectric energy harvesting system using the synchronized switch harvesting on inductor (SSHI) electronic interface. It shows that the electrical response using an ideal SSHI interface is similar to that using the standard interface in a strongly coupled electromechanical system operated at short circuit resonance.

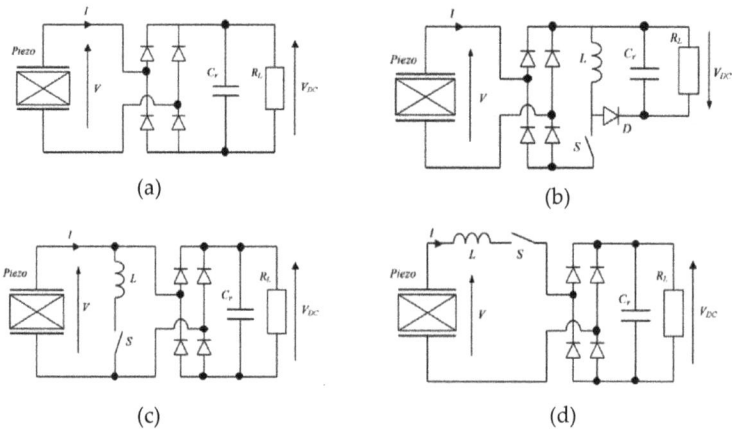

(a)   (b)

(c)   (d)

Figure 4. The interface circuits (a) standard interface (b) Synchronous charge extraction (c) Parallel SSHI (d) Series SSHI

For the development of the MEMS devices, Jeon et al. [30] have successfully developed the first MEMS based micro-scale power generator using d_{33} mode of PZT material. A 170μm × 260μm PZT beam has been fabricated. A maximum output power of 1.01μW across the load of 5.2MΩ at its resonance frequency of 13.9 kHz has been observed. The corresponding energy density is 0.74mWh/cm², which compares favorably to the values of lithium ion batteries.

Figure 5. The first MEMS based micro-scale power generator[30]

Fang et al. [31, 32] successfully developed a PZT MEMS power-generating device based on the d_{31} mode of piezoelectric transducers that uses top and bottom laminated electrodes. The cantilever size is of 12μm thick silicon layer, 2000μm × 500μm cantilever in length and width 500μm × 500μm metal mass (length × height), which generated 1.15μW of effective power when connected to a 20.4kΩ resistance load, leading to a 432mV ac voltage. An improved device was announced later that under the 608Hz resonant frequency, the device generated about 0.89V AC peak–peak voltage output to overcome germanium diode rectifier toward energy storage. The power output obtained was of 2.16μW. Some Other MEMS cantilever piezoelectric power generators examples of different materials and structures can be found in [33] and [34]. Other than single beam structures, Figure 7 [35]shows a MEMS power generator array based on thick-film piezoelectric cantilevers. This device can be tuned to the frequency which expanded the excited frequency bandwidth in ambient low frequency vibration.

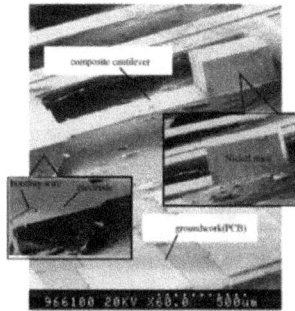

Figure 6. The SEM photo of the fabricated prototype by Fang et al.[32].

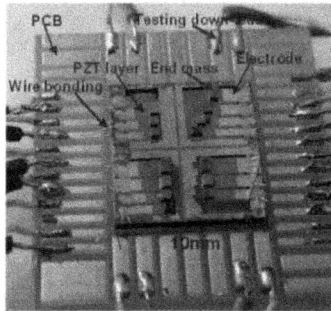

Figure 7. Photograph of power generator array prototype [35]

2. Different types of MEMS power generators and their theoretical models

D_{33} and d_{31} are the two main modes of piezoelectric cantilever beam. In this section, different types of MEMS power generators will be introduced. Readers will be able to see the theoretical models, and the comparison between the experimental results of different modes. The output performances and characteristics for both the d_{33} mode and the d_{31} mode piezoelectric MEMS generators are evaluated using the same dimensions and with the same materials, with the exception of the differing electrode configuration and dimensions of the proof masses. The two devices were then compared for their resonance frequencies, output powers, output voltages and optimal resistive loads.

2.1. Theoretical model and system equations of d_{31} type

In this section, the theoretical model and the development of a d_{31} mode piezoelectric MEMS generator is presented. The d_{31} mode piezoelectric MEMS generator introduced in this chapter is a cantilever type made by using a silicon process which transforms energy by way of the piezoelectric PZT layer. It is laminated with a PZT layer sandwiched between upper and lower electrodes. The PZT sol-gel process that is suitable for fabricating thin film with a thickness of 1~2μm, is often seen in recent researches. But the PZT deposition processes that is applied in the introduced device uses an own developed PZT deposition machine which adopts a "jet-printing" approach based on an aerosol deposition method. This home-made PZT aerosol machine was developed and constructed in order to fabricate a high-quality PZT thin film more efficiently.

For the modeling and analysis of the output performance of the piezoelectric MEMS generator connected with a resistive load, several methods are available. Electrical equivalent circuit model, force equilibrium analysis and energy method are the commonly used methods [36, 37]. The study of the characteristics of a PZT bender utilizing energy method model has been performed in previous studies and the model has shown fair accuracy in various conditions of mechanical stress. Therefore, the analyzing of the output performance of the device in this chapter will be based on the energy method.

Figure 8 shows the configuration of the d_{31} mode piezoelectric MEMS generator. For fabricating the piezoelectric MEMS generator, a beam structure was manufactured and then covered with a PZT layer with a laminated upper and lower electrode. A proof mass was built at the tip of the beam to adjust the structure resonant frequency of the piezoelectric MEMS generator to fit the most adaptable frequency to match the ambient vibration of the surroundings. The beam structure was designed to operate at resonant frequency for maximum stress and strain so as to also maximize electric power output.

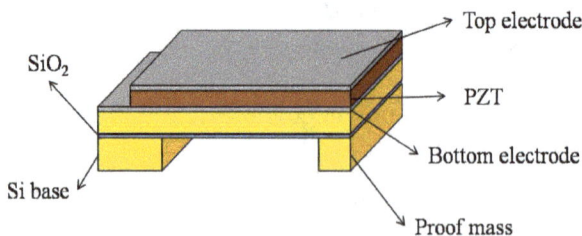

Figure 8. Schematic diagram of the d_{31} mode piezoelectric MEMS generator

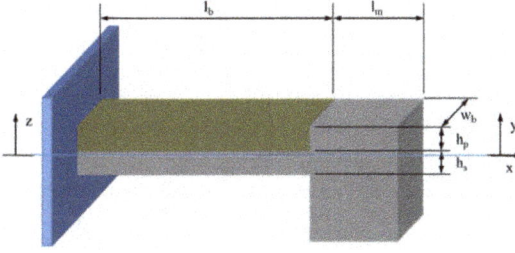

Figure 9. Dimension definitions of the d_{31} mode piezoelectric MEMS generator.

Figure 9 shows the dimension definitions of the d_{31} mode piezoelectric MEMS generator. In the figure, l_b is the length of the beam, l_m the length of the proof mass, h_p the thickness of the piezoelectric material, h_s the thickness of the beam structure (silicon), w_b the width of the beam, z the base vertical displacement and y the distance to the neutral axis of the beam.

The constitutive equations of piezoelectric materials are following the definition in IEEE Standard on Piezoelectric [38]:

$$T_p = c_{pq}^E S_q - e_{kp} E_k \tag{1}$$

$$D_i = e_{iq} S_q + \varepsilon_{ik}^S E_k \tag{2}$$

,where T is the stress (N/m²), S the strain, E the electric field (V/m), D is the electric displacement (Coulomb/m²). "c^E" is the stiffness measured under the constant electric field. "ε^S" is the dielectric constant or permittivity under constant strain. "e" is the piezoelectric constant (Coulomb/m²).

Some other forms of the constitutive equations are:

$$S_q = s_{pq}^E T_p + d_{kq} E_k \tag{3}$$

$$D_i = d_{ip} T_p + \varepsilon_{ik}^T E_k \tag{4}$$

$$S_q = s_{pq}^D T_p + g_{kq} D_k \tag{5}$$

$$E_i = -g_{ip} T_p + \beta_{ik}^T D_k \tag{6}$$

$$T_p = c_{pq}^D S_q - h_{kp} D_k \tag{7}$$

$$E_i = -h_{iq} S_q + \beta_{ik}^S D_k \tag{8}$$

Equation (1) and (2) can be written in a matrix form:

$$\begin{Bmatrix} \mathbf{T} \\ \mathbf{D} \end{Bmatrix} = \begin{bmatrix} \mathbf{c}^E & -\mathbf{e} \\ \mathbf{e} & \varepsilon^S \end{bmatrix} \begin{Bmatrix} \mathbf{S} \\ \mathbf{E} \end{Bmatrix} \tag{9}$$

The model for a d31 type cantilever beam with piezoelectric elements MEMS generator can be obtained with an energy method approach. The generalized form of Hamilton's Principle for an electromechanical system, neglecting the magnetic terms and defining the kinetic (T_k), internal potential (U), and electrical (W_e) energies, as well as the external work (W), is given by:

$$V.I. = \int_{t_1}^{t_2} [\delta(T_k - U + W_e) + \delta W] dt = 0 \tag{10}$$

The individual energy terms are defined as:

$$T_k = \int_{V_s} \frac{1}{2} \rho_s \dot{\mathbf{u}}^t \dot{\mathbf{u}} dV_s + \int_{V_p} \frac{1}{2} \rho_p \dot{\mathbf{u}}^t \dot{\mathbf{u}} dV_p \tag{11}$$

$$U = \int_{V_s} \frac{1}{2} \mathbf{S}^t \mathbf{T} dV_s + \int_{V_p} \frac{1}{2} \mathbf{S}^t \mathbf{T} dV_p \tag{12}$$

$$W_e = \int_{V_p} \frac{1}{2} \mathbf{E}^t \mathbf{D} dV_p + \int_{V_{pe}} \frac{1}{2} \mathbf{E}^t \mathbf{D} dV_{pe} \tag{13}$$

The subscripts s, p and pe indicate the inactive (structural) sections of the beam volume, the piezoelectric element of the beam volume and the piezoelectric element outside the beam structure respectively. The mechanical displacement is denoted by u(x,t) with ρ the density. The contributions to We due to fringing fields in the structure and free space are neglected.

Considering nf discretely applied external point forces, $f_k(t)$, at positions x_k , and nq charges, q_j, applied at discrete electrodes with positions x_j , the external work term is defined in terms of the local mechanical displacement, $u_k = u(x_k, t)$, and the scalar electrical potential, $\phi_j = \phi(\mathbf{x_j}, t)$:

$$\delta W = \sum_{k=1}^{nf} \delta \mathbf{u}_k \mathbf{f}_k(t) - \sum_{j=1}^{nq} \delta \phi_j q_j(t) \tag{14}$$

The above definitions, as well as the constitutive relations of a piezoelectric material (1-9), are used in conjunction with a variational approach to rewrite equation (10):

$$\int_{t_1}^{t_2} \begin{bmatrix} \int_{V_s} \rho_s \delta \dot{\mathbf{u}}^t \dot{\mathbf{u}} dV_s + \int_{V_p} \rho_p \delta \dot{\mathbf{u}}^t \dot{\mathbf{u}} dV_p - \int_{V_s} \delta \mathbf{S}^t \mathbf{c}_s \mathbf{S} dV_s - \int_{V_p} \delta \mathbf{S}^t \mathbf{c}^E \mathbf{S} dV_p \\ + \int_{V_p} \delta \mathbf{S}^t \mathbf{e}^t \mathbf{E} dV_p + \int_{V_p} \delta \mathbf{E}^t \mathbf{e} \mathbf{S} dV_p + \int_{V_p} \delta \mathbf{E}^t \varepsilon^S \mathbf{E} dV_p \\ + \int_{V_{pe}} \delta \mathbf{E}^t \varepsilon^S \mathbf{E} dV_{pe} + \sum_{k=1}^{nf} \delta \mathbf{u}_k \mathbf{f}_k(t) - \sum_{j=1}^{nq} \delta \phi_j q_j(t) \end{bmatrix} dt = 0 \tag{15}$$

Three basic assumptions are introduced: the Rayleigh-Ritz procedure, Euler-Bernoulli beam theory, and that the electrical field across the piezoelectric is constant. In the Rayleigh-Ritz approach, the displacement of a structure can be written as the sum of nr individual modes, $\psi_{ri}(x)$, multiplied by a mechanical temporal coordinate, $r_i(t)$. For a beam in bending status, only the transverse displacement is considered and the mode shape is a function only of the axial position, x. Furthermore, the base excitation is assumed to be in the transverse direction as well:

$$\mathbf{u}(\mathbf{x},t) = \sum_{i=1}^{nr} \psi_{ri}(x) r_i(t) = \psi_r(x)\mathbf{r}(t) \tag{16}$$

Similarly, the electric potential for each of the nq electrode pairs can be written in terms of a potential distribution, ψ_{vj}, and the electrical temporal coordinate, $v_j(t)$.

$$\phi(\mathbf{x},t) = \sum_{j=1}^{nq} \psi_{vj}(\mathbf{x}) v_j(t) = \psi_v(\mathbf{x})\mathbf{v}(t) \tag{17}$$

The Euler-Bernoulli beam theory allows the axial strain in the beam to be written in terms of the beam displacement and the distance from the neutral axis as:

$$\mathbf{S}(\mathbf{x},t) = -y\frac{\partial^2 u(x,t)}{\partial x^2} = -y\psi_r''\,\mathbf{r}(t) \tag{18}$$

Because the MEMS power generator is a composite beam structure, the actual composite beam can be replaced with an equivalent beam made of one material to simplify the analysis. Therefore, the silicon material will be represented by the piezoelectric material in the following derivation. For the composite beam structure, the neutral axis is located at $\bar{y}$ (from the bottom of the beam):

$$\bar{y} = \frac{2c_p h_p h_s + c_p h_p^2 + c_s h_s^2}{2\left(c_p h_p + c_s h_s\right)} \tag{19}$$

, where c_p and c_s are the stiffness of the piezoelectric material and the silicon. Noted that for a special case which the neutral axis is right at the interface of the piezoelectric material and the silicon, the thickness of the piezoelectric material can be obtained from equation (19):

$$h_p = \sqrt{\frac{c_s}{c_p}} h_s \tag{20}$$

The bending rigidity of the composite beam structure could be shown as:

$$\frac{w_b\left(c_p^2 h_p^4 + 4c_p h_p^3 c_s h_s + 6c_p h_p^2 c_s h_s^2 + 4c_p h_p c_s h_s^3 + c_s^2 h_s^4\right)}{12\left(c_p h_p + c_s h_s\right)} \tag{21}$$

In order for replacing the silicon material by the piezoelectric material, the ratio of the elastic constant of the silicon to piezoelectric material, $\eta_s = c_s/c_p$, is used. Then the effective moment of inertia can be obtained from equation (21):

$$I = \frac{w_b\left(h_p^4 + 4h_p^3 h_s \eta_s + 6h_p^2 h_s^2 \eta_s + 4h_p h_s^3 \eta_s + h_s^4 \eta_s^2\right)}{12\left(h_p + h_s \eta_s\right)}$$

$$= \frac{\left(\mu_h^4 + 4\mu_h \eta_s + 6\mu_h^2 \eta_s + 4\mu_h^3 \eta_s + \eta_s^2\right)}{12\left(\mu_h + \eta_s\right)} w_b h_s^3 \tag{22}$$

, where $\mu_h = h_p/h_s$. If the neutral axis is right at the interface of the piezoelectric material and the silicon, the effective moment of inertia can be simplified from equation (22):

$$I = \frac{w_b h_p^2 \left(h_p + h_s\right)}{3} \tag{23}$$

Substituting Equations (16), (17) and (18) into Equation (15), the above equation can be written in terms of mass, M, stiffness, K, coupling, Θ, and capacitive terms, C_p, to obtain the governing equations in Equations bellow:

$$\mathbf{M}\ddot{\mathbf{r}} + \mathbf{K}\mathbf{r} - \mathbf{\Theta}\mathbf{v} = \sum_{k=1}^{nf} \psi_r^t(x_k) \cdot f_k(t) \tag{24}$$

$$\mathbf{\Theta}^t \mathbf{r} + \mathbf{C_p}\mathbf{v} = \sum_{j=1}^{nq} \psi_v(\mathbf{x}_j) \cdot q_j(t) \tag{25}$$

where,

$$\mathbf{M} = \int_{V_s} \psi_r^t \rho_s \psi_r dV_s + \int_{V_p} \psi_r^t \rho_p \psi_r dV_p \tag{26}$$

$$\mathbf{K} = \int_{V_s} (-y\psi_r'')^t \mathbf{c}_s (-y\psi_r'') dV_s + \int_{V_p} (-y\psi_r'')^t \mathbf{c}^E (-y\psi_r'') dV_p \tag{27}$$

$$\mathbf{\Theta} = \int_{V_p} (-y\psi_r'')^t \mathbf{e}^t (-\nabla \cdot \psi_v) dV_p \tag{28}$$

$$\mathbf{C_p} = \int_{V_p} (-\nabla \cdot \psi_v)^t \varepsilon^S (-\nabla \cdot \psi_v) dV_p + \int_{V_{pe}} (-\nabla \cdot \psi_v)^t \varepsilon^S (-\nabla \cdot \psi_v) dV_{pe} \tag{29}$$

The applied external force input to the system is the base excitation is denoted as $\ddot{z}_B$. The loading is summated for all the elements and can be reduced to the integral over the structure length. Assumed that the device is uniform in the axial direction, the right hand side of equation (24) can be written as:

$$\int \psi_r^t(x) \cdot f(t) = \int_0^{l_b} \psi_r^t(x) \cdot (-m\ddot{z}_B) dx = \mathbf{F_B}\ddot{z}_B \tag{30}$$

, where $\mathbf{F_B}$ is the forcing vector for the uniform device in the axial direction. However, the device now consists of two separate sections, the uniform beam and uniform proof mass. Both contribute to the inertial loading of the device. The proof mass displacement is calculated in terms of the displacement and rotation of the tip of the beam. A forcing function is defined in terms of the mass per length of the proof mass, mm, and two additional terms are calculated to make up the modified input matrix [39]:

$$\mathbf{F_B} = -\left(m\int_0^{l_b} \psi_r^t(x)dx + m_m\int_{l_b}^{l_b+l_m} \psi_r^t(l_b)dx + m_m\int_{l_b}^{l_b+l_m} \left(\psi_r'(l_b)x\right)^t dx \right) \tag{31}$$

Mechanical damping can be added through the addition of a damping matrix, C, to equation (24). The right hand side term of equation (25) can be differentiated with respect to time to obtain current. The current can be related to the voltage, assuming that the electrical loading is a resistor, R_l.

$$\mathbf{M\ddot{r} + C\dot{r} + Kr - \Theta v = F_B \ddot{z}_B} \tag{32}$$

$$\mathbf{\Theta^t \dot{r} + C_p \dot{v}} + \frac{1}{R_l}\mathbf{v} = 0 \tag{33}$$

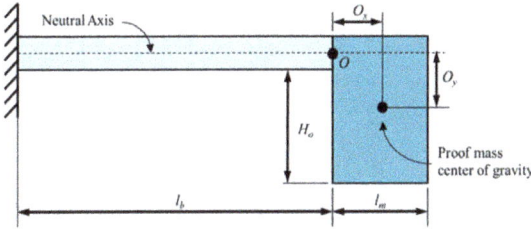

Figure 10. Schematic diagram of the assumed beam configuration

In order to lower the resonance frequency of the piezoelectric energy harvester, it needs to add a proof mass at the tip of the cantilever beam. Figure 10 is the schematic diagram of the beam with tip proof mass. It is assumed that the center of gravity of the mass does not coincide with the end of the beam, O. The Euler-Bernoulli beam theory is used to determine the governing equations in terms of the mechanical displacement:

$$EI\,\psi_{rN}^{(4)} - m\omega^2\psi_{rN} = 0 \tag{34}$$

and can be solved generally for the Nth mode:

$$\psi_{rN} = c\sinh\lambda_N x + d\cosh\lambda_N x + e\sin\lambda_N x + f\cos\lambda_N x \tag{35}$$

The constants (c, d, e, and f) can be solved by using the boundary conditions of the beam with the mass. With a reasonable assumption that the both the beam and the proof mass are uniform in the axial direction with mass per lengths of m and mm, respectively, it is possible to determine the boundary conditions at the point where the beam and the mass are connected, y_{lb}:

$$EIy_{lb}'' - \omega_N^2 I_0 y_{lb}' - \omega_N^2 S_0 y_{lb} = 0 \tag{36}$$

$$EIy_{lb}''' + \omega_N^2 M_0 y_{lb} + \omega_N^2 S_0 y_{lb}' = 0 \tag{37}$$

where: $M_0 = m_m\, l_m$, $S_0 = M_0\, O_x$, $I_0 = I_{yy} + M_0(O_x^2 + O_y^2)$, E is the axial modulus of the beam, I is the second moment of area of the beam, I_{yy} is the moment of inertia of the proof mass around its center of gravity, and ω_N is the natural frequency of the beam. By defining $\bar{\lambda}_N = \lambda_N l_b$, $\bar{M}_0 = M_0 O_x$, $\bar{S}_0 = S_0 / ml_b^2$ and $\bar{I}_0 = I_0 / ml_b^3$, the boundary conditions are used to obtain the matrix equation.

$$\begin{bmatrix} A_{11} & A_{12} \\ A_{21} & A_{22} \end{bmatrix} \begin{bmatrix} e \\ f \end{bmatrix} = 0 \tag{38}$$

$$\begin{aligned} A_{11} &= \left(\sinh\bar{\lambda}_N + \sin\bar{\lambda}_N\right) + \bar{\lambda}_N^3 \bar{I}_0 \left(-\cosh\bar{\lambda}_N + \cos\bar{\lambda}_N\right) \\ &\quad + \bar{\lambda}_N^2 \bar{S}_0 \left(-\sinh\bar{\lambda}_N + \sin\bar{\lambda}_N\right) \end{aligned} \tag{39}$$

$$\begin{aligned} A_{12} &= \left(\cosh\bar{\lambda}_N + \cos\bar{\lambda}_N\right) + \bar{\lambda}_N^3 \bar{I}_0 \left(-\sinh\bar{\lambda}_N - \sin\bar{\lambda}_N\right) \\ &\quad + \bar{\lambda}_N^2 \bar{S}_0 \left(-\cosh\bar{\lambda}_N + \cos\bar{\lambda}_N\right) \end{aligned} \tag{40}$$

$$\begin{aligned} A_{21} &= \left(\cosh\bar{\lambda}_N + \cos\bar{\lambda}_N\right) + \bar{\lambda}_N \bar{M}_0 \left(\sinh\bar{\lambda}_N - \sin\bar{\lambda}_N\right) \\ &\quad + \bar{\lambda}_N^2 \bar{S}_0 \left(\cosh\bar{\lambda}_N - \cos\bar{\lambda}_N\right) \end{aligned} \tag{41}$$

$$\begin{aligned} A_{22} &= \left(\sinh\bar{\lambda}_N - \sin\bar{\lambda}_N\right) + \bar{\lambda}_N \bar{M}_0 \left(\cosh\bar{\lambda}_N - \cos\bar{\lambda}_N\right) \\ &\quad + \bar{\lambda}_N^2 \bar{S}_0 \left(\sinh\bar{\lambda}_N + \sin\bar{\lambda}_N\right) \end{aligned} \tag{42}$$

The mode resonance frequencies can be obtained by solving for $\bar{\lambda}_N$ such that $\begin{vmatrix} A_{11} & A_{12} \\ A_{21} & A_{22} \end{vmatrix} = 0$.

Successive values of $\bar{\lambda}_N$ correspond to the modes of the beam and the natural frequency of each mode can be determined with: $\omega_N^2 = \bar{\lambda}_N^2 \sqrt{\dfrac{EI}{ml_b^4}}$. The solution of equation (35) can be written in terms of a single arbitrary constant, say f:

$$\psi_{rN} = f\left[\left(\cosh\lambda_N x - \cos\lambda_N x\right) - A_{12}/A_{11}\left(\sinh\lambda_N x - \sin\lambda_N x\right)\right] \tag{43}$$

The effective mass of the structure can be obtained from the Lagrange equations of motion and replaces equation (26) when a proof mass is added to a cantilever beam.

$$\mathbf{M} = \int_{V_s} \psi_r^t \rho_s \psi_r dV_s + \int_{V_p} \psi_r^t \rho_p \psi_r dV_p + M_0 \psi_r^t (l_b) \psi_r (l_b)$$
$$+ 2S_0 M_0 \psi_r^t (l_b) \psi_r' (l_b) + I_0 M_0 \psi_r'^t (l_b) \psi_r' (l_b) \tag{44}$$

The governing equation shown in (32) can be written in an alternative form by dividing through by $\mathbf{M}$ and making use of the definitions $\omega_1 = \sqrt{K/M}$ and $\zeta_m = C/2M\omega_1$:

$$\ddot{r} + 2\zeta_m \omega_1 \dot{r} + \omega_1^2 r - \Theta/M v = F_B \ddot{z}_B / M \tag{45}$$

$$\Theta \dot{r} + C_p \dot{v} + \frac{1}{R_l} v = 0 \tag{46}$$

The dimensionless factors $\tau = \omega_1 R_l C_p$, $\kappa^2 = \Theta^2/K C_p$ and $\Omega = \omega/\omega_1$ are introduced, where ω is the base input frequency and the system response is calculated:

$$\left| \frac{r}{F_B \ddot{z}_B} \right| = \frac{1}{K} \frac{\sqrt{1 + (\tau\Omega)^2}}{\sqrt{\left[1 - (1 + 2\zeta_m \tau)\Omega^2\right]^2 + \left[\left(2\zeta_m + (1 + \kappa^2)\tau\right)\Omega - \tau\Omega^3\right]^2}} \tag{47}$$

$$\left| \frac{v}{F_B \ddot{z}_B} \right| = \frac{1}{\Theta} \frac{\kappa^2 \tau\Omega}{\sqrt{\left[1 - (1 + 2\zeta_m \tau)\Omega^2\right]^2 + \left[\left(2\zeta_m + (1 + \kappa^2)\tau\right)\Omega - \tau\Omega^3\right]^2}} \tag{48}$$

$$\left| \frac{P}{(F_B \ddot{z}_B)^2} \right| = \frac{1}{2\sqrt{MK}} \frac{\kappa^2 \tau\Omega^2}{\left[1 - (1 + 2\zeta_m \tau)\Omega^2\right]^2 + \left[\left(2\zeta_m + (1 + \kappa^2)\tau\right)\Omega - \tau\Omega^3\right]^2} \tag{49}$$

Equation (47) gives the generalized mechanical displacement, which can be converted to actual displacements by multiplying it with the mode shape. The system can be analyzed at short-circuit and open-circuit conditions by letting the electrical load resistance tending to zero and infinity, respectively. Two optimal frequency ratios for maximum power generation can be obtained, which correspond to the resonance (subscript sc) and anti-resonance (subscript oc) frequencies of the beam structure:

$$\Omega_{sc} = 1, \qquad \Omega_{oc} = \sqrt{1 + \kappa^2} \tag{50}$$

The power can be optimized with respect to the load resistance to obtain an optimal electrical load. This is achieved by optimizing the power with respect to the dimensionless constant, τ:

$$\tau_{opt} = \frac{1}{\Omega} \sqrt{\frac{(\Omega^2 - 1)^2 + (2\Omega\zeta_m)^2}{(\Omega^2 - (\kappa^2 + 1))^2 + (2\Omega\zeta_m)^2}} \tag{51}$$

Substituting equation (51) into power equation (49) can found that:

$$
\left| \frac{P}{\left(F_B \ddot{z}_B \right)^2 \big/ 2\sqrt{MK}} \right| = \frac{\frac{\Omega}{\kappa^2}\sqrt{\left(\frac{\Omega^2-1}{2\kappa^2\Omega}\right)^2 + \left(\frac{\zeta_m}{\kappa^2}\right)^2}}{\left(8\left(\frac{\zeta_m}{\kappa^2}\right)^2\Omega^2 + 2\left(\frac{\Omega^2-1}{\kappa^2}\right)^2\right)\sqrt{\left(\frac{\Omega^2-\left(1+\kappa^2\right)}{2\kappa^2\Omega}\right)^2 + \left(\frac{\zeta_m}{\kappa}\right)^2 + 4\left(\frac{\zeta_m}{\kappa^2}\right)\Omega\sqrt{\left(\frac{\Omega^2-1}{2\kappa^2\Omega}\right)^2 + \left(\frac{\zeta_m}{\kappa^2}\right)^2}}} \tag{52}
$$

It can be found that except the geometric dimensions, the output power is only the function of Ω, ζ_m and κ. For MEMS-scale devices, ζ_m is generally at least an order of magnitude smaller than κ_2 [40]. With this assumption, the power output at both the resonance and anti-resonance frequencies (under optimal electrical load) is approximated as:

$$
\left| P_{opt} \right| \approx \frac{\left(F_B \ddot{z}_B \right)^2}{16\sqrt{MK}\,\zeta_m} \tag{53}
$$

2.2. Theoretical model and system equations of d33 type

This section presents the theoretical model and the development of the d_{33} mode piezoelectric MEMS generator. It is composed of interdigitated electrodes at the top of the PZT layer. The aerosol deposition method is also adopted to fabricate a high-quality PZT thin film more efficiently.

For piezoelectric elements, the longitudinal piezoelectric effect can be much larger than the traverse effect ($d_{33}/d_{31} \sim 2.4$ for most piezoelectric ceramics [41]). For this reason, it is desirable to operate the device in the d_{33} mode. The d_{33} mode operation occurs when the electric field and the strain direction coincide. Figure 11 shows the configuration of the d_{33} mode piezoelectric MEMS generator. For fabricating the piezoelectric MEMS generator, a beam structure was manufactured and then covered with a PZT layer with a laminated upper electrode. A proof mass was also built at the tip of the beam.

Since the output voltage is a function of the output charge and the capacitance between the interdigitated electrodes, the output voltage can be adjusted by the distance between the interdigitated electrodes. Therefore, the following text will also show readers the relationships between the distance of the interdigitated electrodes with the output voltage and power output performance.

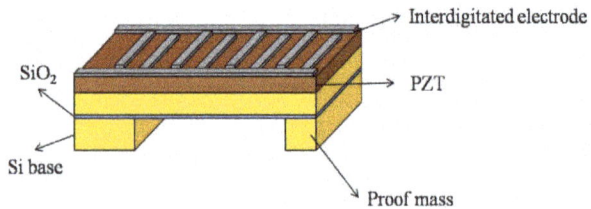

Figure 11. Schematic diagram of the d_{33} mode piezoelectric MEMS generator

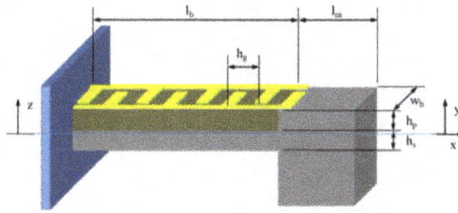

Figure 12. Dimension definitions of the d33 mode piezoelectric MEMS generator.

Figure 12 shows the dimension definitions of the d33 mode piezoelectric MEMS generator. In the figure, l_b is the length of the beam, l_m the length of the proof mass, h_p the thickness of the piezoelectric material, h_s the thickness of the beam structure (silicon), h_g the interval of the interdigitated electrodes, w_b the width the beam, z the base vertical displacement and y the distance to the neutral axis the beam.

Since the electric field is not completely in the axial direction through the thickness of the piezoelectric element, nor is the section of piezoelectric element under the electrode completely inactive, an approximate model for the interdigitated electrode-configuration has been adopted. It is assumed that the region of the piezoelectric element under the electrode is electrically inactive, whereas the section between the electrodes utilizes the full d33 effect. Figure 13 shows the geometry of the approximate model.

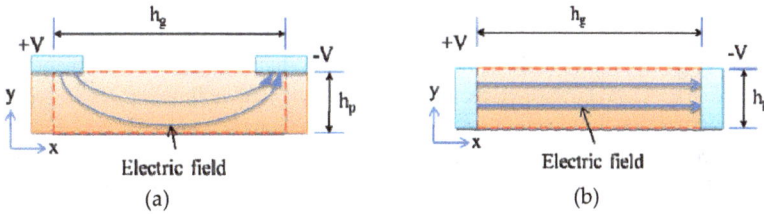

Figure 13. (a) Interdigitated electrode configuration (b) the model approximation

The model for a d33 type cantilever beam with piezoelectric elements MEMS generator can be obtained with an energy method approach. The generalized form of Hamilton's Principle for modeling the electromechanical system is as shown in equation (10). The individual energy terms (the kinetic T_k, internal potential U, and electrical W_e) are defined in equations (11), (12), and (13). It is important to note that although the device is made up of a number of separate piezoelectric regions, there is only one electrode pair and the voltage across all the elements will be the same. Since the strain varies along the length of the beam, different amounts of charge will be generated in each region and the charge sums to give the total charge output of the device. Therefore, the electric potential can be written as:

$$\phi(\mathbf{x},t) = \psi_v(\mathbf{x})\bar{\mathbf{v}}(t) \tag{54}$$

Following the procedure and the assumptions in the previous section and considering only one interdigitated electrode pairs, the governing equations can be rewritten as:

$$\mathbf{M}\ddot{\mathbf{r}} + \mathbf{C}\dot{\mathbf{r}} + \mathbf{K}\mathbf{r} - \mathbf{\Theta}\bar{\mathbf{v}} = \mathbf{F_B}\ddot{z}_B \tag{55}$$

$$\mathbf{\Theta}^t\dot{\mathbf{r}} + \mathbf{C_p}\dot{\bar{\mathbf{v}}} + \frac{1}{R_l}\bar{\mathbf{v}} = 0 \tag{56}$$

where,

$$\mathbf{M} = \int_{V_s} \psi_r^t \rho_s \psi_r dV_s + \int_{V_p} \psi_r^t \rho_p \psi_r dV_p + M_0 \psi_r^t(l_b)\psi_r(l_b)$$
$$+ 2S_0 M_0 \psi_r^t(l_b)\psi_r'(l_b) + I_0 M_0 \psi_r'^t(l_b)\psi_r'(l_b) \tag{57}$$

$$\mathbf{K} = \int_{V_s} (-y\psi_r'')^t \mathbf{c}_s (-y\psi_r'') dV_s + \int_{V_p} (-y\psi_r'')^t \mathbf{c}^E (-y\psi_r'') dV_p \tag{58}$$

$$\mathbf{\Theta} = \int_{V_p} (-y\psi_r'')^t \mathbf{e}^t (-\nabla \cdot \psi_v) dV_p \tag{59}$$

$$\mathbf{C_p} = \int_{V_p} (-\nabla \cdot \psi_v)^t \varepsilon^S (-\nabla \cdot \psi_v) dV_p \tag{60}$$

$$\mathbf{F_B} = -\left(m\int_0^{l_b} \psi_r^t(x)dx + m_m \int_{l_b}^{l_b+l_m} \psi_r^t(l_b)dx + m_m \int_{l_b}^{l_b+l_m} \left(\psi_r'(l_b)x\right)^t dx \right) \tag{61}$$

In order to lower the resonance frequency of the piezoelectric energy harvester, a proof mass was added at the tip of the cantilever beam. The modal shape for a cantilever beam with the addition of the mass is shown as in equation (34). The following electric potential distribution is assumed to give a constant electric field in one piezoelectric element between interdigitated electrode pair. The potential distribution varies from +1 at the electrode on one side to 0 at the electrode on the other side. The function ψ_v can be shown as:

$$\psi_v = \begin{cases} x/h_g, & 2kh_g \le x \le (2k+1)h_g \\ -x/h_g + 2, & (2k+1)h_g \le x \le 2(k+1)h_g \end{cases} \tag{62}$$
$$k = 0,1,2\ldots$$

The governing equation shown in (55) and (56) can be written in an alternative form by dividing through by M and making use of the definitions $\omega_1 = \sqrt{K/M}$ and $\zeta_m = C/2M\omega_1$:

$$\ddot{r} + 2\zeta_m\omega_1\dot{r} + \omega_1^2 r - \Theta/M\bar{v} = F_B\ddot{z}_B/M \tag{63}$$

$$\Theta\dot{r} + C_p\dot{\bar{v}} + \frac{1}{R_l}\bar{v} = 0 \tag{64}$$

The dimensionless factors $\tau = \omega_1 R_l C_p$, $\kappa^2 = \Theta^2/KC_p$ and $\Omega = \omega/\omega_1$ are introduced, where ω is the base input frequency and the system response is calculated:

$$\left|\frac{r}{F_B \ddot{z}_B}\right| = \frac{1}{K} \frac{\sqrt{1+\left(\tau\Omega\right)^2}}{\sqrt{\left[1-\left(1+2\zeta_m\tau\right)\Omega^2\right]^2 + \left[\left(2\zeta_m+\left(1+\kappa^2\right)\tau\right)\Omega - \tau\Omega^3\right]^2}} \tag{65}$$

$$\left|\frac{\bar{v}}{F_B \ddot{z}_B}\right| = \frac{1}{\Theta} \frac{\kappa^2\tau\Omega}{\sqrt{\left[1-\left(1+2\zeta_m\tau\right)\Omega^2\right]^2 + \left[\left(2\zeta_m+\left(1+\kappa^2\right)\tau\right)\Omega - \tau\Omega^3\right]^2}} \tag{66}$$

$$\left|\frac{P}{\left(F_B\ddot{z}_B\right)^2}\right| = \frac{1}{2\sqrt{MK}} \frac{\kappa^2\tau\Omega^2}{\left[1-\left(1+2\zeta_m\tau\right)\Omega^2\right]^2 + \left[\left(2\zeta_m+\left(1+\kappa^2\right)\tau\right)\Omega - \tau\Omega^3\right]^2} \tag{67}$$

The results are identical to the d$_{31}$ mode piezoelectric MEMS generator as shown in (47), (48), and (49).

The system can be analyzed at short-circuit and open-circuit conditions by letting the electrical load resistance tending to zero and infinity, respectively. Two optimal frequency ratios for maximum power generation can be obtained, which correspond to the resonance and anti-resonance frequencies of the beam structure:

$$\Omega_{sc} = 1, \qquad \Omega_{oc} = \sqrt{1+\kappa^2} \tag{68}$$

The power can be optimized with respect to the load resistance to obtain an optimal electrical load. This is achieved by optimizing the power with respect to the dimensionless constant, τ:

$$\tau_{opt} = \frac{1}{\Omega} \sqrt{\frac{\left(\Omega^2-1\right)^2 + \left(2\Omega\zeta_m\right)^2}{\left(\Omega^2-(\kappa^2+1)\right)^2 + \left(2\Omega\zeta_m\right)^2}} \tag{69}$$

This is the same as the results of the d$_{31}$ mode piezoelectric MEMS generator as shown in (51). Substituting equation (69) into power equation (67) can found that:

$$\left|\frac{P}{\left(F_B\ddot{z}_B\right)^2 / 2\sqrt{MK}}\right| = \frac{\frac{\Omega}{\kappa^2}\sqrt{\left(\frac{\Omega^2-1}{2\kappa^2\Omega}\right)^2 + \left(\frac{\zeta_m}{\kappa^2}\right)^2}}{\left(8\left(\frac{\zeta_m}{\kappa^2}\right)^2\Omega^2 + 2\left(\frac{\Omega^2-1}{\kappa^2}\right)^2\right)\sqrt{\left(\frac{\Omega^2-\left(1+\kappa^2\right)}{2\kappa^2\Omega}\right)^2 + \left(\frac{\zeta_m}{\kappa}\right)^2} + 4\left(\frac{\zeta_m}{\kappa^2}\right)\Omega\sqrt{\left(\frac{\Omega^2-1}{2\kappa^2\Omega}\right)^2 + \left(\frac{\zeta_m}{\kappa^2}\right)^2}} \tag{70}$$

It can be found that except the geometric dimensions, the output power is only the function of Ω, ζ_m and κ. With the assumption that for MEMS-scale devices, ζ_m is generally at least an order of magnitude smaller than κ^2 [40], the power output at both the resonance and anti-resonance frequencies (under optimal electrical load) is approximated as:

$$\left|P_{opt}\right| \approx \frac{\left(F_B \ddot{z}_B\right)^2}{16\sqrt{MK}\zeta_m} \tag{71}$$

3. Fabrication of piezoelectric MEMS power generators

3.1. PZT deposition method

Fabricating the PZT layer using an aerosol deposition method has been proven to be a quick, efficient and easy-to-pattern MEMS process [42, 43]. The aerosol deposition equipment deposited PZT film up to 0.1 micrometer per minute. Figure 14 shows the schematic diagram of the aerosol deposition equipment. The PZT powder with a particle size smaller than 1μm in diameter was put in a continuously vibrating powder chamber in order to suspend the PZT particles. Nitrogen or Helium gas was connected to the powder chamber with gas flow rate of 4~6 liters per minute so as to bring the PZT particles through the nozzle and into the deposition chamber. With the deposition chamber in a vacuum, the pressure difference between the power chamber and the deposition chamber accelerated the PZT particles and forced them to jet out from the nozzle inside the deposition chamber and deposit onto the wafer surface with high speed. The wafer substrate was then carried by an X-Y moving stage so that deposition over the entire area of the PZT took place. Both the flow rate of the inlet gas and the scan speed of the X-Y moving stage were then used to control the deposition rate and the roughness of the deposited PZT layer.

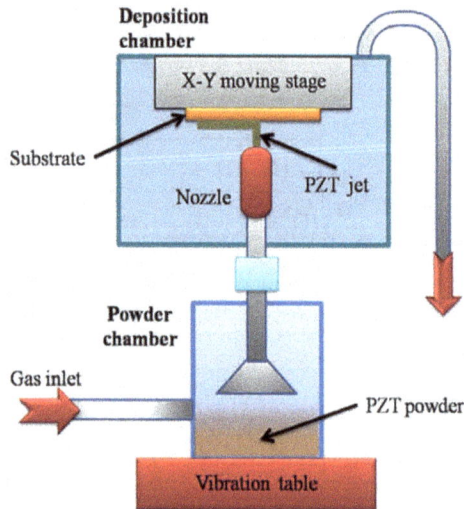

Figure 14. Schematic diagram of the aerosol deposition machine

Figure 15 shows the SEM photography of the PZT layer as deposited by aerosol deposition with a thickness of up to 28μm. A lift-off method was adopted to pattern the PZT layer that was deposited by the aerosol deposition machine. A photoresist[4] with suitable hardness and adhesion between the photoresist and PZT powder was needed for the lift-off process to prevent damage to the photoresist during processing and to limit accumulation of the PZT powder at the sidewall. Figure 16 shows the SEM photograph of the sidewall of the PZT layer patterned by the lift-off method.

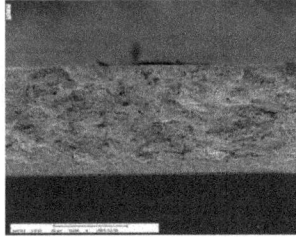

Figure 15. SEM Photograph of the cross-sectional view of 28μm thickness PZT layer after deposition

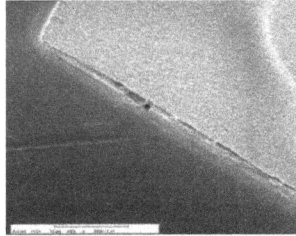

Figure 16. SEM Photograph of a patterned PZT layer by lift-off method

An annealing process was required to improve the characteristics of the material. To investigate the effects at different annealing temperatures, the relationship between polarization and the electric field of the annealed PZT film with 5μm in thickness at different annealing temperatures were undertaken using a ferroelectric analyzer (TF ANALYZER 2000). Figure 17 shows the measured P-E hysteresis curves. The applied electrical field was 75MV/m at 100Hz. The remnant polarizations were 7~9.3μC/cm2 after annealing above 450°C, which shows much improvement when compared to non-annealed PZT layers. The measurement results show that the coercive field decreased with respect to an increase in annealing temperature.

The crystalline phase of the deposited PZT layer associated with the different annealed temperatures can be characterized by XRD (x-ray diffraction). The non-annealed crystalline phase was used as a reference point. (See figure 18) The findings indicate that a perovskite phase in the PZT powder remains after a 650°C annealing process. Therefore, after the PZT film was deposited, it was then annealed at 650°C for 3 hours in a furnace and then cooled to room temperature. It should be noticed that PZT microstructures will crack easily when

---

[4] A photoresist KMPR-1050 (MicroChem Corp.) or THB-151N (JSR Micro Inc.) was used in this work.

the annealing temperature is higher than 700°C. Similarly, acceptable piezoelectric constants cannot be obtained for annealing temperature lower than 450°C.

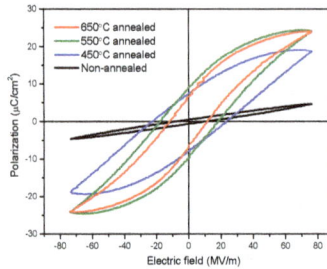

Figure 17. P-E hysteresis curve of a 5μm PZT layer at different annealing temperatures

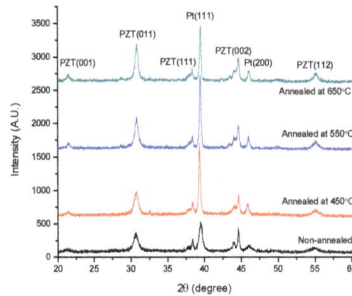

Figure 18. XRD scan of the PZT layers at different annealing temperatures

3.2. MEMS fabrication process of the device

The piezoelectric MEMS generator was a laminated cantilever structure which was composed of a supporting silicon membrane, a piezoelectric layer and laminated electrodes. Both the d_{31} and d_{33} mode piezoelectric MEMS generator introduced in this chapter were designed to incorporate a $3000 \times 1500 \mu m^2$ size cantilever beam structure with an 11μm thickness comprised of a 5μm piezoelectric PZT layer and a 1μm SiO_2 at the bottom of the beam structure. For the d_{33} mode device, the interdigitated electrodes were fabricated with 30μm widths and 30μm gaps. The proof mass for the d_{31} mode piezoelectric MEMS generator was fabricated under the beam structure with dimensions of $500 \times 1500 \times 500 \mu m^3$, and $750 \times 1500 \times 500 \mu m^3$ for the d_{33} mode. A different proof mass dimension comparing to the d_{31} mode piezoelectric MEMS generator was used to show readers how the proof masses influence the resonance frequency. Most of the process steps were undertaken in a standard clean room environment. The piezoelectric material PZT thin film deposition was deposited using aerosol deposition machine.

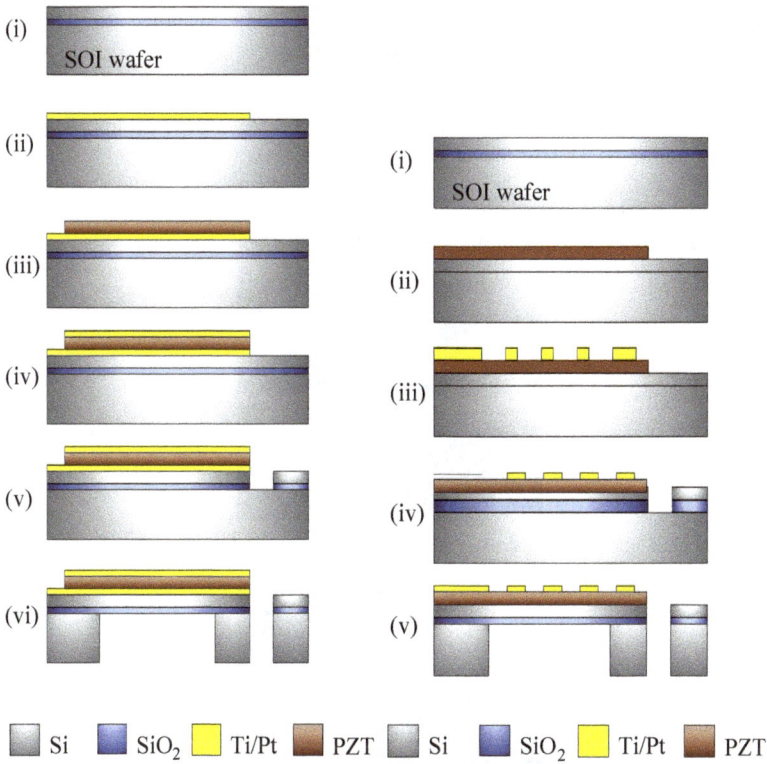

(i) SOI wafer

(ii)

(iii)

(iv)

(v)

(vi)

(i) SOI wafer

(ii)

(iii)

(iv)

(v)

Si   SiO₂   Ti/Pt   PZT   Si   SiO₂   Ti/Pt   PZT

Figure 19. Fabrication processes of d₃₁ (left) and d₃₃ (right) devices

Figure 19 shows the fabrication process of both mode of the piezoelectric MEMS generator. SOI wafers with a 5μm device layer and a 1μm buried oxide layer was used in the process. The processes are similar to one another except for the second step, where the d₃₁ generator has the bottom electrode deposited with a 30nm Ti and 220nm Pt on the top-side of the SOI wafer using an e-beam evaporator. PZT layers of 5μm were then deposited onto the bottom electrode of the d₃₁ device, using the aerosol deposition method described above. For the d₃₃ device the PZT layer is directly deposited upon the SOI wafer. The patterning processes required in the previous steps were done by lift-off processes. Then, the annealing process was taken place at a furnace of 650°C for 3 hours. Afterwards, an e-beam evaporator was used to deposit the top electrode with 30nm Ti and 220nm Pt and then patterned by lifting-off. The beam shape was defined and etched on the top side DRIE. The buried oxide layer was etched out using RIE at the same time. Finally a DRIE process was then used to etch the wafer from the back side until the beam was released. The proof mass was made at the same time and its size adjusted during the etching to the back side. The PZT layer was then poled under a high electric field. For the poling process, the device was heated up to 160°C using a hot plate, followed by poling under 100V for 30 minutes, and then allowed to cool slowly to room temperature with the electric field applied through continuously during the entire heating and cooling process.

The SEM of the finished d_{31} and d_{33} modes piezoelectric MEMS generator are shown in figure 20 and figure 21. The cantilever beams were covered with laminated electrode and the proof mass at the tips can be seen. The beam structures could be seen to be bent upwards due to the thermal expansion difference for PZT and to the silicon wafer after the PZT cooled down to room temperature from 650°C.

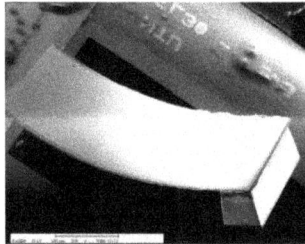

Figure 20. SEM photograph of a finished d_{31} mode device

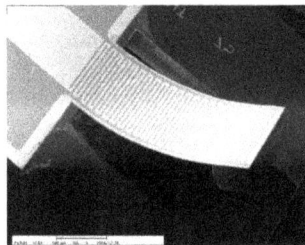

Figure 21. SEM photograph of a finished d_{33} mode device

4. Discussion on different types of MEMS power generators

4.1. Comparison between d31 and d33 mode piezoelectric MEMS generators

The d_{31} mode and the d_{33} mode piezoelectric MEMS generators were both excited at a 2g acceleration level. The measurement results are summarized in Table 3. The optimal load was found to be inversely proportional to the capacitance of the piezoelectric material [44, 45]. For the same dimensions of the beam shape of the d_{31} and d_{33} mode devices, it was obvious that the capacitance of the d_{31} mode device was larger than the d_{33} mode device. Therefore, the optimal resistive load for the d_{31} mode device was smaller than that of the d_{33} mode device. The output power for the d_{33} mode piezoelectric MEMS generator was smaller than that for the d_{31} mode piezoelectric MEMS generator. This was due to the PZT material of the d_{33} mode device which was poled by the interdigitated electrodes and which results in a non-uniform poling direction. The material under the electrodes was not used because it was not poled correctly. Furthermore, the further the distance from the surface of the PZT material, the less effective the poling electric field strength will be. This causes an efficiency drop for the d_{33} mode piezoelectric MEMS generator when compared to the d_{31} mode piezoelectric MEMS generator. Nevertheless, the output voltage of the d_{33} mode piezoelectric MEMS generator was higher than that of the d_{31} mode piezoelectric MEMS generator and easily adjusted by the gap of the interdigitated electrodes under the same dimensions of the beam shape.

Mode	Resonant Frequency	Optimal Load	Power Output	Voltage Output (open circuit)	Voltage Output (with load)
d_{31}	255.9 Hz	150kΩ	2.099μW	2.415V$_{P-P}$	1.587V$_{P-P}$
d_{33}	214.0 Hz	510kΩ	1.288μW	4.127V$_{P-P}$	2.292V$_{P-P}$

Table 3. The Output performance of the d_{31} and d_{33} mode piezoelectric MEMS generators at 2g acceleration

5. Conclusion

In this chapter, the theoretical analysis, design and manufacture methods of two basic piezoelectric MEMS generators were introduced. For these piezoelectric MEMS generators, we investigated the relationship between output voltage and output power at different resistive loads.

The measurement results show that the d_{31} mode piezoelectric MEMS generator had a maximum open circuit output voltage of 2.675V$_{P-P}$ and a maximum output power of 2.765μW with a 1.792V$_{P-P}$ output voltage at resonant frequency of 255.9Hz at a 2.5g acceleration level. The d_{33} mode piezoelectric MEMS generator showed a maximum open circuit output voltage of 4.127V$_{P-P}$ and a maximum output power of 1.288μW with a 2.292V$_{P-P}$ output voltage at resonant frequency of 214Hz at a 2g acceleration level. The output power and the output voltage are also influenced by the driven acceleration intensely.

When comparing the output characteristics of both the d_{31} mode and the d_{33} mode piezoelectric MEMS generators, the results showed that the d_{31} mode device made of a PZT sandwiched between laminated electrodes was better in output power performance than the d_{33} mode device that composed of interdigitated electrodes at the top.

Author details

Wen Jong Wu* and Bor Shiun Lee
Department of Engineering Science and Ocean Engineering, National Taiwan University, Taipei, Taiwan

6. References

* Corresponding Author

[1]   N. G. Elvin, A. A. Elvin and M. Spector. A self-powered mechanical strain energy sensor. Smart Materials & Structures. 2001;10 293-299

[2]   J. M. Rabaey, M. J. Ammer, J. L. da Silva, D. Patel and S. Roundy. PicoRadio supports ad hoc ultra-low power wireless networking. Computer. 2000;33 42-48

[3]   S. Roundy, P. K. Wright and J. Rabaey. A study of low level vibrations as a power source for wireless sensor nodes. Computer Communications. 2003;26 1131-1144

[4]   R. Amirtharajah and A. P. Chandrakasan. Self-powered signal processing using vibration-based power generation. Ieee Journal of Solid-State Circuits. 1998;33 687-695

[5]   S. Roundy, D. Steingart, L. Frechette, P. Wright and J. Rabaey. Power sources for wireless sensor networks. Wireless Sensor Networks, Proceedings. 2004;2920 1-17

[6]   S. Roundy, E. S. Leland, J. Baker, E. Carleton, E. Reilly, E. Lai, B. Otis, J. M. Rabaey, P. K. Wright and V. Sundararajan. Improving power output for vibration-based energy scavengers. Ieee Pervasive Computing. 2005;4 28-36

[7]   H. A. Sodano. A Review of Power Harvesting from Vibration Using Piezoelectric Materials. The Shock and Vibration Digest. 2004;36 197

[8]   N. S. Shenck and J. A. Paradiso. Energy scavenging with shoe-mounted piezoelectrics. Ieee Micro. 2001;21 30-42

[9]   G. K. Ottman, H. F. Hofmann and G. A. Lesieutre. Optimized piezoelectric energy harvesting circuit using step-down converter in discontinuous conduction mode. Ieee Transactions on Power Electronics. 2003;18 696-703

[10] N. N. H. Ching, H. Y. Wong, W. J. Li, P. H. W. Leong and Z. Y. Wen. A laser-micromachined multi-modal resonating power transducer for wireless sensing systems. Sensors and Actuators a-Physical. 2002;97-8 685-690

[11] M. El-hami, R. Glynne-Jones, N. M. White, M. Hill, S. Beeby, E. James, A. D. Brown and J. N. Ross. Design and fabrication of a new vibration-based electromechanical power generator. Sensors and Actuators a-Physical. 2001;92 335-342

[12] S. Meninger, J. O. Mur-Miranda, R. Amirtharajah, A. Chandrakasan and J. H. Lang. Vibration-to-electric energy conversion. Very Large Scale Integration (VLSI) Systems, IEEE Transactions on. 2001;9 64-76

[13] Y. Chiu and V. F. G. Tseng. A capacitive vibration-to-electricity energy converter with integrated mechanical switches. Journal of Micromechanics and Microengineering. 2008;18, 104004

[14] M. Miyazaki. Electric-Energy Generation through Variable-Capacitive Resonator for Power-Free LSI. IEICE - Transactions on Communications. 2004;87 549

[15] S. Roundy and P. K. Wright. A piezoelectric vibration based generator for wireless electronics. Smart Materials & Structures. 2004;13 1131-1142

[16] T. Starner. Human-powered wearable computing. Ibm Systems Journal. 1996;35 618-629

[17] L. Mateu and F. Moll. Optimum piezoelectric bending beam structures for energy harvesting using shoe inserts. Journal of Intelligent Material Systems and Structures. 2005;16 835-845

[18] L. Mateu and F. Moll. Appropriate charge control of the storage capacitor in a piezoelectric energy harvesting device for discontinuous load operation. Sensors and Actuators a-Physical. 2006;132 302-310

[19] S. Xu, Y. G. Wei, J. Liu, R. Yang and Z. L. Wang. Integrated Multilayer Nanogenerator Fabricated Using Paired Nanotip-to-Nanowire Brushes. Nano Letters. 2008;8 4027-4032

[20] P. J. Cornwell, J. Goethal, J. Kowko and M. Damianakis. Enhancing power harvesting using a tuned auxiliary structure. Journal of Intelligent Material Systems and Structures. 2005;16 825-834

[21] K. Mossi, C. Green, Z. Ounaies and E. Hughes. Harvesting energy using a thin unimorph prestressed bender: Geometrical effects. Journal of Intelligent Material Systems and Structures. 2005;16 249-261

[22] N. M. White, P. Glynne-Jones and S. P. Beeby. A novel thick-film piezoelectric micro-generator. Smart Materials & Structures. 2001;10 850-852

[23] H. S. Yoon, G. Washington and A. Danak. Modeling, optimization, and design of efficient initially curved piezoceramic unimorphs for energy harvesting applications. Journal of Intelligent Material Systems and Structures. 2005;16 877-888

[24] D. Guyomar, A. Badel, E. Lefeuvre and C. Richard. Toward energy harvesting using active materials and conversion improvement by nonlinear processing. Ultrasonics, Ferroelectrics and Frequency Control, IEEE Transactions on. 2005;52 584-595

[25] E. Lefeuvre, A. Badel, A. Benayad, L. Lebrun, C. Richard and D. Guyomar. A comparison between several approaches of piezoelectric energy harvesting. Journal De Physique Iv. 2005;128 177-186

[26] E. Lefeuvre, A. Badel, C. Richard, L. Petit and D. Guyomar. A comparison between several vibration-powered piezoelectric generators for standalone systems. Sensors and Actuators a-Physical. 2006;126 405-416

[27] E. Lefeuvre, A. Badel, C. Richard and D. Guyomar. Piezoelectric energy harvesting device optimization by synchronous electric charge extraction. Journal of Intelligent Material Systems and Structures. 2005;16 865-876

[28] A. Badel, D. Guyomar, E. Lefeuvre and C. Richard. Efficiency enhancement of a piezoelectric energy harvesting device in pulsed operation by synchronous charge inversion. Journal of Intelligent Material Systems and Structures. 2005;16 889-901

[29] Y. C. Shu, I. C. Lien and W. J. Wu. An improved analysis of the SSHI interface in piezoelectric energy harvesting. Smart Materials & Structures. 2007;16 2253-2264

[30] Y. B. Jeon, R. Sood, J. H. Jeong and S. G. Kim. MEMS power generator with transverse mode thin film PZT. Sensors and Actuators a-Physical. 2005;122 16-22

[31] H. B. Fang, J. Q. Liu, Z. Y. Xu, L. Dong, D. Chen, B. C. Cai and Y. Liu. A MEMS-based piezoelectric power generator for low frequency vibration energy harvesting. Chinese Physics Letters. 2006;23 732-734

[32] H. B. Fang, J. Q. Liu, Z. Y. Xu, L. Dong, L. Wang, D. Chen, B. C. Cai and Y. Liu. Fabrication and performance of MEMS-based piezoelectric power generator for vibration energy harvesting. Microelectronics Journal. 2006;37 1280-1284

[33] D. Shen, J. H. Park, J. Ajitsaria, S. Y. Choe, H. C. Wikle and D. J. Kim. The design, fabrication and evaluation of a MEMS PZT cantilever with an integrated Si proof mass for vibration energy harvesting. Journal of Micromechanics and Microengineering. 2008;18, 055017

[34] M. Marzencki, Y. Ammar and S. Basrour. Integrated power harvesting system including a MEMS generator and a power management circuit. Sensors and Actuators a-Physical. 2008;145 363-370

[35] J.-Q. Liu, H.-B. Fang, Z.-Y. Xu, X.-H. Mao, X.-C. Shen, D. Chen, H. Liao and B.-C. Cai. A MEMS-based piezoelectric power generator array for vibration energy harvesting. Microelectronics Journal. 2008;39 802-806

[36] J. Ajitsaria, S. Y. Choe, D. Shen and D. J. Kim. Modeling and analysis of a bimorph piezoelectric cantilever beam for voltage generation. Smart Materials & Structures. 2007;16 447-454

[37] A. Erturk and D. J. Inman. Comment on 'modeling and analysis of a bimorph piezoelectric cantilever beam for voltage generation'. Smart Materials & Structures. 2008;17, 058001

[38] N. E. duToit, B. L. Wardle and S. G. Kim. Design considerations for MEMS-scale piezoelectric mechanical vibration energy harvesters. Integrated Ferroelectrics. 2005;71 121-160

[39] N. E. duToit and B. L. Wardle. Performance of microfabricated piezoelectric vibration energy harvesters. Integrated Ferroelectrics. 2006;83 13-32

[40] B. M. Xu, Y. H. Ye, L. E. Cross, J. J. Bernstein and R. Miller. Dielectric hysteresis from transverse electric fields in lead zirconate titanate thin films. Applied Physics Letters. 1999;74 3549-3551

[41] X. Y. Wang, C. Y. Lee, Y. C. Hu, W. P. Shih, C. C. Lee, J. T. Huang and P. Z. Chang. The fabrication of silicon-based PZT microstructures using an aerosol deposition method. Journal of Micromechanics and Microengineering. 2008;18, 055034

[42] X.-Y. Wang, C.-Y. Lee, C.-J. Peng, P.-Y. Chen and P.-Z. Chang. A micrometer scale and low temperature PZT thick film MEMS process utilizing an aerosol deposition method. Sensors and Actuators A: Physical. 2008;143 469-474

[43] Y. C. Shu and I. C. Lien. Efficiency of energy conversion for a piezoelectric power harvesting system. Journal of Micromechanics and Microengineering. 2006;16 2429-2438

[44] Y. C. Shu and I. C. Lien. Analysis of power output for piezoelectric energy harvesting systems. Smart Materials & Structures. 2006;15 1499-1512

Chapter 8

Wideband Electromagnetic Energy Harvesting from a Rotating Wheel

Yu-Jen Wang, Sheng-Chih Shen and Chung-De Chen

Additional information is available at the end of the chapter

1. Introduction

Harvesting kinetic energy from the environment is an ideal alternative to traditional electro-chemical batteries as a power source for low-power electronics such as pressure sensors and thermometers. A tire pressure monitoring system (TPMS) is a key device in preventing a flat tire, which is one of the most frequent causes of car accidents. The U.S.A., European Union, and a number of Asian countries are currently considering the implementation of TPMS-related legislation. Existing TPMS are powered by a battery, but there are drawbacks to these batteries, such as their limited durability and environmental pollution upon disposal. For this reason, this chapter is aimed at developing a TPMS energy harvester in a rotating wheel.

Researches on energy harvesting from the environment for TPMS have grown in the recent years. The first part of this chapter introduces some of the representative ways to scavenge energy from a rotating wheel or tire. A wide bandwidth or similar frequency-adjusting energy harvesting device is necessary to power a TPMS. This is an emerging research topic with great challenge and the need for further improvement. This chapter presents a well-weighted pendulum with nonlinear effects to help the pendulum adjust its natural frequency to match the wheel rotation frequency. This chapter also formulates mathematical models of the weighted pendulum using the Euler-Lagrange formulation according to arbitrary configurations. This well-weighted pendulum oscillates at various wheel speeds with a larger swing angle and angular velocity than an ill-weighted pendulum design. Electromagnetic induction converts the kinetic energy produced by the pendulum into electrical energy. Numerical analysis reveals that a well-weighted pendulum generates hundreds of micro-Watts. This chapter establishes and analyzes dynamic models with electromechanical couplings, and introduces a novel circular Halbach array to augment magnetic strength on the coil side of the array. At an optimum external resistance, the power output of the well-weighted pendulum was approximately 300 to 550 micro-Watts at

200-500 rpm. This performance demonstrated that this wideband electromagnetic energy harvester has the potential to replace traditional batteries in a TPMS.

2. Energy harvester for a tire pressure monitor system

A tire-pressure monitoring system (TPMS) is an electronic system designed to monitor the air pressure inside pneumatic tires on automobiles, airplanes, straddle-lift carriers, forklifts, and other vehicles (Fig. 1). A TPMS provides the driver with real-time tire pressure information and gives a warning message if the tire pressure falls below the recommended value. In 2000, the U.S. government issued the TREAD act, which forces car manufacturers to install TPMS in all new vehicles sold in the U.S. The European Union and a number of Asian countries are currently considering the implementation of TPMS-related legislation. The TPMS is expected to become one of the most popular electronic products in vehicle applications.

Grand Cherokee WK
Tire pressure sensor

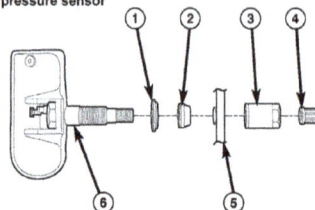

1. Metal washer
2. Sensor-To-Wheel Seal
3. Valve Stem Nut (with pressed-in washer)
4. Valve stem cap
5. Wheel
6. Sensor

(a)   (b)

Figure 1. (a) TPMS module installed in a tire (b) Detailed components of TPMS by BMW.

A TPMS consists of a wireless pressure sensor inside each tire and a receiver in the car. Existing systems are powered by lithium batteries, but there are drawbacks to these batteries: limited durability, relatively large size and weight, and environmental pollution upon disposal. As a result, they have gradually become less favored by the public. For this reason, research on harvesting energy from the environment for TPMS has grown over the past few years (Hatipoglu & Urey, 2009; Matsuzaki & Todoroki, 2008). This chapter discusses a TPMS energy harvester device to satisfy the critical demand for a sustainable TPMS power supply.

2.1. A review of previous research on wide-bandwidth energy-harvesting methods

Instead of relying on embedded batteries, an energy harvesting device converts mechanical vibration energy into electricity. Possible methods of powering wireless devices installed on a vehicle include harvesting energy from a rotating tire or road vibrations. Previous reports regarding energy harvesting from a rotating wheel or tire (Hatipoglu & Urey, 2009; Lohndorf et al., 2007; Matsuzaki & Todoroki, 2008; Roundy 2008) have utilized vibration energy. One of the critical issues is that the maximal frequency band of the energy harvesting device is limited and cannot be chosen arbitrarily (Roundy 2008).

(Leland et al., 2006) presented a tuneable-resonance vibration energy scavenger that axially compresses a piezoelectric bimorph to decrease its resonance frequency. They showed that an axial preload can adjust the resonance frequency of a simply-supported bimorph to 24% below its unloaded resonance frequency. The power output to a resistive load was 65–90% of the nominal value at frequencies 19–24% below the unloaded resonance frequency. Prototypes produced 300–400 μW of power at driving frequencies of 200-250 Hz. However, the compressive axial preload of the piezoelectric device cannot automatically adjust itself. The other method of ensuring that the energy harvester has a self-tuning resonance frequency capability to match the base motion frequency is integrating the harvester with an accelerometer, a displacement sensor, and an actuator to tune the stiffness of the harvester based on the base vibration frequency (Leland & Wright, 2006). Passive frequency-adjusting energy-harvesting mechanisms have grown in use over the past few years (Daminakis et al., 2005; Gu & Livermore, 2010; Lee et al., 2007; Mansour et al., 2010; Marzencki et al., 2009). (Youngsman et al., 2005) presented a model for a frequency-adjustable vibration energy harvester.

(Sari et al., 2008) developed a wideband electromagnetic vibration-to-electrical micro power generator. This generator covers a wide band of external vibration frequencies by implementing a number of serially connected cantilevers of different lengths and varying natural frequencies. This device generated 0.4 μW of continuous power with 10mV voltage at an external vibration frequency range of 4.2–5 kHz, covering a band of 800 Hz. Possible disadvantages of this approach are that the maximum power is too small to power a TPMS, and the suitable vibration frequency is too high for a rotating wheel. Other researchers have attempted to develop wide-band structures to achieve high-efficiency energy-scavenging devices (Marinkovic & Koser, 2009; Zhang & Chen, 2010).

(Stanton et al., 2009) validated a nonlinear energy harvester capable of bidirectional hysteresis. This design invokes hardening and softening responses within the quadratic potential field of a power generating piezoelectric beam (with a permanent magnet end mass) by tuning nonlinear magnetic interactions. This technique increases the bandwidth of the device from 12±0.5 Hz to 11~14 Hz. To consider the nonlinear effects in electromagnetic harvest mechanisms, Dallago et al., 2010 presented an analytical model that considers the non-linear electromagnetic repulsion force and the flux linkage of the coil.

Bower & Arnold, 2009 developed a non-resonant energy harvester that employs a spherical magnet ball that moves arbitrarily in a cavity wrapped with copper coil windings. This harvester demonstrated root-mean-square voltages ranging from 80 to 700 mV. Platt et al., 2005, developed a low-frequency electric power generator using PZT ceramics for the same purpose.

A wide bandwidth or similar frequency-adjusting energy harvesting device is necessary for a TPMS. This is an emerging research topic with great challenges and the need for further improvement. This chapter proposes the use of a novel well-weighted pendulum method to harvest kinetic energy from a rotating wheel. In this design, the pendulum oscillates in response to periodic changes in the tangential component of gravity. The greatest advantage of this approach is that the rotating frequency of the wheel increases linearly as the car speeds up. Unlike traditional energy-harvesting devices, which have a fixed natural frequency, the radial acceleration of the rotating wheel will raise the pendulum's natural frequency of oscillation as the car increases its speed.

2.2. TPMS power consumption

Knowing the power consumption of TPMS is a prerequisite to designing a suitable energy harvester. Fig. 2 shows the all-in-one package block diagram of the TPMS produced by Orange Electronics, which includes a pressure sensor, an 8-bit microcontroller (MCU), an RF transmitter, and a 2-axis accelerometer sensor. This TPMS is installed on the wheel rims to provide independent, real-time air pressure measurements for each tire that can be transmitted to the vehicle instrument cluster to instantly inform the driver.

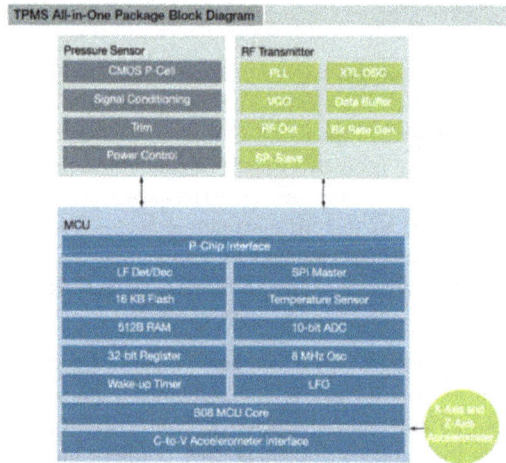

Figure 2. TPMS block diagram by Orange Electronics.

Fig. 3 depicts the Orange Electronics P409S TPMS, P409S, employed in the experiment. The low-current measurement circuit is in-series connected with the TPMS (Fig. 4). From LF411, the output voltage V_o provides 2 mA/V exchange rate. Table 1 lists the operation modes and power consumption of the TPMS. The whole operation, which consists of pressure measurement, RF transmitter activation, temperature measurement, and idle mode, lasts 130.2 seconds and consumes average power 48.1 μw.

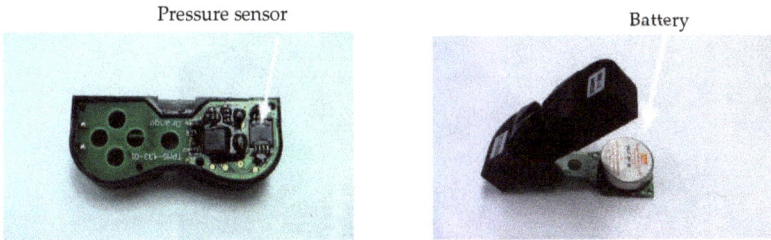

Figure 3. Orange Electronics P409S TPMS.

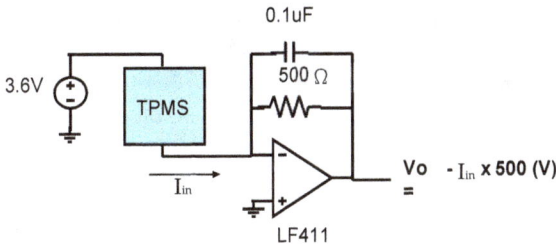

Figure 4. Low current measurement circuit.

Item	Action	Number of times	Duration (sec)	Peak current (mA)	Average current (mA)	Average power (mW)
A	Pressure sensor active	36	0.01	2.56	1.20	4.32
B	Idle mode	42	3.1	0.00260	0.00260	0.00936
C	RF signal	5	0.042	4.75	4.33	15.6
D	Pressure and Temp. measurement	1	0.01	2.56	1.75	6.30

Table 1. Operation mode and power consumption of TPMS

3. Kinetic energy of the wheel

The pressure sensing module of TPMS was mounted under a rubber tire to measure the tire pressure. The first step in designing the energy harvesting device for TPMS was to determine the kinetic energy of a wheel. A 3-axis accelerometer and a radio frequency (RF) transmission module with a sampling rate 50 Hz were mounted on a wheel rim to collect kinematic data when the car was driving at any constant speed Fig. 5. The acceleration data were then transferred to a receiver connected to a laptop computer in the test vehicle.

Figure 5. 3-axis accelerometer with the RF module mounted on the wheel.

Fig. 6 plots the normal and tangential accelerations of the accelerometer in red and blue lines, respectively, when the car travelled at a constant speed of 70 km/hr (619 rpm) on an asphalt road. The normal acceleration resulted from both centripetal and gravitational forces during wheel rotation. Even at low speeds, such as 20 km/hr (177 rpm), the centripetal force was 7 times larger than gravity. The tangential acceleration experienced only the tangential component of gravity. Fig. 7 shows the energy spectrum of the tangential acceleration, the peak of which is located at approximately 10 Hz and is very close to the wheel rotation frequency. This result was similar even when the test car was driving on a macadam road at various speeds. This indicates that other acceleration terms, such as that caused by tire vibration from the road, are insignificant. Therefore, the tangential acceleration at each point of the rim caused by the variability of the tangential component of gravity during rotation is a reliable energy source for a rotating wheel under any road conditions.

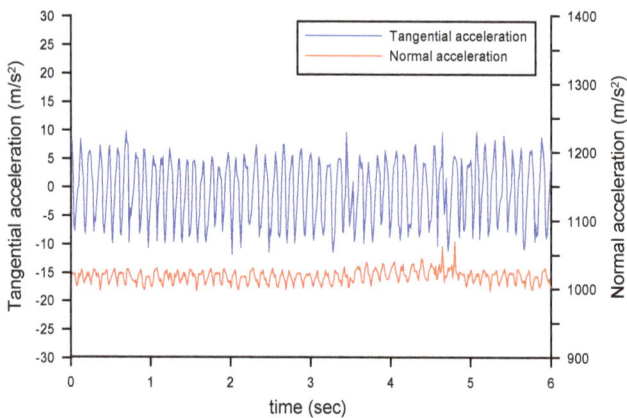

Figure 6. Normal and tangential acceleration data at 70 km/hr (619 rpm).

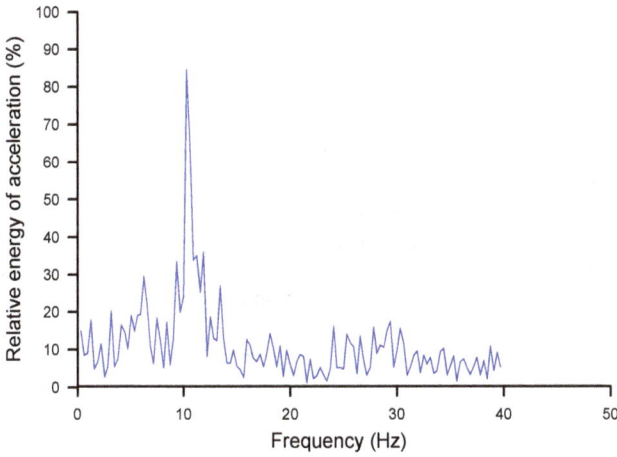

Figure 7. Energy spectrum of tangential acceleration at 70 km/hr.

4. Weighted pendulum by arbitrary configurations

This section derives the governing equation of the weighted pendulum from the arbitrary configuration to show the generality (Wang et al., 2012a). Consider a weighted pendulum mounted on a wheel (Fig. 8). The locations of the center and the pivot with reference to a fixed frame (x, y) are $(x_0, 0)$ and (x_1, y_1), respectively. Introducing the wheel rotation angle Θ in Fig. 8 gives

$$x_1 = x_0 + R_2 \sin\Theta$$
$$y_1 = R_2 \cos\Theta$$

(1)

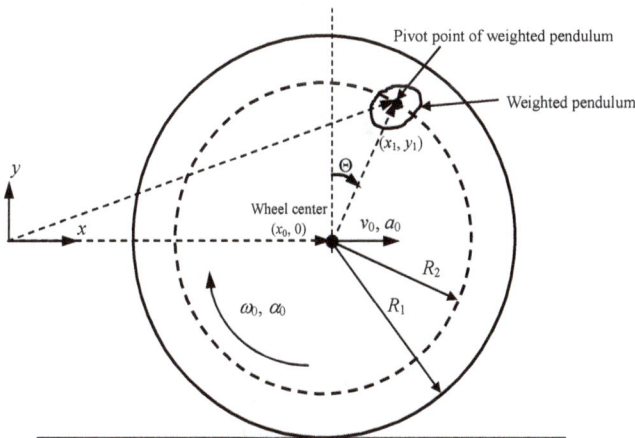

Figure 8. Schematic diagram of a rolling wheel.

Fig. 9 shows the detailed configuration of the weighted pendulum, which can be an arbitrary shape but whose rotation axis is not located at the center of mass. This study defines a local coordinate $(\bar{x}, \bar{y})$ attached to the weighted pendulum such that its origin is located at the pivot and its center of mass is located at the extended line of the $\bar{y}$-axis. Considering an infinitesimal element dm on the weighted pendulum, the position of dm can be defined by a distance r and an angle ϕ (Fig. 9). θ is the swing angle of the weighted pendulum. Note that the profile of the weighted pendulum represented by the dashed line in Fig. 9 denotes an arbitrary swing angle θ with reference to the pivot point while the solid line denotes a specific position $\theta = 0$. The swing angle can be considered as the relative position between the weighted pendulum and the wheel. Further define an angle φ such that

$$\phi + \theta + \varphi - \Theta = \frac{\pi}{2} \tag{2}$$

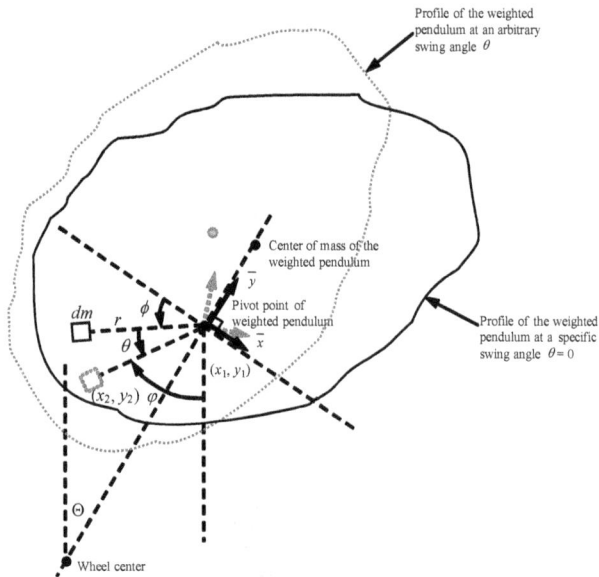

Figure 9. Diagram of a weighted pendulum.

For an arbitrary swing angle θ, the location of dm with respect to (x, y) is

$$\begin{aligned} x_2 &= x_0 + R_2 \sin\Theta - r\sin\varphi \\ y_2 &= R_2 \cos\Theta - r\cos\varphi \end{aligned} \tag{3}$$

where R_2 is the distance between the pivot and the wheel center. By differentiating Eq. (3) with respect to time t, the velocity of dm can be obtained as follows:

$$\dot{x}_2 = v_0 + R_2\dot{\Theta}\cos\Theta - r\dot{\varphi}\cos\varphi$$
$$\dot{y}_2 = -R_2\dot{\Theta}\sin\Theta + r\dot{\varphi}\sin\varphi \tag{4}$$

where v_0 is the speed of the wheel center and $\dot{\Theta}$ is the wheel rotation speed. The equations of motion of the weighted pendulum can be derived using the Lagrange equation. Based on Eq. (4), the kinetic energy of the weighted pendulum can be obtained by integrating over the volume $\bar{V}$ with respect to the local coordinate $(\bar{x}, \bar{y})$:

$$
\begin{aligned}
T &= \frac{1}{2}\int(\dot{x}_2^2 + \dot{y}_2^2)dm \\
&= \frac{1}{2}\int\Big[v_0^2 + R_2^2\dot{\Theta}^2 + r^2\dot{\varphi}^2 + 2v_0(R_2\dot{\Theta}\cos\Theta - r\dot{\varphi}\cos\varphi) \\
&\quad - 2R_2r\dot{\Theta}\dot{\varphi}\cos(\Theta - \varphi)\Big]\rho d\bar{V}
\end{aligned}
\tag{5}
$$

Eq. (5) uses the relation

$$dm = \rho d\bar{V} \tag{6}$$

where ϱ is density of the weighted pendulum and $d\bar{V}$ is an infinitesimal volume. The potential energy is

$$V = g\int\big[R_2\cos\Theta - r\cos\varphi\big]\rho d\bar{V} \tag{7}$$

where g is the gravitational acceleration. According to the Lagrangian function L:

$$
\begin{aligned}
L &= T - V \\
&= \frac{1}{2}\int\Big[v_0^2 + R_2^2\dot{\Theta}^2 + r^2\dot{\varphi}^2 + 2v_0(R_2\dot{\Theta}\cos\Theta - r\dot{\varphi}\cos\varphi) - 2R_2r\dot{\Theta}\dot{\varphi}\cos(\Theta - \varphi) \\
&\quad - 2g(R_2\cos\Theta - r\cos\varphi)\Big]\rho d\bar{V}
\end{aligned}
\tag{8}
$$

According to the Euler–Lagrange formulation, the equation of motion can be written as:

$$\frac{d}{dt}\left(\frac{\partial L}{\partial \dot{\varphi}}\right) - \frac{\partial L}{\partial \varphi} = 0 \tag{9}$$

Substituting Eq. (8) into Eq. (9) yields the following equation:

$$
\begin{aligned}
&\int\rho r^2\ddot{\varphi}d\bar{V} - a_0\int\rho r\cos\varphi d\bar{V} - R_2\ddot{\Theta}\int\rho r\cos(\Theta - \varphi)d\bar{V} \\
&\quad + R_2\dot{\Theta}^2\int\rho r\sin(\Theta - \varphi)d\bar{V} + g\int\rho r\sin\varphi d\bar{V} = 0
\end{aligned}
\tag{10}
$$

in which $a_0 = \ddot{x}_0$. Substituting Eq. (9) into Eq. (10) yields

$$
\begin{aligned}
&\left(\int\rho r^2 d\bar{V}\right)(\ddot{\Theta} - \ddot{\theta}) - a_0\int\rho r\sin(\phi + \theta - \Theta)d\bar{V} \\
&\quad - R_2\ddot{\Theta}\int\rho r\sin(\phi + \theta)d\bar{V} - R_2\dot{\Theta}^2\int\rho r\cos(\phi + \theta)d\bar{V} + g\int\rho r\cos(\phi + \theta - \Theta)d\bar{V} = 0
\end{aligned}
\tag{11}
$$

Recall that the $\bar{y}$-axis passes through the center of mass of the weighted pendulum. Therefore, the mass distribution is symmetric with respect to the $\bar{y} - \bar{z}$ plane, which gives $\int \rho r \cos \phi d\bar{V} = 0$. Eq. (11) then becomes

$$\left(\int \rho r^2 d\bar{V} \right) \ddot{\theta} - R_2 \dot{\Theta}^2 \left(\int \rho r \sin \phi d\bar{V} \right) \sin \theta$$
$$= \left(\int \rho r^2 d\bar{V} \right) \ddot{\Theta} - a_0 \left(\int \rho r \sin \phi d\bar{V} \right) \cos(\theta - \Theta) - \qquad (12)$$
$$R_2 \ddot{\Theta} \left(\int \rho r \sin \phi d\bar{V} \right) \cos \theta - g \left(\int \rho r \sin \phi d\bar{V} \right) \sin(\theta - \Theta)$$

where $\int \rho r^2 d\bar{V} = I$ is the mass moment of inertia. At this point, and without loss of generality, assume a constant wheel speed and a small swing angle of the pendulum to facilitate the study of the natural frequency of the weighted pendulum. After linearization, the equation of motion can be written as

$$\left(\int \rho r^2 d\bar{V} \right) \ddot{\theta} - R_2 \dot{\Theta}^2 \left(\int \rho r \sin \phi d\bar{V} \right) \theta = -g \left(\int \rho r \sin \phi d\bar{V} \right) \sin \Theta \qquad (13)$$

The natural frequency ω_n of the weighted pendulum is

$$\omega_n = \left(-R_2 \dot{\Theta}^2 \frac{\int \rho r \sin \phi d\bar{V}}{\int \rho r^2 d\bar{V}} \right)^{1/2} \qquad (14)$$

This study defines the characteristic length, L^*, as

$$L^* = -\frac{\int \rho r^2 d\bar{V}}{\int \rho r \sin \phi d\bar{V}} \qquad (15)$$

Because of the definition of the local coordinate $(\bar{x}, \bar{y})$, $\sin \phi$ is a negative number. Therefore, the characteristic length L^* is a positive number. After linearization, the natural frequency at a constant wheel speed can be written as

$$\omega_n = \dot{\Theta} \sqrt{\frac{R_2}{L^*}} \qquad (16)$$

Note that ω_n is proportional to the wheel rotation frequency $\dot{\Theta}$. If the weighted pendulum is designed to make

$$L^* = R_2 \qquad (17)$$

then the natural frequency of the weighted pendulum is equal to the wheel rotation frequency at any car speed to achieve high-efficiency energy harvesting. By integrating the weighted pendulum with magnets and coils, kinetic energy can be transformed into electrical energy during wheel rotation.

By adding the damping term, Eq. (12) can be written as

$$
\left(\int \rho r^2 d\overline{V} \right) \ddot{\theta} + C_T \dot{\theta} - R_2 \dot{\Theta}^2 \left(\int \rho r \sin \phi d\overline{V} \right) \sin \theta
$$
$$
= \left(\int \rho r^2 d\overline{V} \right) \ddot{\Theta} - a_0 \left(\int \rho r \sin \phi d\overline{V} \right) \cos(\theta - \Theta) - \tag{18}
$$
$$
R_2 \ddot{\Theta} \left(\int \rho r \sin \phi d\overline{V} \right) \cos \theta - g \left(\int \rho r \sin \phi d\overline{V} \right) \sin(\theta - \Theta)
$$

where C_T^* is the total generalized damping constant. Eq. (18) can then be rewritten as

$$
\ddot{\theta} + C_T^* \dot{\theta} - \frac{R_2 \dot{\Theta}^2}{L^*} \sin \theta = \ddot{\Theta} - \frac{a_0}{L^*} \cos(\theta - \Theta) - \frac{R_2 \ddot{\Theta}}{L^*} \cos \theta - \frac{g}{L^*} \sin(\theta - \Theta) \tag{19}
$$

in which $C_T^* = C_e^* + C_m^*$ and $C_T^* = \dfrac{C_T}{\left(\int \rho r^2 d\overline{V} \right)}$. C_e^* and C_m^* are the generalized

electromagnetic damping and the generalized mechanical damping caused by the electromagnetic force and the friction of motion, respectively. By assuming a constant car speed, Eq. (19) becomes

$$
\ddot{\theta} + C_T^* \dot{\theta} - \frac{R_2 \dot{\Theta}^2}{L^*} \sin \theta = -\frac{g}{L^*} \sin(\theta - \Theta) \tag{20}
$$

This section discusses the dynamic behavior of the weighted pendulum by solving the Eq. (19) using Runge-Kutta method. Eq. (19) shows that the dynamic behavior can be determined by two independent parameters: the characteristic length L^* and the total generalized damping constant C_T^*, if the dimensions of the wheel (R_1 and R_2), the velocity and acceleration of the car are known. The following parameters are fixed in all of the following numerical studies: $R_1 = 0.300$ m, $R_2 = 0.203$ m, g = 9.81 m/s², I = 1.847×10⁻⁶ kg-m². I is the mass moment of inertia of the rotor. As an illustrative example, the characteristic length and generalized damping constant are 0.192 m and 0.70 N-s/kg/m, respectively. Fig. 10 shows the swing angle θ driven by gravity in the time domain when the wheel speed is 350 rpm with the initial conditions $\theta(0) = \pi / 3$ and $\dot{\theta}(0) = 0$. To simplify the description in the following figures and discussions, the maximum swing angle, θ_{max}, is defined by the largest swing angle value in the positive direction.

Fig. 11 shows the maximum swing angle at the steady state, θ_{max}, at a wheel speed of 350 rpm and the initial conditions $\theta = 0$ and $\dot{\theta} = 0$ under various damping constants. The swing angle strongly depends on L^* when damping is small. For large damping, the swing angle becomes small and the dynamic system becomes linear. The validity of the numerical results of swing angle for large damping was confirmed by the linear vibration theory. Fig. 11 shows that small damping yields a large swing angle. However, a large swing angle does not guarantee a large amount of power converted.

Fig. 12 shows the steady-state average power for a wheel speed 350 rpm and initial conditions $\theta = 0$ and $\dot{\theta} = 0$ under various damping constants. This figure reveals a compromise between swing angle and damping constant. The optimum total generalized damping to obtain a large amount of power converted is approximately 1.6 N-s/kg/m to 3.0 N-s/kg/m. Fig. 11 and Fig. 12 indicate that the best L^* ranges from 0.190 m to 0.203 m for a specific damping constant.

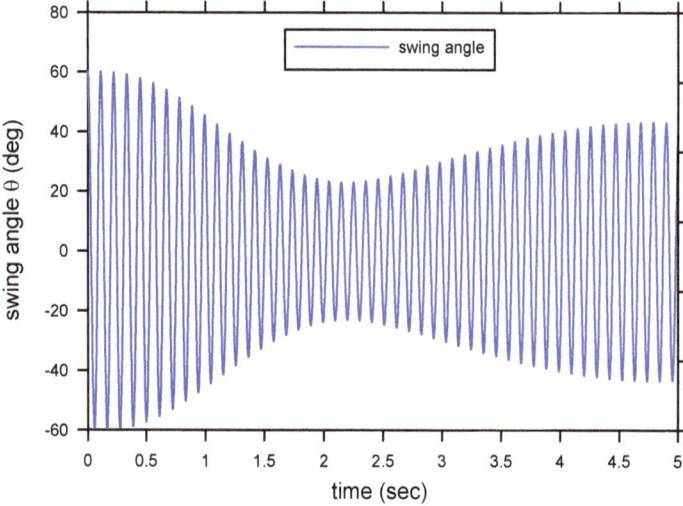

Figure 10. Transient response of the swing angle at the wheel speed 350 rpm and initial conditions $\theta(0) = \pi / 3$ and $\dot{\theta}(0) = 0$.

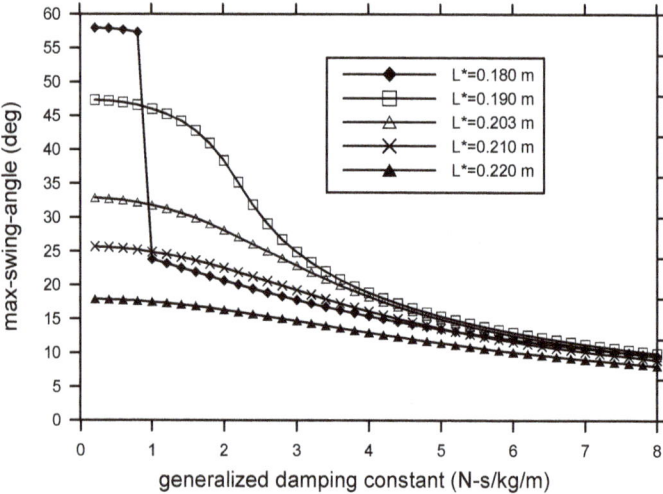

Figure 11. Variation of the steady-state swing angle for the wheel speed 350 rpm and the initial conditions $\theta = 0$ and $\dot{\theta} = 0$ under various total generalized damping constants.

The swing motion of a weighted pendulum driven by gravity exhibits nonlinear behavior. Eq. (19) describes the dynamic behavior of the weighted pendulum, and the requirements of Eq. (17) cause the system to oscillate at resonance at any wheel speed. Fig. 13 shows the variation of maximum swing angle at the steady state under various L^* and wheel rotation speeds with total generalized damping constant $C_T^* = 0.70$ N-s/kg/m. When $L^* = 0.203$ m and the condition $L^* = R_2$ is met, the swing angle varies continuously at different wheel speeds. Compared with with other situations, the maximum swing angle when $L^* > R_2$ is smaller than that when $L^* = R_2$. This means that the power generated decreases when L^* exceeds R_2. When $L^* < R_2$, the maximum swing angle exhibits a sudden discontinuous jump above the critical wheel rotation speed. Using the case $L^* = 0.190$ m as an example, the amplitude gradually decreases with an increasing frequency of wheel rotation until point a is reached, and then suddenly jumps to a smaller value with an increasingly small rotation frequency, as indicated by point b. The narrow region between point a and point b in Fig. 13 is unstable. Beyond point b, the amplitude of the maximum swing angle keeps continuously decreasing. The results above indicate that when L^* is exactly equal to R_2, a steady output power is provided at any speed. This provides evidence of the advantage of using a well-weighted pendulum.

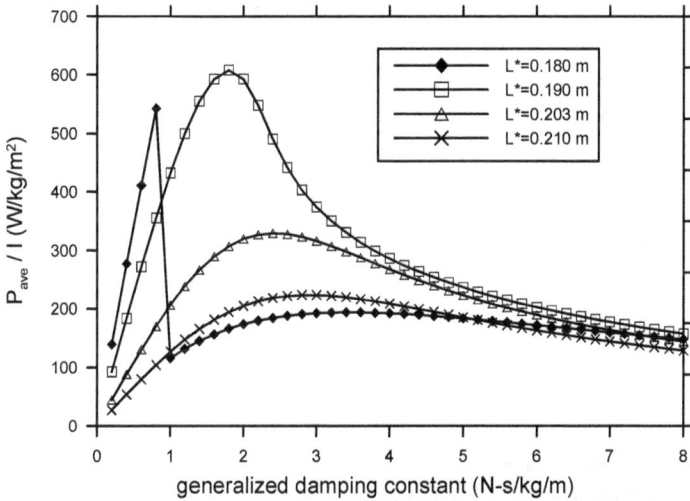

Figure 12. Variation of the steady-state average power for the wheel speed 350 rpm and the initial conditions $\theta = 0$ and $\dot{\theta} = 0$ under various total generalized damping constants.

The discussion above assumes that the car velocity is constant. The following discussion considers two non-constant car velocity modes, acceleration and deceleration (Fig. 14). The characteristic length and total generalized damping constant are 0.203 m and 0.70 N-s/kg/m, respectively, and the initial conditions are $\theta(0) = 0$ and $\dot{\theta}(0) = 0$. In the first 10 seconds, the constant speed mode causes the steady-state value of the maximum angular velocity to reach approximately 17 rad/s (Fig. 14 (a)). In the time interval from 10 to 25 seconds, the car accelerates at a rate of 1.48 m/s², during which the angular velocity increases at a rate of approximately 0.87 rad/s², and the high wheel speed increases the maximum power to approximately 800 W/Kg/m². In the time interval from 25 to 35 seconds, the variation of maximum power and maximum angular velocity is small because the car speed remains constant. In the period from 35 to 45 seconds, the car decelerates at 2.22 m/s², causing the maximum angular velocity to drop at a rate of approximately 0.95 rad/s² and the maximum power to drop at a rate of 43 W/Kg/m² per second. Fig. 14 (b) shows that the maximum power and maximum angular velocity increases linearly with constant acceleration or deceleration. When the well-weighted pendulum satisfies the requirement of Eq. (17), the variation in maximum power and maximum angular velocity is continuous, and no jump phenomenon is apparent.

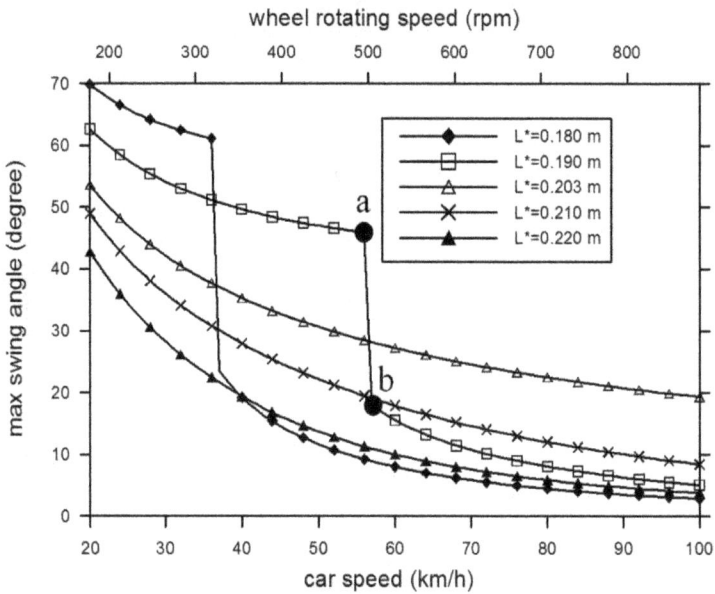

Figure 13. Variation of the steady-state swing angle under various L^* and wheel speeds with initial conditions $\theta = 0$, $\dot{\theta} = 0$ and $Cr^* = 0.70$ N-s/kg/m.

The transient response analysis of the weighted pendulum after large car acceleration is also necessary. To give an example, the total generalized damping constant is 0.5 N-s/Kg/m and the initial conditions are $\theta(0) = 0$ and $\dot{\theta}(0) = 0$. In the first 10 seconds, the wheel rotation speed is constant at 200 rpm, which equals 22.6 km/hr for $R_1 = 0.300$ m. In the time interval of acceleration from 10 to 20 seconds, the wheel speeds up with an angular acceleration $\alpha = 73.7$ rad/s² to reach a constant rotation speed of 710 rpm. This is equal to a 80.2 km/hr car speed, and Fig. 15 shows the transient response of the weighted pendulum. The maximum swing angles of L^* from 0.180 to 0.203 after α = 73.7 rad/s^2 are very close to the results of a small α = 0.105 rad/s^2. The maximum swing angles for L^* = 0.180 and 0.190 m after α = 148 rad/s^2 are far smaller than after α = 73.7 rad/s^2 at the same final constant speed of 710 rpm (Fig. 16). These results show that a huge angular acceleration can cause the jump phenomenon to occur for a critical value of L^*. Fig. 16 shows that the well-weighted pendulum of L^* = R_2 = 0.203 m will not encounter the jump phenomenon when the angular acceleration speed is 148 rad/s^2. The acceleration mode is equal to a car accelerating from 22.6 to 80.2 km/hr in 5 seconds, which is the acceleration limit of a normal car on a highway.

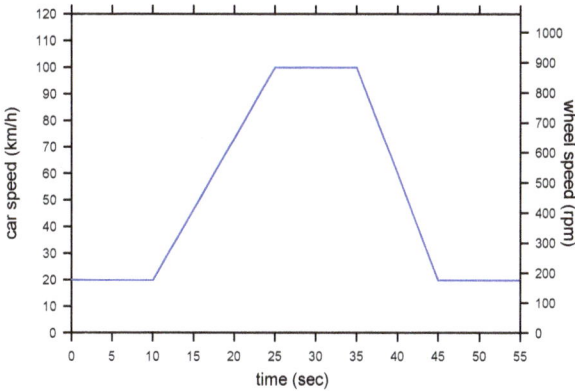

(a) Car speed mode

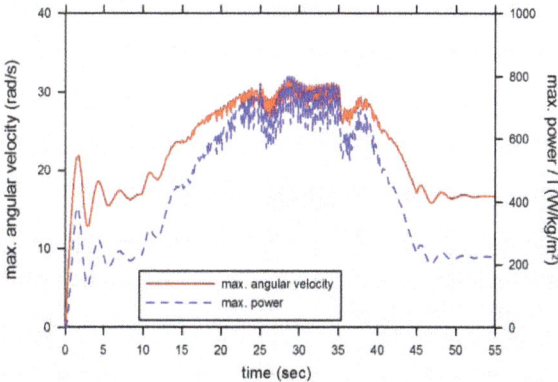

(b) Maximum power and angular velocity, $\dot{\theta}_{max}$, under different car speeds.

Figure 14. Time domain responses of the well-weighted pendulum under car acceleration and deceleration. Characteristic length and total generalized damping constant are 0.203 m and 0.70 N-s/kg/m, respectively.

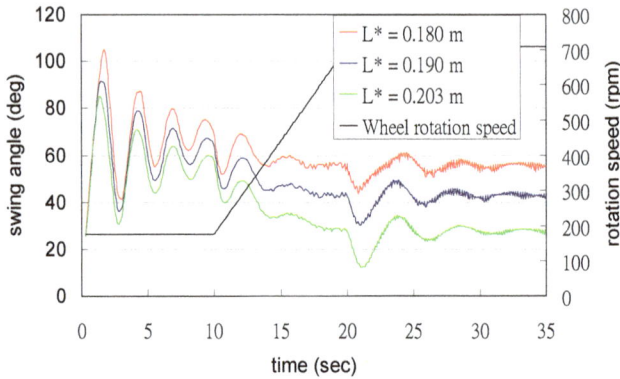

Figure 15. Maximum swing angle under various L^* after angular acceleration of 73.7 rad/s².

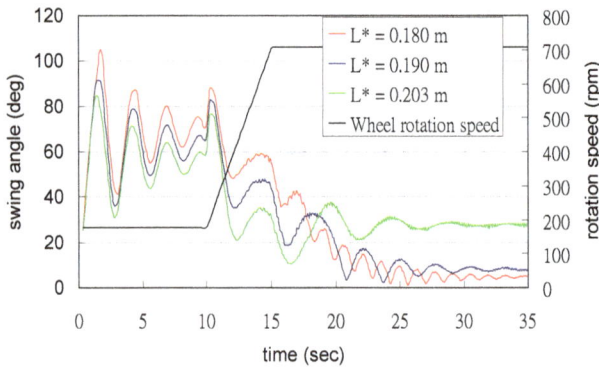

Figure 16. Maximum swing angle under various L^* after angular acceleration of 148 rad/s².

5. Magnetic circuit and power generation analysis

Electromagnetic generation is caused by a coil with a dynamically changing magnetic flux density B. To achieve greater power generation at the same oscillation speed with the same number of turns of the coil, a magnetic circuit design for a dense magnetic flux is necessary. Fig. 17 (a) illustrates a novel arrangement of permanent magnets designed using the Halbach array concept. This arrangement consists of 16 individual sector-shaped permanent Nd-Fe-B magnets with an outer radius r_{out} of 13.0 mm, an inner radius r_{in} of 7.0 mm, and a thickness t_m of 3.0 mm. The red arrows and symbols represent the flux orientation of each magnet. This circular Halbach array augments the magnetic strength on the coil side of the array. Because of the oscillation motion of the weighted pendulum, a disk-type magnet with a periodic arrangement is considered.

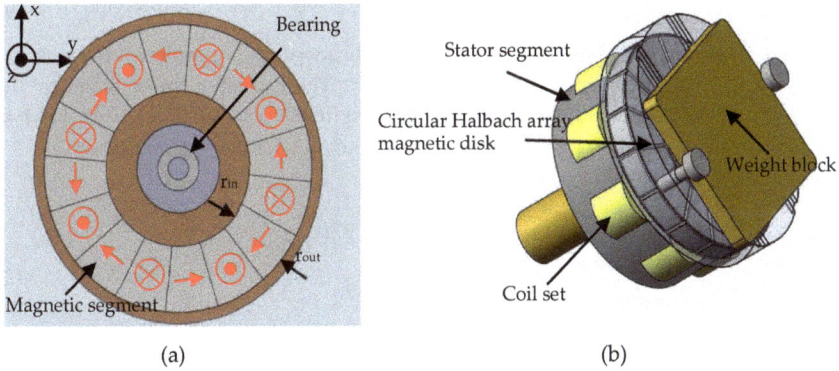

(a)   (b)

Figure 17. Schematic illustrations of circular Halbach array magnetic disk.

Fig. 18 presents the 3D magnetic field distribution of the circular Halbach array disk in the A–A cross section plane located between r_{in} and r_{out} calculated using COMSOL V3.5 software. One period consists of a four-magnet arrangement, and the circular Halbach array disk has four periods. The residual magnetism of each magnet is 1.4 T. The simulation results shown in Fig. 18 indicate that the magnetic flux densities reinforce one another above the array where the coils are placed, but cancel one another below it. This kind of magnetic field distribution increases the induced electromotive force (EMF) ε. Considering the coil motion in the x–y plane, the z component of the magnetic flux density B_z above the circular Halbach array was calculated. The gradient change and the strength of the z component of the average magnetic flux density for the circular Halbach array disk exceed those of normal multipolar magnetic disk. For detailed information on the magnetic circuit analysis, please refer to (Wang et al., 2012b).

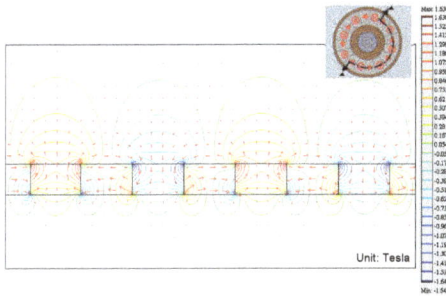

Figure 18. Magnetic field distribution of the circular Halbach array disk.

According to Faraday's law, when a magnet and a coil loop have a relative motion, the changing flux inside the loop can generate an EMF. Considering a coil set in a hollow cylinder shape by spiral winding, the coil set can be divided into N_t layers along the axial direction, and each layer has N_r loops. The magnetic flux through each coil loop moving in the x–y plane can be expressed as $uD[B_z(x_c + \frac{D}{2}) - B_z(x_c - \frac{D}{2})]$, where D, u, x_c, and B_z are the diameter of the loop, linear velocity of the center of the loop, center position of the loop, and z component of magnetic flux density, respectively. The z component of magnetic flux density of each loop in the different layers is acquired from the simulation results. The total EMF ε_{total} can be obtained by superposing $\Phi_{ij}(\theta)$ of each coil for the different layers:

$$\Phi_{ij}(\theta) = uD_{ij}[B_{zi}(x_{c_{ij}} + \frac{D_{ij}}{2}) - B_{zi}(x_{c_{ij}} - \frac{D_{ij}}{2})]$$

$$\varepsilon_{total} = \frac{d\Phi_{total}}{dt} = \frac{d(n\sum_{i=1}^{Nt}\sum_{j=1}^{Nr}\Phi_{ij}(\theta))}{dt} \tag{21}$$

where n is the total number of coil sets. The numerical EMF can be obtained according to the magnetic flux density from above simulation and Eq. (21). The numerical EMF of the circular Halbach array disk was approximately 4.162 V. Applying the chain rule to Faraday's law, the EMF can be expressed as

$$\varepsilon = \frac{d\Phi}{dt} = \frac{d\Phi}{d\theta}\theta \tag{22}$$

With EMF and the angular velocity $\dot{\theta}$ acquired from the numerical results, $d\Phi/d\theta$ can be calculated by Eq. (22). The torque M_m caused by the electromagnetic damping to the rotating system can be represented as

$$M_m(t) = \frac{(d\Phi/d\theta)^2}{R_c + R_L}\dot{\theta} \tag{23}$$

where R_c is the resistance of the coil and R_L is the external resistance. Consequently, the electromagnetic damping C_e can be written as

$$C_e = \frac{(d\Phi/d\theta)^2}{R_c + R_L} \tag{24}$$

The voltage across the resistor was measured using an oscilloscope, and the power consumed by the external resistor was calculated as the square of the root mean square of the voltage (V_{rms}) divided by the external resistance value (i.e., $P_{exp} = (V_{rms})^2/R_L$). Using the electromagnetic damping C_e^* obtained from Eq. (24) and integrating into the following equation, the numerical power consumption of the external resistor ($P_{num.}$) was calculated as

$$P_{num.} = \frac{R_L}{\tau(R_c + R_L)}\int_{t_0}^{t_0+\tau} C_e^* \dot{\theta}^2 dt \tag{25}$$

where τ is the swing period of the weighted pendulum and t_0 is any given time at a steady state. For the proposed design, $d\Phi/d\theta$ was approximately 0.066 V-s, as calculated by Eq. (22). Table 2 lists the magnet and coil parameters of the circular Halbach array disk and Fig. 19 shows the prototype of the energy harvester integrating the weighted pendulum and the circular Halbach array disk. Eqs. (19), (22), and (25) yield the numerical results for power generation and voltage output. Fig. 20 presents the experimental and numerical results for power of the energy harvester (Wang et al., 2012b). The experimental results were averaged from ten sets of data.

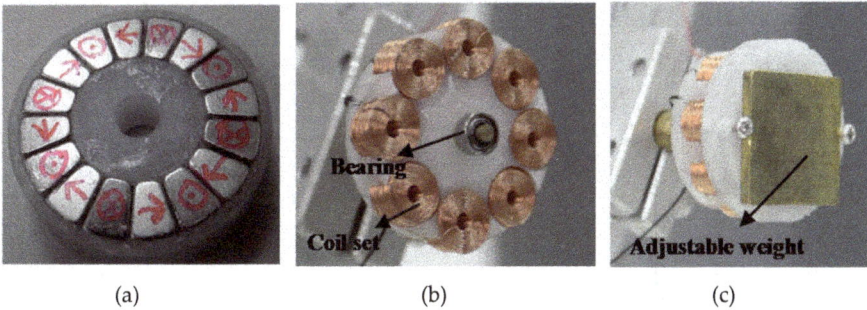

(a)   (b)   (c)

Figure 19. Prototype of the energy harvester. (a) Rear view of the rotor with the circular Halbach array magnetic disk. (b) Front view of the stator. (c) The energy harvester mounted on the rotation plate.

Items	Quantification
Dimensions of individual magnet	r_{out} = 13.0 mm, r_{in} = 7.0 mm, t_m = 3.0 mm
Residual magnetism of magnet by axial magnetization	B_r = 1.4 T
Coil wire diameter	d = 0.1 mm
Coil turns in a layer and coil layers in a set	N_r = 20, N_t = 30
Thickness of coil set	t_c = 3.0 mm
Inside and outside diameter of the coil set	D_{in} = 2.0 mm, D_{out} = 6.0 mm
Air gap between magnetic disk and coil sets	t_p = 1.0 mm
Mass Moment of inertia of the rotor	I = 1.611×10^{-6} kg-m^2
$d\Phi/d\theta$	0.066 V-s
Coil resistance	R_c = 156 Ω

Table 2. Parameters of the circular Halbach array disk for the power generation experiments.

Numerical and experimental results both demonstrate that the electrical power obtained from the weighted pendulum increased monotonically with wheel speed because of the high angular velocity $\dot{\theta}$ of the magnets, which induced a large EMF ε. The difference between the numerical and experimental results in terms of the voltage across the resistor was smaller than that observed in the average power output because power output contains the squared term of voltage. In the power output experiments, the optimum external resistance required to obtain the largest power output ranged from 330 Ω to 510 Ω, which is close to the internal resistance of the coil, but not constant. These results confirm that maximum power is delivered to an external load when the external load resistance equals the sum of the coil resistance and the electrical analogue of the friction damping coefficient. Although a high external resistance can produce a large output voltage for greater efficiency in the pumping circuit, a large power output occurs at an optimum resistance. Hence, there must be a trade-off between circuit design and power transformation.

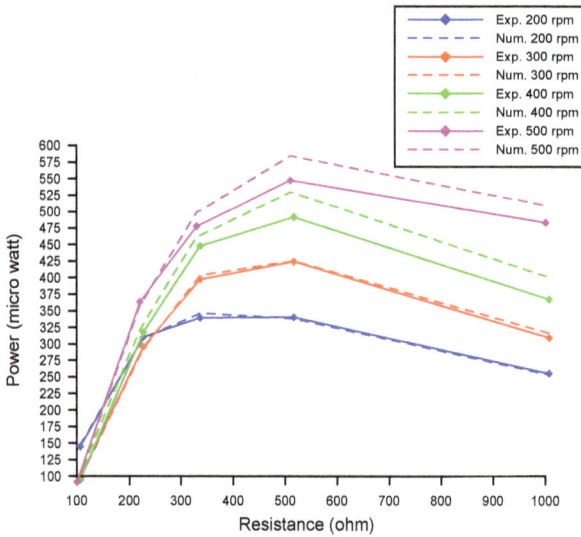

Figure 20. Experimental and numerical results for the circular Halbach array disk. Average power output (consumption) of the external resistor (unit, μW).

6. Conclusion

This chapter proposes a novel well-weighted pendulum design. Numerical analysis and experimental results reveal that several hundred micro-Watts of power, which is sufficient to drive a TPMS, can be harvested from this well-weighted pendulum. If the characteristic length L^* is equal to R_2, then the natural frequency of oscillation of the well-weighted pendulum matches the rotational frequency of wheel at any wheel speed. Consequently, resonance will occur at any wheel speed. Therefore, oscillation at a large angle and velocity for large power generation is feasible.

The magnetic field simulation results in this study show that the novel circular Halbach array disk exhibits a high gradient change and large strength of the z component of the average magnetic flux density because it reinforces the magnetic flux density inside the area in which the coils are placed. This in turn leads to a large power generation. The simulation models of the power generation were built well on the basis of the 3D magnetic field distribution by COMSOL V3.5, the numerical EMF equations, and Faraday's law.

The analytical model of power generation employed for numerical simulation, which showed good agreement with the experimental results, was developed based on models presented in this study. At an optimum external resistance, the power output of the energy harvester integrating the well-weighted pendulum and the multipolar magnetic disk was approximately 300-550 micro-Watts at 200-500 rpm. These results demonstrate that this has the potential to replace a TPMS battery.

Author details

Yu-Jen Wang, Sheng-Chih Shen and Chung-De Chen
National Formosa University, National Cheng Kung University,
Industrial Technology Research Institute, Taiwan

Acknowledgement

The authors would like to thank National Science Council (NSC), and College of Engineering and Research Center for Energy Technology and strategy, National Cheng Kung University for their financial supports to the project (granted number: NSC 98-2221-E-006-260-MY3 and 100-2628-E-006-019-MY3). Moreover, the authors wish to thank South-Industrial Technology Research Institute (S-ITRI) and National Formosa University (NFU) for their related supports.

7. References

Bowers, B. J. & Arnold, D. P. (2009). Spherical, rolling magnet generators for passive energy harvesting from human motion, *J. Micromech. Microeng.*, Vol. 19, No. 9, 2009, 094008, ISSN 0960-1317.

Dallago, E.; Marchesi, M. & Venchi, G. (2010). Analytical model of a vibrating electromagnetic harvester considering nonlinear effects, *IEEE Transactions on Power Electronics*, Vol. 25, No. 8, Aug. 2010, pp. 1989-1997, ISSN 0885-8993.

Daminakis, M.; Goethals, J. & Kowtke, J. (2005). Enhancing Power Harvesting using a Tuned Auxiliary Structure, *Journal of Intelligent Material Systems and Structures*, Vol. 16, No. 10, (2005) 825-834, ISSN 1045-389X.

Gu, L. & Livermore, C. (2010). Passive self-tuning energy harvester for extracting energy from rotational motion, *Applied Physics Letters*, 97, 2010, 081904, ISSN 0003-6951.

Hatipoglu, G. & Urey, H. (2009). FR4-based electromagnetic energy harvester for wireless tire sensor nodes, *Proceedings of the Eurosensors XXIII conference*, Procedia Chemistry 1, 2009, pp. 1211-1214, ISSN 1876-6196.

Lallart, M.; Anton, S. R. & Inman, D. J. (2010). Frequency Self-tuning Scheme for Broadband Vibration Energy Harvesting, *Journal of Intelligent Material Systems and Structures*, Vol. 21, No. 9, 2010, pp. 897-906, ISSN 1045-389X.

Lee, D.G.; Carman, G.P.; Murphy, D. & Schulenburg, C. (June 2007). Novel micro vibration energy harvesting device using frequency up conversion, TRANSDUCER &

EUROSENSORS '07, *The 14th International Conference on Solid-state Sensors, Actuators and Microsystems*, 10-14.

Leland, E.S. & Wright, P.K. (2006). Resonance tuning of piezoelectric vibration energy scavenging generators using compressive axial preload, *Smart Mater. Struct.*, 15 (2006) 1413–1420, ISSN 0964-1726.

Lohndorf, M.; Kvisterøy, T.; Westby, E. & Halvorsen, E. (2007). Evaluation of energy harvesting concepts for tire pressure monitoring systems, *Technical Digest PowerMEMS 2007*, Freiburg, Germany, (2007) 331-334.

Mansour, M. O.; Arafa, M. H. & Megahed, M. (2010). Resonator with magnetically adjustable natural frequency for vibration energy harvesting, *Sensors and Actuators A*, 163, 2010, pp. 297-303, ISSN 0924-4247.

Marinkovic, B. & Koser, H. (2009). Smart sand - a wide bandwidth vibration energy harvesting platform, *Applied Physics Letters*, 94, 103505, 2009, ISSN 0003-6951.

Marzencki, M.; Defosseux, M. & Basrour, S. (2009). MEMS Vibration Energy Harvesting Devices With Passive Resonance Frequency Adaptation Capability, *Journal of Microelectromechanical Systems*, Vol. 18, No. 6, 2009, pp. 1444-1453, ISSN 1057-7157.

Matsuzaki, R. & Todoroki, A. (2008). Wireless monitoring of automobile tires for intelligent tires, *Sensors*, 8 (2008) 8123-8138, ISSN 1424-8220.

Platt, S. R.; Farritor, S. & Haider, H. (2005). On low-frequency electric power generation with PZT ceramics, *IEEE/ASME Transactions on Mechatronics*, Vol. 10, No. 2, April 2005, pp. 240-252, ISSN 1083-4435.

Roundy, S. (2008). Energy harvesting for tire pressure monitoring systems: design considerations, *Technical Digest PowerMEMS 2008*, Sendai, Japan, (2008) 1-6.

Sari, I.; Balkan, T. & Kulah, H. (2008). An electromagnetic micro power generator for wideband environmental vibrations, *Sensors and Actuators A.*, 145-146, 405-413, ISSN 0924-4247.

Shahruz, S.M. (2006). Design of mechanical band-pass filters for energy scavenging, *Journal of Sound and Vibration*, 292 (2006) 987–998, ISSN 0022-460X.

Stanton, S.; McGehee, C. & Mann, B. (2009). Reversible hysteresis for broadband magnetopiezoelastic energy harvesting, *Applied Physics Letters*, 95, 174103, ISSN 0003-6951.

Wang, Y. J.; Chen, C. D. & Sung, C. K. (2012). System design of a weighted-pendulum type electromagnetic generator for harvesting energy from a rotating wheel, *IEEE/ASME Transactions on Mechatronics*, ID: 10.1109/TMECH.2012.2183640, Vol. 99, 2012, pp. 1-10, ISSN 1083-4435.

Wang, Y. J.; Chen, C. D.; Sung, C. K. & Li, C. (2012). Natural Frequency Self-tuning Energy Harvester using a Circular Halbach Array Magnetic Disk, *Journal of Intelligent Material Systems and Structures*, 23, 8, 933-943, ISSN 1045-389X.

Youngsman, J. M.; Luedeman, T.; Morris, D. J.; Anderson, M. J. & Bahr, D. F. (2010). A model for an extensional mode resonator used as a frequency-adjustable vibration energy harvester, *Journal of Sound and Vibration*, 329, 2010, pp. 277-288, ISSN 0022-460X.

Zhang, C. L. & Chen, W. Q. (2010). A wideband magnetic energy harvester, *Applied Physics Letters*, 96, 123507, 2010, ISSN 0003-6951.

Vibrations: Techniques

Chapter 9

Strategies for Wideband Mechanical Energy Harvester

B. Ahmed Seddik, G. Despesse, S. Boisseau and E. Defay

Additional information is available at the end of the chapter

1. Introduction

The energy harvesting market expands day by day, this is mainly due to the number of the implemented low power sensors in different fields, such as: human body, building, car engine...etc. These sensors are in most cases powered by batteries, but the main drawback of this technique is the need of a continuous control of their state of charge, the recharge and the replacement which is in most cases expensive. Thus, in order to overcome these limitations, one of the most promising solutions is to harvest the surrounding energy beside the system to power. In our environment, we can find many types of recoverable energy, for example: mechanical energy, thermal energy and radiative energy (solar, infra-red, radio-frequency). This chapter is dedicated to mechanical energy and more particularly to mechanical vibration energy produced by cars, fridges, mechanical engines and so on. The mechanical to electrical converter can be electromagnetic, electrostatic or piezoelectric. In case of an electromagnetic conversion, the vibrations are used to create a relative movement between a coil and a permanent magnet. In case of an electrostatic transduction, the vibrations are used to create a variable capacitance. In case of a piezoelectric transduction, the vibrations are used to apply a mechanical stress on a piezoelectric material. Actually, a vibration energy harvester (VEH) features 3 main components as presented in the Figure 1.

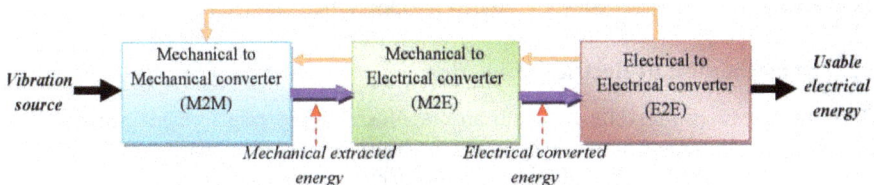

Figure 1. Vibration energy harvester conversion chain

The first stage of the conversion chain is a Mechanical to Mechanical (M2M) converter usually based on a mechanical resonator system. This converter translates the input vibration into a relative displacement between the resonator seismic mass and the vibration source. In addition, by using a resonant mechanism, the relative displacement amplitude can be larger compared to the vibration source displacement amplitude, increasing then the extracted mechanical power from the vibration source. Then, thanks to a dedicated Mechanical to Electrical (M2E) converter, which could be electromagnetic, electrostatic or piezoelectric, the amplified relative displacement is converted into electrical energy. Finally, an Electrical to Electrical (E2E) converter translates this electrical energy into a usable energy with a stable direct voltage able to supply an electrical circuit (3V for example). The efficiency of the harvester is tightly related to each stage of this chain. Moreover, as it can be noted from Figure 1, each stage has an effect on the other stages. Thus, the improvement of the VEH efficiency should take into account all the stages and also the relations between them. In what follows, more details are given for each stage.

1.1. The mechanical to mechanical converter

The first aim of this converter is to translate a vibration into a relative displacement able to actuate the mechanical input of the M2E converter. To make that, a seismic mass is required and the mechanical work that can be produced from the vibration is proportional to this mass and then to its size. This seismic mass is the main limitation in terms of power density capability for the main developed systems. In order to amplify the inertial effect of this seismic mass, it is necessary to use the resonance effect, which means using resonators. Actually, such devices are commonly modeled by a mass spring system connected to the vibration source and damped by the M2E converter and the mechanical losses. The efficiency of the M2M converter could be measured by the amplification gain of the vibration displacement amplitude (Q factor). However, the amplification gain is inversely proportional to the frequency bandwidth of the resonator making the system very sensitive to any change of the input vibration. This shift is commonly occurred especially in vibrations produced by car engine, in which case the frequency depends on the motor speed which is susceptible to change over time. In addition, the VEH resonant frequency is susceptible to change over time because of the aging of the materials. In fact, as the material of the harvester is subjected to a continuous mechanical stress, the mechanical stiffness will be altered during time and so the resonant frequency. To overcome this limitation many solutions have been proposed in literature, most of them are summarized in [1-2]. Actually, a few of these solutions allow an automatic adaptation without comprising the efficiency of the harvester neither the power balance of the VEH.

1.2. The mechanical to electrical converter

Once the vibrations are converted into an amplified relative displacement between two elements, this displacement is converted into electricity using electromagnetic, electrostatic or piezoelectric principle. The efficiency of this converter depends on its mechanical and electrical losses and also its good impedance matching with the mechanical source and electrical load. Many approaches have been developed in order to improve the efficiency of this converter, these approaches depend mainly on the type of the converter, most of them are linked to the system size and the vibration source characteristics [3-5].

1.3. The electrical to electrical converter

The maximum of the extracted electrical power is achieved when the electrical converted power is equal to the mechanical dissipated power in the mechanical structure. However, the mechanical damping depends on the used material, while the electrical damping depends on the converted power (electromechanical coupling of the structure) and the output electrical impedance. When the vibration frequency and amplitude are known and fixed, one can design a harvester to fit the optimized conditions (in terms of resonance and damping forces). However, when the vibration magnitude changes, this equality cannot be satisfied any more since the damping forces (mechanical and electrical) have different variation profile when the vibration amplitude changes. Consequently, the VEH efficiency is decreased. At the present time, only one study has been done in this perspective, which means adapting in real time the damping forces in order to maintain an optimum point of electrical energy extraction [6].

This brief introduction highlights two improvement areas. The first one consists to ensure the tracking of the vibration frequency, while the second one consists to adjust in real time the electrical damping force with respect to the mechanical one. This chapter covers these areas of VEH efficiency improvement. At the present time, more works have been done to cover the first area of investigation than the second one.

The next part of this chapter gives an overview of the main approaches developed in the state of the art to ensure a vibration frequency tracking. These approaches are classified according to the type of vibration source: mixed frequencies vibration or vibration with a main frequency that changes over time.

2. State of the art

Many solutions have been proposed in the state of the art in order to overcome the system degradation related to the shift between the resonant frequency and the vibration one. The best way to make comparison between these solutions is to classify them according to the type of input vibration signal to which they could be subjected. Actually, two main types of vibration signals exist:

2.1. Vibration with multiple of harmonic at different frequencies

Basically, for such signals, the energy of the vibration signal is spread over a wide bandwidth. Hence, using a one degree of freedom resonator will not harvest the energy efficiently even if the resonant frequency is included in the bandwidth of the vibration signal. This kind of signal exists in staircases, buildings, train rails...etc. Solutions developed to extract the maximum of energy from such vibrations spectrum are based on systems with a wide bandwidth. Hereafter the main techniques developed in this issue:

2.1.1. High electrical damping systems

Despesse et al., [7] proposed an electrostatic converter with a high electrical coupling coefficient in order to broaden the resonance peak. The fabricated prototype is able to recover mechanical vibration below 100 Hz, with a global conversion efficiency of 60% at 50Hz. In fact, the main disadvantage of this structure is the quality factor of the converter, the resonance peak is broadened by increasing the electrical damping coefficient, the quality factor is then decreased and therefore, the quantity of the scavenged energy is relatively decreased when the input vibration frequency reaches the resonant frequency compared to a system with a high quality factor. To take advantage from this solution without decreasing permanently the quality factor and keep the same efficiency as a high quality factor system when the resonant frequency is equal to the vibration one, we should adjust in real time the electrical coupling and then the electrical damping. When the vibration input frequency is equal to the resonant frequency, the electrical coupling could be very low enabling a full resonant effect and when the vibration frequency shifts, the coupling could be increased to reach a higher output power. In fact, if the mechanical to electrical converter can reach a high electromechanical coupling, it is easy to temporarily decrease this coupling by mismatching the output electrical impedance.

2.1.2. Multi-modes systems

Shahruz et al., [8] proposed to expand the bandwidth using a multi modal structure. This structure is composed of several cantilevers. Each one has a defined resonance frequency. However, for a given vibration frequency, there is only one cantilever excited at its resonance frequency and all the others generate only a few amount of energy which limits the power density of the whole system. Another solution has been proposed by Roundy et al., [9] similar to the previous one, which consists on using a mechanical resonator composed of 3 different proof masses and four springs, they predict that the bandwidth could be multiplied by 3; nevertheless, the functionality of this system has never been experimentally verified. However, in both cases the power density of the converter is reduced since the harvested energy is proportional to the seismic mass, and in such cases only one cantilever works efficiently. Nevertheless, it can be interesting to use this technique to harvest a main vibration frequency and its harmonics that can be significantly separated in frequency but well known in advance.

2.1.3. Non-resonant system

Yang et al., [10] proposed another idea to broaden the VEH bandwidth, this idea consists to design a harvester where the effect of the air damping can be controlled. The idea is quite similar to the first one, except that for the present approach, the mechanical damping is increased instead of the electrical one, decreasing then significantly the mechanical extracted power. Actually, this solution is always less interesting than a high quality factor VEH, the bandwidth is in fact just increased because the mechanical damping limits significantly the output power when the vibration frequency fits the resonant frequency and not because it increases the output power outside the resonant frequency.

2.2. Harmonic vibrations

This is the most common type of vibrations, the energy of vibration is mostly concentrated around one determined frequency. However, there are two origins to the shift between the resonant frequency and the vibration one for such type of vibration:

- The vibration frequency is susceptible to change overtime.
- The change of the VEH materials properties because of the aging.

Hence, solutions developed in this issue trend to minimize the shift between the resonant frequency and the vibration one by tuning the resonant frequency. To do so, two types of approaches have been developed:

2.2.1. Passive tuning of the resonant frequency:

The passive way for tuning the resonant frequency means that the adaptation system does not need to be supplied by an external source of energy. Hereafter some techniques developed in this perspective:

- *Using non linear spring*

Marzencki et al., [11] has proposed an energy harvesting device employing the mechanical non linear strain stiffness using a clamped-clamped beam. For such systems the stiffness depends on the amplitude and the frequency of the vibration source. Hence, a well design of the structure could allow an adaptation between the vibration frequency and the resonant one. In this work, they report a tuning ratio of resonance frequency of over 36% for a clamped-clamped beam device with an input acceleration of 2g. This solution allows an efficient way to make a passive dynamic adaptation of the resonant frequency. However, there are some limits of this technique: a high resonant frequency tuning ratio requires a high acceleration. In addition, the frequency response of a non linear VEH has a hysteresis aspect, which means that when the frequency exceeds a specific value, the output power drops off dramatically, and it is not possible to come back to the previous point unless the vibration frequency decreases to the start frequency of the VEH. Furthermore, the principle efficiency is very dependent on the input vibration amplitude. For low amplitudes, there is a limited non-linear effect. Inversely, for large input displacements, the non-linear effect limits the relative displacement and then the output power.

- *Manuel tuning of the resonant frequency*

Leland et al., [12] suggests another technique for tuning the resonance frequency, it consists on applying *manually* an axial preload in order to change the resonance frequency and match the frequency of the vibration source. Using this technique, the developed system is able to adjust the resonance frequency of about 24% below its unloaded resonance frequency. This solution is viable only when the vibration frequency is known a priori and is not susceptible to change over time, since it allows fitting the right resonance frequency before implementation of the VEH.

2.2.2. Active tuning of the resonant frequency:

The active tuning means that the resonant frequency is adjusted in real time using an active process. The drawback of this technology is the power required to make the dynamic tuning of the resonant frequency. Hereafter, the main active tuning techniques:

- *Application of a force*

The application of a force means that the VEH is equipped with an actuator, magnetic [13], piezoelectric [14] or electrostatic actuator [15]. The main task of this actuator is to apply an additional force on the seismic mass in the same direction as the vibration one in order to affect the mechanical stiffness of the VEH (added or substituted a force to the spring force of the resonator). This force induces a change in the effective mechanical stiffness and then the resonant frequency. Among all the works made in this perspective only a few of them report a positive power balance between the output power delivered by the VEH and the energy required to adjust in real time the resonant frequency. Eichhorn et al., [16] have presented a smart and self-sufficient frequency tunable vibration energy harvester, they report a resonant frequency tuning ratio up to 26% and a consumption of the tuning system about10% of the VEH output power. However, the system could perform the frequency adjustment only once a 22s. Another system proposed by Lallart et al., [17] based on an original approach that permits increasing the effective bandwidth by a factor of 4 in terms of mechanical vibration with a positive power balance.

- *Electrical load adaptation*

The idea of this technique is to adjust the resonant frequency by adjusting the value of the electrical load coupled with the VEH. This technique has been used for electromagnetic harvester. Since the stiffness is related to the electrical current in the coil, by adjusting the electrical load, the current could be changed and then the resonant frequency as shown in Figure 2, [18]. However, this approach requires too much power for the implementation, about 1000 times the power generated by the VEH.

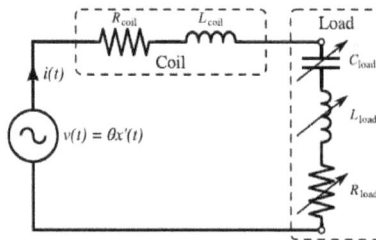

Figure 2. Method for adjusting the resonant frequency by adapting the electrical load principal scheme [18]

As it can be noted from this overview of the existed solutions, the main limit is either the lack of dynamic adaptability or the tuning ratio. To have a good dynamic of resonant frequency it is necessary to use active solutions. However, the main drawback of such techniques is the negative power balance between the output power from the converter part and the power required to make the tuning of the resonant frequency. The works of Eichorn [16] and Lallart [17] show the possibility to develop wideband system with a positive power balance, their results are for great interest especially for autonomous system development. However, in both cases the positive balance is achieved by making a compromise with either the frequency of tuning adjustment or the tuning ratio. This brings us to conclude that it remains a large quantity of work to perform in this perspective in order to achieve a complete wideband harvester, able to be implemented in environment where the frequency vibration frequency is susceptible to change over time regardless the decrease of the output power due to the shift between the input frequency and the resonant one. Among laboratories interested by developing a viable solution using active technique there is *CEA-Leti*. Three different solutions have been developed in this laboratory. The next part gives details of each of these developed solutions, the first solution is based on the amplification of the generated relative displacement at off resonance, this solution is applicable for both types of vibration described before. The second and the third solutions are based on active tuning of the resonant frequency and are more applicable for the second type of vibration.

3. Solutions developed by *CEA-Leti*

3.1. Amplification of relative displacement at off resonance (rebound technique) [19]

In this part, a new approach for amplifying the relative movement of a cantilever system at off-resonance is presented. The aim is to broaden the resonance peak of resonators without compromising the quality factor of the system. The idea is to gather the resonance phenomenon conditions at off resonance. This could be done by adding a *rebound* mechanism to the VEH, when the speed of the vibration source reaches an extremum, the seismic mass is mechanically connected to the vibration source via a high stiffness spring, the movement of the seismic mass is then inverted and its speed is increased, which means a transfer of energy from the source to the VEH. This operation is called *"rebound"*. This approach is useful for VEH operating in environments where vibrations could be spread over a wide bandwidth or characterized by one main frequency susceptible to change over time. In the following, the principle is presented in details by giving the modeling and optimization approaches.

3.1.1. Description of the approach

The original idea of this approach is based on the principal of the elastic collision theory. When two solid bodies enter into collision there is an exchange of mechanical energy between these two bodies. After collision, the small mass goes in the opposite direction with a higher speed as shown in Figure 3. In the case of VEH, the small mass will represent the seismic mass of the harvester, while the big one will represent the mass of the vibration source.

The following calculations are intended to estimate the final energy of the seismic mass m_2 after the collision in order to compare it to its initial energy and the speed of m_1 before the collision. The objective is to deduce the energy gain of m_2 and the way to maximize it.

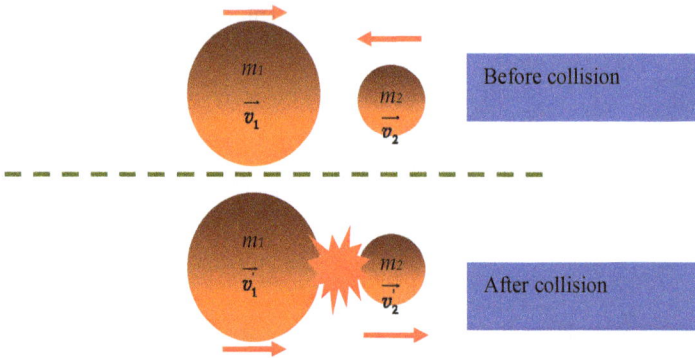

Figure 3. Illustration of the direct collision between two masses

The conservation of the kinetic energy before and after the collision leads to the following equations:

$$\begin{cases} m_1\vec{p}_1 + m_2\vec{p}_2 = m_1\vec{p}\,'_1 + m_2\vec{p}_2\,' \\ m_1v_1^2 + m_2v_2^2 = m_1v\,'^2_1 + m_2v\,'^2_2 \end{cases} \tag{1}$$

Where: v_1 and v_2 are the speed of the mass m_1 and m_2, respectively, p_1 and p_2 are the quantity of movement ($m_i.v_i$),. This leads to the final speed v'_2 of the seismic mass:

$$\vec{v}'_2 = -\vec{v}_2 + 2\vec{V}_I \tag{2}$$

Where V_I is the speed of the center of inertia of the whole moving system giving by the following equation:

$$V_I = \frac{m_1v_1 + m_2v_2}{m_1 + m_2} \tag{3}$$

By considering a mass m_2 much smaller than the mass m_1 the center of inertia speed becomes: $V_I = v_1$

It can be seen that the mass m_2 changes its movement direction after the collision and goes back with a higher speed (v'_2). A speed gain of twice the speed of the mass m_1 before the collision is reached. Considering the mass m_1 much greater than the mass m_2, the achieved gain in terms of kinetic energy E_1-E'_1 of the seismic mass m_2 is as follows:

Before collision:

$$E_1 = \frac{1}{2}m_2v_2^2 \tag{4}$$

After collision

$$E_1' = \frac{1}{2}m_2 v'^2_2 = \frac{1}{2}m_2 v_2^2 + \boxed{2m_2 v_1^2 + 2m_2 v_1 v_2}$$ (5)

This energy gain is proportional to the square of the vibration source speed v_1 and to the product between the vibration source speed v_1 and the seismic mass speed v_2. As soon as the seismic mass reaches a speed higher than the vibration source one, the energy gain becomes mainly related to the last term. Higher the initial speed (or initial energy) of the seismic mass is, higher the energy gain is, like a resonant mechanism on a half period.

This is the basic idea of the present approach. Let us investigate in more detail how it could be possible to implement this approach with a real VEH to amplify the movement of the seismic mass on a random type vibration source.

3.1.2. Application of the rebound mechanism for VEH

The process of rebound mechanism to implement with the VEH will help to extract more energy from the environment at off resonance for harmonic signals and even for random vibration. To understand the operating principle of this technique, we will look in more detail at the mechanical behavior of the VEH at or close to the resonance. Consider the equivalent model of a converter consisting of a spring with a stiffness k, attached from one side to a seismic mass m, and from the other side to the vibration source. The mass is also attached to the vibration source via damper as shown in Figure 4. At the resonance, the speed of the source and the effort imposed by the spring on the vibration source are in phase opposition. In other words, the source provides a mechanical work to cantilever and not the reverse. Hence, maximum of mechanical energy is transferred from the source to the resonant system (resonance phenomenon). The system is then able to extract more energy from the vibration source such that the relative displacement is larger (the force exerted by the spring is important), which means that the quality factor is high. However, at off resonance, the vibration source displacement and the effort imposed by the spring on the support are not synchronized, reducing then the average power transferred to the seismic mass. Hence, only a small amount of energy is absorbed by the mass spring system from the vibration source.

By considering the previous idea, more absorption of mechanical work from the environment at off resonance can be ensured by synchronizing a rebound of the seismic mass on the vibration source when the vibration source speed is in opposite direction with the seismic mass speed, the kinetic energy gain of the seismic mass is tightly related to the vibration source speed and to the initial seismic mass speed (as given by the equation (5)), it is more convenient to synchronize the rebound to a speed extremum of the vibration source. This will amplify the relative speed of the seismic mass and then the absorbed energy from the vibration source. The energy absorbed from the vibration source increases gradually, higher the seismic mass speed becomes, higher the extracted mechanical energy is. When the speed of the seismic mass is greater, the force exerted on the source during the rebound is higher and then mechanical work extracted from the vibration source is higher too. Obviously, there is a physical limit to this amplification mechanism, this limit is fixed by the damping coefficient of the structure; a structure with a high quality factor will allow more amplification and vice versa, like for a resonance mechanism. To validate the principle, the rebound mechanism is introduced to a mass spring harvester system as shown in Figure 4. The spring k_1 represents the system guidance that has minimal spring stiffness. The rebound is applied by connecting mechanically the vibration source to a the seismic mass via a large stiffness spring k_2, This connection is ensured by using two actuators Act_1 and Act_2, as shown in the figure below. When these actuators are activated, the mechanical stiffness of the resonant system is modified (from k_1 to $k_1+k_2 \sim k_2$). Hence, the system will feature two natural frequencies: f_{r1}, related to the stiffness k_1 when the spring k_2 is not connected and f_{r2}, related to the stiffness k_2.when the spring k_2 is connected.

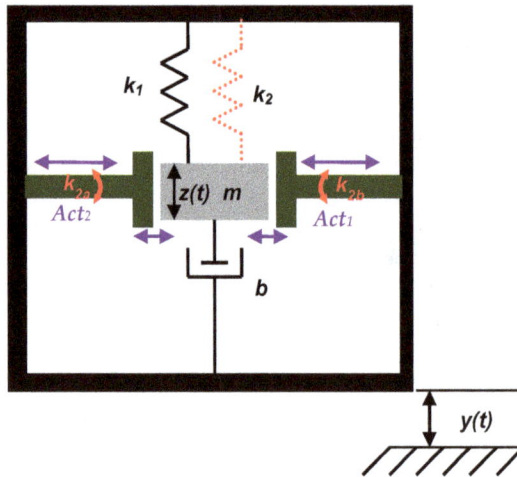

Figure 4. Equivalent VEH model with the rebound process

For a non damped collision, one can obtain a theoretical speed gain of the seismic mass twice the speed of the vibration source during the rebound, (as explained above). The next section gives an over view about the different techniques that could be used for actuating the rebound.

3.1.3. Rebound mechanism choice

Many types of mechanisms could be used for applying the rebound. This mechanism should be able to apply the rebound at any time by connecting on demand the seismic mass to the vibration source via a spring of stiffness k_2 significantly higher than k_1. Hereafter the different types of mechanism that could be used:

- *Thermal actuation*

A current is applied in a thermal resistance when the vibration source speed reaches its maximum value; the thermal material expands and then makes a connection with the seismic mass. The main disadvantage of this technique is the reaction time of the material and its power consumption.

- *Electromagnetic actuation*

It is also possible to actuate the rebound by using an electromagnetic actuator composed of a coil and a core; such a method is used for breaking motors. When the coil is powered, the magnetic core moves and blocks the seismic mass. The main disadvantage of this approach is the power consumption which is relatively high compared to the power that can be scavenged (<1 mW for a centimeter scale device).

- *Piezoelectric actuation*

A third solution is to use a piezoelectric actuator enabling a short displacement with a high effort in a good agreement with the need to efficiently pinch the seismic mass. Furthermore the reaction time is very short compared to the blocking time (100 µs to few ms) and its consumption is relatively low (capacitive mechanism).

The piezoelectric actuation has been adopted to implement the present idea thanks to its accuracy, time of response, low power consumption and its compatibility with the application. The piezoelectric actuators chosen are a linear actuators developed by CEDRAT Technology (APA400M) placed on each side of the seismic mass.

3.1.4. Simulations results

The time simulation diagram is presented in Figure 5. This diagram is composed of two working phases. The first one is used when the resonant frequency of the system is f_{r1} (rebound process deactivated), and the second phase is used when the resonant frequency is f_{r2} (the rebound process activated).

First, it is supposed that the system starts oscillating at a random frequency, different than the resonant one. The algorithm starts computing the speed and the displacement of the seismic mass. When a maximum speed of the vibration source is detected, the rebound is activated, the simulation phase is then changed, the simulation jumps to the second phase. This latter remains a certain period of time (the maintain of the connection of k_2 with the seismic mass). After that, the system goes back to the first phase (seismic mass connected to the spring k_1 only). This operation is repeated as often as the vibration source reaches some maximum speed, which means twice a period of vibration. The transition from one phase to another updates the initial condition of the new phase from the final position at the previous phase (initial displacement and speed, of both the seismic mass and the vibration source).

This simulation process has been used in different conditions in terms of vibration frequency, of rebound time duration and mechanical quality factor. After a deep investigation, some criteria have been established allowing an optimal amplification of the seismic mass's relative displacement:

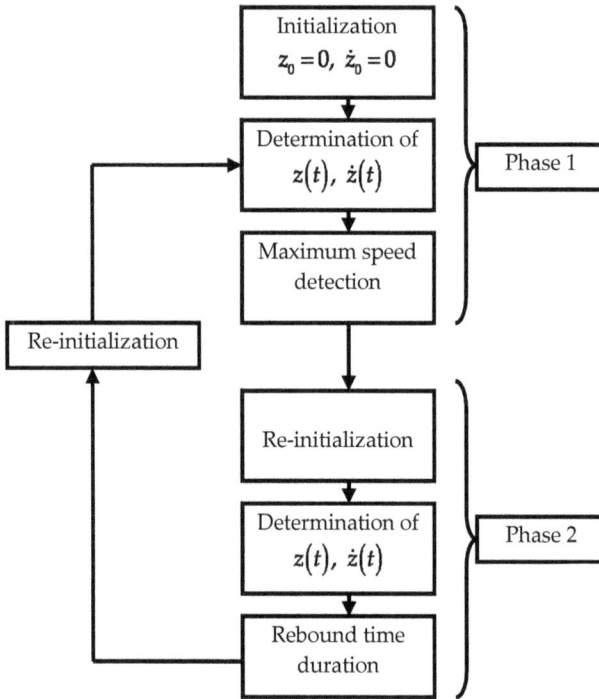

Figure 5. Rebound simulation diagram

- *The rebound time duration*

The simulation results show that the optimal rebound time duration is related to the resonant frequency f_{r2} occurring during the rebound time:

$$\Delta t = \frac{1}{2f_{r2}} \tag{6}$$

In fact, the rebound is a compression/decompression cycle of the spring k_2, corresponding to the half of the resonant period occurring during the rebound time.

- *The second resonant frequency*

The simulation shows that the operating frequency bandwidth where there is a positive gain in terms of relative displacement is limited between f_{r1} and $f_{r2}/2$. Hence, the higher the distance between f_{r1} and f_{r2} is, the higher the bandwidth of the harvester is. To enlarge the operating frequency bandwidth, it is interesting to choose a large resonant frequency f_{r2}. However, a high resonant frequency f_{r2} implies short rebound duration, this will conduct to more difficulties to actuate the rebound. Thus, a tradeoff has been made between the bandwidth and the mechanical challenges to reduce the rebound time.

The first resonant frequency fr_1, while k_2 is open, has been fixed at 50Hz and the second one, during the rebound, at 200Hz.

The Figure 6 below presents the seismic mass speed amplification reached at the following conditions:

Parameter	Value
Input frequency	80Hz
Input acceleration	1g
The first natural resonant frequency (fr_1)	50Hz
The second natural resonant frequency (fr_2)	200Hz
Time rebound duration	2.5ms
The mechanical quality factor of the structure when the rebound process is activated	20, 30 and 40

Table 1. Simulation parameters for rebound system

The figure below shows the transient behavior of the seismic mass speed for different quality factor values. The maximal amplification is related to the mechanical quality factor. Hence, higher the quality factor is, higher the amplification gain is.

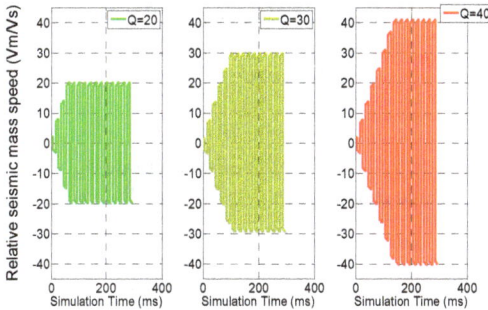

Figure 6. Seismic mass speed amplification (*Vm*: Seismic mass speed and *Vs*: vibration source speed)

The next section gives some details about the electronic attended to ensure the control of the actuators.

3.1.5. Drive electronic

The aim of this electronic is to deliver the command signals to activate and deactivate the actuators (act1 and act2). These signals are provided after measuring and processing the acceleration signal of the vibration source. Hence, the setting of electronic components is based on the following specifications:

- Output signal should be square, with an optimal duration time,
- The input vibration spectrum is comprised inside the system bandwidth (fr_1 and $fr_2/2$).

The selected electronic architecture is composed of 5 stages placed in series:

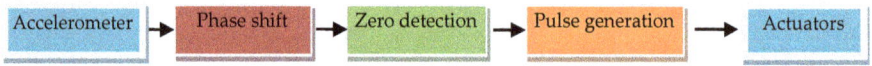

Figure 7. Synoptic scheme of the selected electronic

- *Phase shift*

It is worth to remind that the rebound time duration is extremely low compared to the vibration period. The present approach needs then an extreme accuracy in terms time of activating/deactivating the actuators. However, the rebound is actuated after processing the acceleration signal. Hence, this operation will introduce a delay on the command signals. In order to overcome this limitation, a phase shift is added in order to make compensation to the delay introduced by the electronic. Hence, the information arrives to the last stage at the right moment.

- *Zero detection*

Detecting a vibration source speed extremum is equivalent to detect the zero acceleration of the input vibration. The second stage is then attended to detect zero acceleration on the vibration source. This is done by comparing the acceleration signal with zero using a comparator, the output voltage of the comparator changes from 0 to V_{cc} (for positive speed maximum) or from V_{cc} to 0 Volt (for negative speed maximum).

- *Pulse generation*

The third stage is attended to generate a pulse of an accurate duration each time the signal delivered by the previous stage change of state (rising or falling edge).

- *Power circuit*

The power circuit contains the switches to power from an external source to the actuators used for processing to the rebound.

The presented electronic was developed and tested with the mechanical system; the next part presents the experimental results

3.1.6. Experimental validation:

The experimental setup is shown in Figure 8 below. The cantilever is represented by its equivalent model, which is composed of a mass, a spring and a damper system. All these components are enclosed in the casing which is mechanically connected to the vibration source.

The device *(a)* is a Laser vibrometer (type: LSV250) connected to a computer in order to measure the displacement magnitude of the seismic mass. The acceleration of the vibration source is measured by an accelerometer *(e)*, the measured signal is provided to the electronic

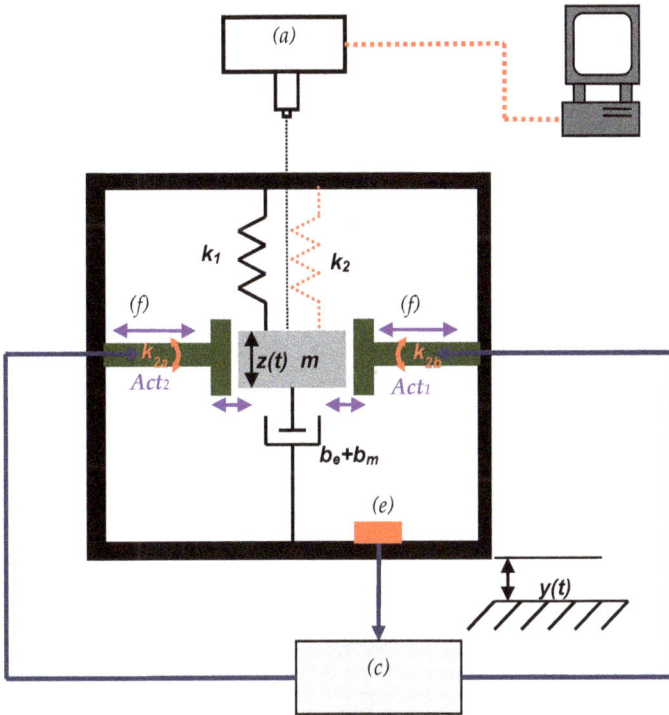

Figure 8. Experimental setup scheme for the rebound technique validation

(c) described in the previous section. This electronic processes the measured signal and generates the driving signals to the actuators (APA400M) (f) for connecting or disconnecting the spring k_2 relaying the seismic mass to the vibration source.

The manufactured structure is shown in the Figure 9 followed by a brief definition of the different components.

Figure 9. Picture of the fabricated structure

Components	Definition
1	Cantilever made of stainless k_1
2	Tip mass
3	Piezoelectric actuators APA 400M
4	The casing/support connected to the vibration source
5	Acceleration sensor

Table 2. Structure components definition

The Figure 10 shows the displacement gain achieved by the present approach as a function of the input vibration frequency compared to same system without the rebound mechanism. This gain is defined as the ratio between the relative displacement at off resonance obtained by using the present approach over the displacement obtained without activating the rebound when the maximum of speed is occurred. As expected by theory, the gain depends effectively on the input frequency, the gain is more important for frequencies much higher than the first resonant frequency because the amplitude of the relative displacement is close to the physical maximum since the vibration frequency is close the resonant one. This figure shows a difference between the theoretical expectation and the experimental results in terms of displacement gain due to the fact that in the theoretical study the damping induced by the actuators themselves was not taken into account.

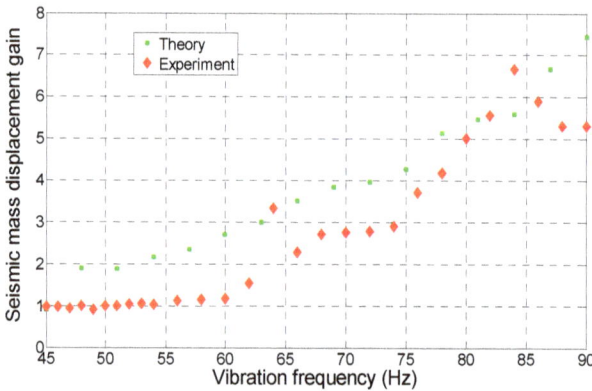

Figure 10. Experimental and theoretical results of the relative displacement gain achieved by the rebound technique

These results present a great advantage of the rebound technique for the increase of the VEH efficiency over a wide frequency band. The amplification of the seismic mass displacement allows more mechanical energy extraction when the resonant frequency is not equal to the input one. Nevertheless, the drawback that remains for the present approach is a large power consumption to actuate the piezoelectric actuators. The energy required for each rebound is estimated at 200µJ. Further works are under investigation in order to reduce as low as possible this consumption.

3.1.7. Conclusions

In this part of the chapter, a new approach for amplifying the movement of the seismic mass at off resonance have been shown by theory and experiments. A relative displacement seven times higher than the one achieved with a single resonator at off resonant frequency (90Hz) was shown. This best gain occurs at twice the natural frequency of the structure. This relative displacement gain corresponds to an output electrical power gain equal to 49, which represents a good prospect in the field of energy harvesting. This technique ensures a dynamic amplification of the relative displacement in real time, with a high efficiency without the need of a control loop, a simple measure of the sign of the acceleration is sufficient to control the whole system. Nevertheless, the electrical consumption of the actuator applying the rebound is still too large to make the system completely autonomous. This first demonstrator validate the principle with an actuator over-sized, a significant reduce of its consumption promises a good perspective in this rebound technique.

If this rebound mechanism can be applied for a large number of vibration types (up to random vibrations), it is nevertheless interesting to inspect if other techniques, with narrower applications but easier to use, can be used to enlarge the frequency response. The next part presents two ways to follow a main vibration frequency that moves during the time.

3.2. Active tuning of the resonant frequency:

In the present section, two approaches for a dynamic tuning of the resonant frequency are given. These techniques could be applicable where the main vibration frequency is susceptible to change over time. It could be used for vibrations spread over a wide bandwidth as well, except that in this case the system will track only one main frequency. For both techniques given in what follows, the idea is to make a tuning of the resonant frequency by changing the stiffness of a piezoelectric material.

The most used structure shape for piezoelectric transduction in case of harvesting mechanical vibration is presented by the figure below:

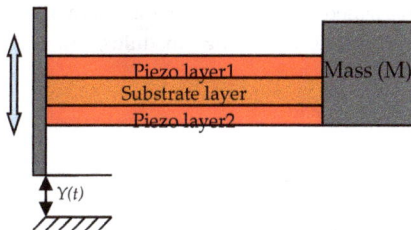

Figure 11. Piezoelectric cantilever shape

The structure is composed of three main components:

- *The substrate*: the substrate is usually added to piezoelectric harvester for two aims: enhancing the effective mechanical quality factor of the structure, removing the stress neutral line of the whole structure outside the symmetry axes of the piezoelectric part in order not to reduce the output generated power by electrical charges compensation.
- *The piezoelectric part*: stressed under mechanical stress, it converts the mechanical extracted energy into electrical one.
- *The seismic mass*: used to adjust the resonant frequency and enhance the amplitude of the relative displacement.

All these components are bounded together as shown in the previous figure.

One way to quantify the structure capability to change its resonant frequency is to estimate resonant frequency tuning ratio. For a cantilever based piezoelectric structure, the tuning ratio is given by the equation below:

$$\frac{f_{max} - f_{min}}{f_{min}} = \sqrt{\frac{I_b + 2x_1x_0I_p}{I_b + 2x_0I_p} \cdot \frac{I_b\frac{L^3}{3} - 2x_0I_p\left(L^2L_p + \frac{L_p^3 - L^3}{3} - LL_p^2\right)}{I_b\frac{L^3}{3} - 2x_1x_0I_p\left(L^2L_p + \frac{L_p^3 - L^3}{3} - LL_p^2\right)}} - 1 \tag{7}$$

I_b and I_p represent the moment of inertia of the substrate and the piezoelectric part, respectively.

L and Lp represent the length of the beam and the piezoelectric layers, respectively.

$$x_0 = \frac{Y_{p\,min}}{Y_b}, \quad x_1 = \frac{Y_{p\,max}}{Y_{p\,min}}$$

$Y_{p\,min}$ and $Y_{p\,max}$: the minimal and maximal value of the piezoelectric Young's modulus.

Y_b: the Young's modulus of the substrate

Equation (7) shows that the tuning ratio depends on the sizes of the structure, on mechanical material properties (x_0) and electromechanical properties (x_1). The choice of the piezoelectric material is based on its Young's modulus sensibility to the external conditions, a high sensibility will allow more change of the piezoelectric Young's modulus, and then a high tuning ratio.

3.2.1. Application of a DC electric field

3.2.1.1. Introduction

Mechanical and electromechanical properties of piezoelectric materials depend on external constraints. Among these properties, there is the effect of an applied electric field on the stiffness of the piezoelectric material. Thus, using a good piezoelectric material, in terms of stiffness variation under a static electric field, a high resonant frequency tuning ratio could be achieved. Adjusting the level of the applied electric field will adjust the resonant frequency and hence a controllable resonant frequency VEH could be achieved.

3.2.1.2. Theory of the approach

The dependence that exists between the stiffness of the piezoelectric material and the strength of the applied electric field could be noted from the complete equation of piezoelectricity given by the following expression:

$$\varepsilon_{ij} = s^E_{ijlm}\sigma_{lm} + d_{ijn}E_n + \frac{1}{2}\tau^E_{ijlmpq}\sigma_{lm}\sigma_{pq} + \frac{1}{2}a_{ijnr}E_n E_r + \kappa_{ijlmn}\sigma_{lm}E_n \qquad (8)$$

This equation features two different parts, the first one relies the deflection to the stress by a constant parameter (s^E), while the second one shows a non constant coefficient of proportionality between the deflection and the mechanical stress, this non constant parameters depends on the applied electric field (E_n). This effect reflects the non linear behavior of piezoelectricity when it is subjected to a DC electric field. This effect varies from one type of piezoelectric material to another; it depends also on how the electric field is applied on the material in terms of strength and direction.

Equation (8) is the general piezoelectric equation, but considering our cantilever design, some assumptions can be taken into account: the mechanical behavior of the cantilever is elastic, only one component of the electric field vector of the stress and strain tensors are taken into account. Hence, this leads to the following simplified equation relating the stress to the strain and the applied electric field:

$$\varepsilon_x = s_{11}\sigma_x + d_{31}E_3 + \frac{1}{2}a_{113}E_3^2 + \kappa_{113}\sigma_x E_3 \qquad (9)$$

This leads to the following expression of the mechanical stress:

$$\sigma_x = \frac{1}{s_{11} + \kappa_{113}E_3}\varepsilon_x - \frac{d_{31}E_3 + \frac{1}{2}a_{113}E_3^2}{s_{11} + \kappa_{113}E_3} \qquad (10)$$

The expression relaying the stress to the strain shows clearly the dependence that exists between the applied electric field and the stiffness of the piezoelectric material. However, it is difficult to determine the parameters appearing in this model, in most cases they are not provided by suppliers. Another simple equivalent model has been proposed by Thornburgh et al [20], reflecting the same effect by using a simple relation. This model is based on the linear constitutive equations of the piezoelectric material, we keep only the first and the second terms of the equation (9). Except that the effect of the DC electric field is reflected on the piezoelectric coefficient (d_{31}). The value of this one depends on the level of the applied electric field. The equation showing this dependence is given by (11), where the d_{31} is the piezoelectric strain coefficient at $0kV/cm$ and q_{31} is called piezoelasticity coefficient:

$$d_{31}^{*} = d_{31} + q_{31}\varepsilon_x \tag{11}$$

Using this relation, the stress function of the strain can be expressed as:

$$\sigma_x = Y_p'\varepsilon_x - \beta \tag{12}$$

Y_p': is the expression of the piezoelectric effective Young's modulus as a function of the applied electric field and is expressed as follows:

With $Y_p' = \frac{1 - q_{31}E_3}{s_{11}}$ and $\beta = d_{13}E_3$

For a bimorph cantilever shape, the resonant frequency is then expressed as follows:

$$f_r \frac{1}{2\pi} = \sqrt{\frac{w}{4(M + 0.24M_b)L^3}\left(Y_p(1 - q_{31}E_3)\left(6t_s^2t_p + 12t_st_p^2 + 8t_p^3\right) + Y_st_s^3\right)} \tag{13}$$

The optimization of the tuning ratio involves in first step the choice of a piezoelectric material. This material should allow a high resonant frequency shift under a low applied DC electric field without compromising the efficiency of the mechanical-to-electrical conversion. Finding a piezoelectric material having a high electric coupling can induce a material with a low quality factor reducing the harvesting power, it is then important to find a material with a good compromise, we introduce then a new figure of merit for choosing the piezoelectric material taking into account the following parameters:

- *The electromechanical coupling coefficient k31:*

It is necessary to choose a material with a high electromechanical coupling coefficient, this will improve the efficiency of the electromechanical conversion.

- *The maximum electrical field supported by the piezoelectric material (Emax and Emin):*

As it can be seen from the expression of the resonant frequency (13), the highest the supported electric field is, the highest and the tuning ratio is. It is then important to take into account the limits of the applied electric field imposed by the piezoelectric material.

- *The coefficient of piezo-elasticity:*

The effect of the applied electric field on the stiffness of the piezoelectric material is described by the coefficient of the piezo-elasticity *q31* as shown by equation (11). This means that a material with a high coefficient of piezoelasticity will provide a high resonant frequency tuning ratio.

- *Dielectric losses coefficient:*

The limiting factor of the present approach is the dielectric losses of the material. A material with a high dielectric losses coefficient presents more leakage current. This will induce higher power consumption for the management electronic of the system. It is then important to choose a material with low losses.

The figure of merit taking into account the different constraints above, can be expressed as:

$$\lambda_{p-1} = \frac{k_{31}^2 \left(E_{max} - E_{min} \right)}{\tan(\delta)} q_{31} \tag{14}$$

After an overview of the most used piezoelectric material, it was found that the best material for this application is the PZN-PT, allowing the best compromise between the resonant frequency tuning and the electromechanical conversion. Despite the best performance of this material, it remains actually quite expensive compared to others.

3.2.1.3. The experimental validation of the approach:

- *The manufactured device*

The Figure 12 is a picture of the fabricated structure, it consists on a bimorph cantilever shape. The structure sizes and the main electromechanical properties are presented in the table below:

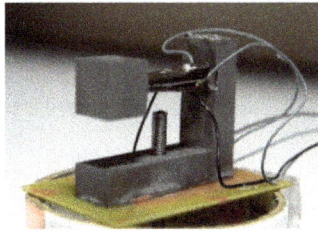

Figure 12. Picture of the fabricated structure

Parameter	Value
Substrate material:	
Y_b (GPa)	210
Density (kg.m^{-3})	7500
Length x width x thickness (mm)	30x10x0.8
Piezoelectric material (PZN-PT):	
Y_{p-max} (GPa)	100
Density (kg.m^{-3})	8800
k_{31}	0.9
relative dielectric constant (ε)	7500
Length x width x thickness (mm)	30x10x1

Table 3. Characteristics of the fabricated structure

- *The experimental results:*

The Figure 13 presents the relative displacement amplitude as a function of the input frequency for different applied DC voltage on the piezoelectric layers. It presents also the structure resonant frequency as a function of the applied DC voltage. A tuning ratio of the resonant frequency up to 20% has been obtained by applying an electric field from -1 to +6kV/cm. Theory shows that a tuning ratio of 26% should be reached in these conditions, this discrepancy is due to the errors introduced during the fabrication process. The bounding of the different layers has been done using an adhesive. The given theoretical model does not take into account the parameters of this adhesive, which explain for the main part the difference between the experimental and theoretical results.

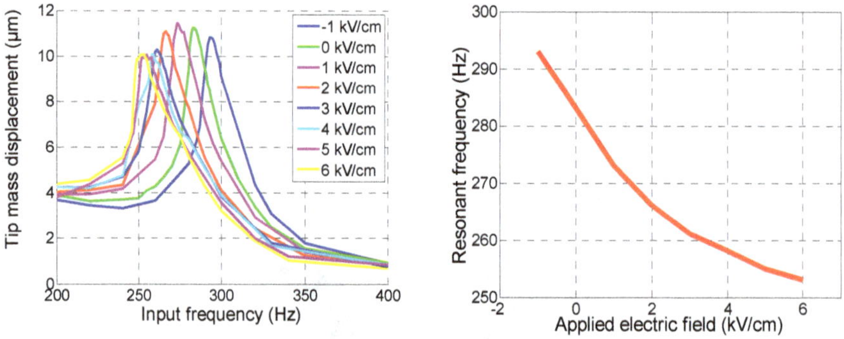

Figure 13. The effect of the applied DC electric field on the resonant frequency

The most important challenge while designing a VEH able to adjust its resonance frequency automatically, is the power balance between the converter output power and the power required to drive the frequency tuning. As explained before, most of developed wideband VEH have a negative power balance. In the next section, a new low power consumption electronic is proposed, this electronic is under development within the *CEA-Leti*, it allows a dynamic tuning of the resonant frequency by tracking the maximum output power point.

3.2.1.4. The resonant frequency tuning electrical circuit

The drive electrical circuit is composed of two principal parts, the power circuit and the control circuit. The first one allows the flow of the power between an electrical energy storage element and the piezoelectric material in order to apply a DC electric field, while the second one controls the level of the applied electric field according to the shift that exists between the vibration frequency and the resonant one.

The block diagram of the control circuit is given by Figure 14 below. The aim of this electronic is to determine how much is the resonant frequency lower (or higher) than the vibration one. First of all, this shift is determined by measuring the phase difference between the acceleration of the vibration source and the piezoelectric voltage. At resonance the phase shift between these two signals is equal to a quarter of a period, when the resonant frequency is higher than the vibration one, this phase shift is higher than this quarter of a period and inversely when the resonant frequency is lower than the vibration one. Thus, after measuring this phase shift, two cases may occur: (i) the phase shift is lower than a quarter of a period, this means that the voltage already applied on the piezoelectric material is higher than the voltage that should be applied across the material ($f<f_{vib}$). (ii) the phase shift is higher than a quarter of a period, this means that the voltage already applied on the piezoelectric material is lower than the voltage which should be applied across the material ($f>f_{vib}$). In the first case, energy is transferred from the electrical energy storage element into the piezoelectric material, the switch k_1 is closed first, the close time of both switches should correspond exactly to the energy expected to be injected into the piezoelectric capacitance to reach the right DC voltage across the piezoelectric material. After switching on k_1 during the right time, the expected energy is stored into the magnetic core, it is then switched off and k_2 is switched on, the stored energy in the magnetic core is then injected in the piezoelectric capacitance and the voltage applied across the piezoelectric material attends its intended value. The process for the second case is the same as the first one, except in this case the energy transfer is made in the other direction in order to decrease the voltage across the piezoelectric material by closing first k_2 and then k_1. The energy is then restored to the electrical energy storage. The power used to maintain the right voltage across the piezoelectric material is just the losses that occur during the power transfer in the fly-back converter.

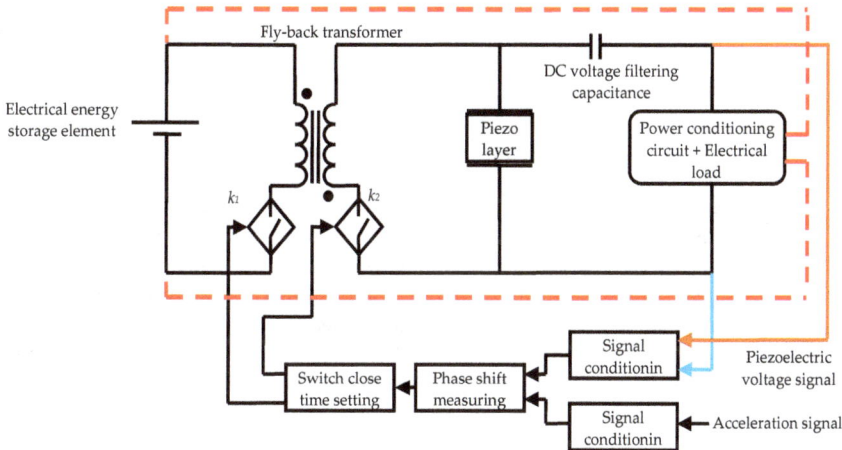

Figure 14. Resonant frequency adaptation circuit

3.2.1.5. The power balance

The Figure 15 below shows the power balance of the whole VEH including the resonant frequency tuning system. It is assumed that the resonant frequency is adjusted by step of 0.1Hz and it can be changed up to 10 times a second. The accelerometer consumption is not taken into account.

The mechanical, dielectric and electrical losses are investable, these losses are common for all VEH systems. The most important point shown in this power balance is the positive net output power. The whole system is self sufficient including the power required by the electronic for racking the vibration frequency and also the power needed to apply the right DC voltage across the piezoelectric material.

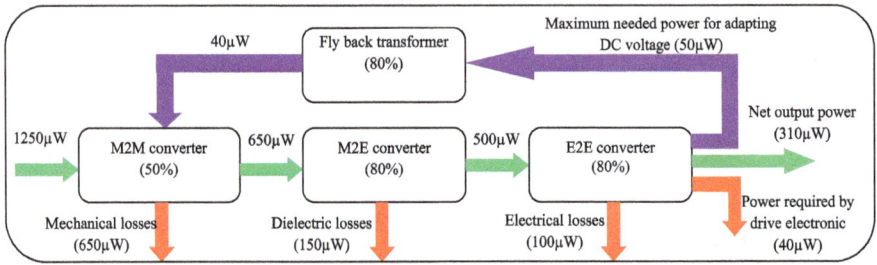

Figure 15. Complete power balance of the whole system

This result is very promising for the next generation of piezoelectric VEH, it shows that the real time resonant frequency tuning can be energetically positive. Nevertheless, the actual piezoelectric material which is in good agreement with this technique is still expensive (few 100€ per cm³) and their quality factor is quite limited (< 100).

In the next part of this section, another solution for a dynamic resonant frequency adjustment is given. This solution is quite similar to the previous one, it consists to adjust the electrical load connected to the piezoelectric material in order to adjust the electrical stiffness of the piezoelectric material.

3.2.2. Adaptation of the electrical load [21]

This subpart presents a declination of the previous solution for dynamic resonant frequency tuning. This one is based on the dependence that exists between the mechanical stiffness of a piezoelectric material and the electrical conditions to which it is subjected. In fact, the deflection of a piezoelectric bulk material under an applied mechanical force is higher when it is placed in short circuit (zero electric field) than in open circuit (no charge displacement), this means that the stiffness is lower in case of short circuit conditions than in open circuit conditions. The relation between the stiffness at short and open circuit is given by (14).

$$s^D = s^E - d^2 / \varepsilon^T \Leftrightarrow s^E = \frac{s^D}{1 - k^2} \tag{15}$$

with

- s^D: Mechanical compliance of piezoelectric at open circuit conditions
- s^E: Mechanical compliance of piezoelectric at short circuit conditions
- d: Piezoelectric coefficient
- ε: Dielectric permittivity
- k: Electromechanical coupling coefficient.

One way to obtain a variable stiffness between these two limits (s^D and s^E) is to connect the piezoelectric material with an adjustable non-dissipative electrical load. However, this electrical load should not affect the quality factor of the structure. Thus, connecting a variable capacitor seems to be a good compromise, able to change the resonant frequency with less power losses. The value of the capacitance set the effective dielectric permittivity of the piezoelectric material and then the piezoelectric material stiffness as it is shown by the equation (16), where Y_p: is the effective Young's modulus for the piezoelectric, C_p is the piezoelectric capacitance, C_{sh} the capacitance connected with the piezoelectric material, and A is the effective section of the piezoelectric material. Hence, by adjusting the value of the connected capacitance (C_{sh}), it is possible to adjust the value of the piezoelectric stiffness and then the resonant frequency.

$$Y_p = \left(s^E - \frac{d_{33}^2 A}{t_p \left(C_p + C_{sh} \right)} \right)^{-1} \tag{16}$$

In case of a cantilever shape like in the Figure 11, the resonant frequency of the harvester becomes:

$$f = \frac{1}{2\pi} \sqrt{ \frac{3 \left[2. \left(s^E - \frac{d_{31}^2 A}{t_p \left(C_p + C_s \right)} \right)^{-1} I_p + Y_b I_b \right]}{L^3 M} } \tag{17}$$

With M: the effective mass, t_p: the thickness of the piezoelectric layers ans L : represents the beam length.

3.2.2.1. Choice of the piezoelectric material

As for the previous technique, to choose the suitable material we define a new figure of merit. This one is based on a compromise between the following parameters:

- High electromechanical coupling coefficient k_{ij}:

For the present approach, the electromechanical coupling has a significant effect on the resonant frequency tuning ratio as shown by the equation (14). It is better to choose a material with a high electromechanical coupling.

- Low dielectric losses:

In order to reduce as much as possible the power losses, it is necessary to choose a material with a low dielectric losses. But, unlike the previous approach, the coefficient of dielectric losses will have no effect on the power consumption of the power management electronic.

- The effective permittivity:

For the present approach, it is better to have a material with a high electrical capacitance, which means high dielectric permittivity because it minimizes the effect of the parasitic capacitances on the adjustment of the resonant frequency. If the parasitic capacitance is at the same order as the piezoelectric one, the variation of the shunt capacitance will have a minor effect on the tuning ratio of resonant frequency.

- *The coupling mode:*

As the vibrations are supposed to be straight and unidirectional, two modes could be used for ensuring an efficient electromechanical coupling, the longitudinal mode (polarization and mechanical stress axes are collinear), or the transverse mode where the polarization and mechanical stress axes are perpendicular. As the electromechanical coupling is higher in the longitudinal mode, this one enables a better tuning ratio.

Finally, by taking into account all these parameters and their effect on resonant frequency tuning, the figure of merit is:

$$FOM = \frac{k_{3j}^2 . \varepsilon_{33}^T}{\tan(\delta)}, j = 1,3 \tag{18}$$

F.O.M(x1e3)	PZT	PMN-PT	PZN-PT
Transversal mode (31)	39.85	104.54	204.12
Longitudinal Mode (33)	136.96	447.7	618.52

Table 4. Comparison between the different piezoelectric materials

The table above shows the *FOM* of three different piezoelectric materials in two different modes:

It can be noted from this table that the PZN-PT in longitudinal coupling mode presents the best figure of merit and seems to be the suitable material for this method of resonant frequency tuning.

The next part presents the experimental results obtained with this technique on a structure prototype.

3.2.2.2. The experimental validation:

- *The manufactured structure*

The piezoelectric prototype developed by CEA-Leti is presented Figure 16. It consists on a bimorph piezoelectric cantilever, each piezoelectric layer is composed of a number of subparts mechanically bounded together in series. The polarization axis of each subpart is oriented on the direction of the resulted mechanical stress in order to work in the longitudinal mode. The electrodes of the subparts are connected in parallel, in order to obtain the highest equivalent capacitance.

This structure has been mounted on a shaker, the piezoelectric part has been coupled with a variable capacitance, and the obtained experimental results are presented in the next section.

- *The experimental results*

The first measurements show that the resonant frequency is equal to 208 Hz at short circuit condition and 294 Hz at open circuit condition, which represents **41% of tuning ratio**. The

Parameter	Value
Substrate material:	
Y_b (GPa)	210
Density (kg.m^{-3})	7500
Length x width x thickness (mm)	30x10x1
Piezoelectric material (PZN-PT):	
Y_{p-max} (GPa)	100
Density (kg.m^{-3})	8800
k_{33}	0.94
relative dielectric constant (ε)	7500
Length x width x thickness (mm)	30x10x1

Table 5. Characteristics of the fabricated structure

Figure 16. Picture of the fabricated structure

Figure 17 below presents the resonant frequency as a function of the shunt capacitance. The capacitance value is normalized to the blocked capacitance of the piezoelectric material (C_s/C_p). The blocked piezoelectric capacitance C_p is equal to 1nF.

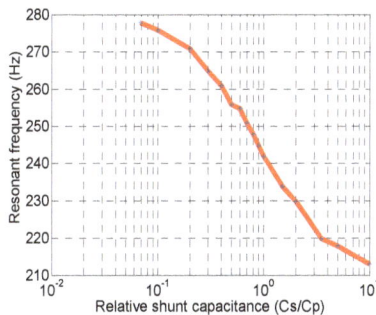

Figure 17. Resonant frequency as a function of the shunt capacitance

A tuning ratio up to 31% is noted when the shunt capacitance varies between $0.07C_p$ and 10 C_p. The power extracted using an optimal load is equal to 320µW for 0.1g@250Hz.

3.2.2.3. Resonant frequency tuning drive electronic

As mentioned before, the ultimate goal of the work is to develop an automatic system able to adjust in real time the harvester resonant frequency to the main frequency of the vibration source. The idea here is to couple the VEH with an adjustable capacitive load. The block diagram of the drive electronic is given by the following figure. The VEH is first connected to the adjustable capacitive load composed of two capacitances, C_1 and C_2. It is supposed that the power conditioning circuit requires a very low voltage ($\approx 3V$) compared to the piezoelectric output voltage ($>10V$). Hence, the equivalent shunt capacitance that affects the piezoelectric stiffness is the sum of C1 with C_2. Nevertheless, the capacitance C_2 enables a power transfer from the piezoelectric material to the power conditioning circuit and has more effect on the extracted electrical energy and then it enables an adjustment of the electrical damping.

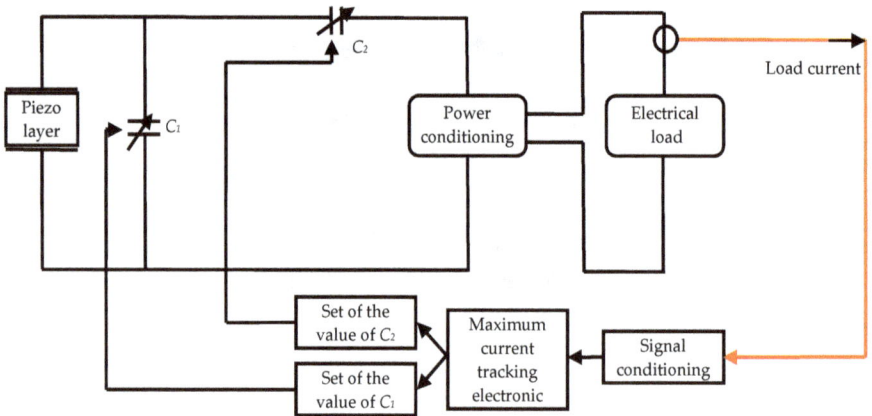

Figure 18. Block diagram circuit for the resonant frequency tuning by adapting the electrical load

The objective of the drive electronic is to track the maximum output power flowing through the electrical load by adjusting the electrical damping (adjusting C_2) and the resonant frequency (adjusting C_1 and C_2). Hence, the reaction time of a change of C_2 is shorter than a change of C_1, since the first has an effect of the transfer of energy while the second one on the resonant frequency. This electronic is composed of two loops, a slower one for adjusting the capacitance C_1 and a faster one to adjust the capacitance C_2.

The first measurements show that the whole electronic consumes about $30\mu W$ which represents only 10% of the $300\mu W$ generated power.

4. Conclusions

Through this chapter, it has been demonstrated that there is a real need to ensure a real time tracking of the vibration frequency. The issue of the adaptation between the input vibration characteristic and the mechanical characteristics of the VEH is capital for the development of robust and efficient VEH. Within this chapter, a special focus has been made on the solutions developed in *CEA-Leti* laboratory. Three solutions have been presented, the first one expects to amplify the relative displacement of any vibration type (random) occurring on a wide width frequency bandwidth, the second one consists to apply a DC electric field on the piezoelectric layer in order to adjust its stiffness and then the resonant frequency of the structure. The third solution consists to adjust the electrical load coupled with the piezoelectric material in order to adjust its stiffness and then the resonant frequency of the structure. Each solution has been validated experimentally, the first one enables an operation over one octave of frequency (50Hz to 100Hz), the second one a frequency tuning from (250Hz to 300Hz) and the last one from (210Hz to 280Hz). This chapter introduces also the drive electronic for each strategy, the drive electronic part is the most critical point for active techniques since it requires in most cases a huge amount of energy. First results obtained in CEA-Leti laboratory show that the resonant frequency tuning can significantly increase the net output power without consuming too much power to be managed (about 10% of the converter output power). Anyway, the successful of VEH systems is tightly related to the frequency bandwidth. So, the commercialization of such systems at large scale could not be done until overcoming efficiently the limits imposed by the bandwidth issue. Active techniques allowing positive power balance let us hope a large flexibility in the near future of VEH.

Author details

B. Ahmed Seddik, G. Despesse, S. Boisseau and E. Defay
LETI, CEA, Minatec Campus, Grenoble, France

5. References

[1] Zhu D, Tudor M J, Beeby S P. Strategies for increasing the operating frequency range of vibration energy harvesters. A review, Meas. Sci. Technol. Vol. 21, 2010.

[2] Yang Y, Soh C K. Toward Wideband Vibration Based Energy Harvesting. J. Intelligent Material and Structures, Vol 21, n°18, (2010), pp 1867-1897.

[3] Zhou W, Yong X. Vibration energy harvesting device based on air spaced piezoelectric cantilever, Applied Physics Letters, vol 90, n°26, (2007).

[4] Paquin S, St-Amant Y. Improving the performance of a piezoelectric energy harvester using a variable thickness beam. Smart Mater. Struct. Vol 19, (2010), 105020.

[5] Zheng Q, Xu. Y. Asymmetric air-spaced cantilevers for vibration energy harvesting. Smart Mater. Struct, Vol 17, (2008).

[6] Mitcheson P D, Toh TT, Wong KH, Burrow S G, Holmes A S. Tuning the Resonant Frequency and Damping of an Electromagnetic Energy Harvester Using Power Electronics, Circuits and Systems II: Express Briefs, IEEE Transactions , vol 58, n°12, (2011), pp792-796.

[7] Despesse G, Chaillout J-J, Jager T, Léger J-M, Vassilev A, Basrour S, Charlot B. High damping electrostatic system for vibration energy scavenging. Proc. SoC-EUSAI, (2005), pp283-286.

[8] Shahruz S. Design of mechanical band-pass filters for energy scavenging, J. Sound. Vib. Vol 292, (2006), pp987–998.

[9] Roundy S, Leland S E, Baker J, Carleton E, Reilly E, Lai E, Otis B, Rabaey J M, Sundararajan V, Wright P K. Improving power output for vibration-based energy scavengers, Pervasive. Comput. 4 (1) (2005), pp28–36.

[10] Yang B, Lee C. Non-resonant electromagnetic wideband energy harvesting mechanism for low frequency vibrations. Microsyst Technol, Vol 16,(2010) pp961-966.

[11] Marzencki M, Defosseux M and Basrour S. MEMS Vibration Energy Harvesting Devices with Passive Resonance Frequency Adaptation Capability, J. Michroelectromech syst. Vol 18, n°6, (2009).

[12] Leland E S and Wright P K. Resonance tuning of piezoelectric vibration energy scavenging generators using compressive axial preload, Smart Mater. Struct. Vol 15, n°5, (2006), pp1413–1420.

[13] Chella V-R, Prasad M G, Shi Y and Fisher F T. A vibration energy harvesting device with bidirectional resonance frequency tenability, Smart. Mater. Struct. Vol17, (2008).

[14] Peters C, Maurath D, Schock W, Mezger F, Manoli Y. A closed-loop wide-range tunable mechanical resonator for energy harvesting systems. J. Micromech. Microeng. Vol 19, (2009).

[15] Morgan B, Ghodssi R. Vertically-Shaped Tunable MEMS Resonators. J. Micrelectromech. Syst. Vol 17, n°1, (2008), pp85-92.

[16] Eichhorn C, Tchagsim R, Wihelm N, Woias P. Smart Self sufficient frequency tunable vibration energy harvester. J. Micromech. Microeng, Vol 21, n°10, (2011).

[17] Lallart M, Anton S R, Inman D J. Frquency self tuning scheme for broadband vibrations energy harvesting. J. Intelligent Material and Structures, June 2010, Vol 21, n°9, pp897-906.

[18] Cammarano A, Burrow S G, Barton D A, Carrella A, Clarella L R. Tuning a resonant energy harvester using a generalized electrical load. Smart Mater. Struct. Vol 19, (2010).

[19] Ahmed Seddik B, Despesse G, Defay E, Boisseau S. Increased bandwidth of mechanical energy harvester. J. Sensors & Transducers, Vol. 13, Special Issue, (2011), pp.62-72

[20] Thornbourgh R P, Chattopadhyay A. Nonlinear actuation of smart composites using a coupled piezoelectric-mechanical model, Smart Mate. Struct. Vol 10, (2001), pp743-749.

[21] Ahmed Seddik B, Despesse G, Defay G. Autonomous Wideband Mechanical Energy Harvester. Porc, IEEE ISIE, 2012.

Chapter 10

Analysis of Energy Harvesting Using Frequency Up-Conversion by Analytic Approximations

Adam Wickenheiser

Additional information is available at the end of the chapter

1. Introduction

Energy harvesting is the process of capturing energy existing in the environment of a wireless device in order to power its electronics without the need to manually recharge the battery. By replenishing on-board energy storage autonomously, the need to recharge or replace the battery can be eliminated altogether, enabling devices to be placed in difficult-to-reach areas. Vibration-based energy harvesting in particular has garnered much attention due to the ubiquity of vibrational energy in the environment, especially around machinery and vehicles (Roundy et al., 2003). Although several methods of electromechanical transduction from vibrations have been investigated, this chapter focuses on utilizing the piezoelectric effect.

Piezoelectric energy harvesters convert mechanical energy into electrical through the strain induced in the material by inertial loads. Typically, piezoelectric material is mounted on a structure that oscillates due to excitation of the host structure to which it is affixed. If a natural frequency of the structure is matched to the predominant excitation frequency, resonance occurs, where large strains in the piezoelectric material are induced by relatively small excitations. In order to take advantage of resonance, the natural frequency of the device must be matched to the predominant frequency component of the base excitation (Anderson & Wickenheiser, 2012). For many potential applications, ambient vibrations are low frequency, requiring longer length scales or a larger mass to match the resonance frequency to the excitation frequency (Roundy et al., 2003; Wickenheiser & Garcia, 2010a; Wickenheiser, 2011). In order to shrink the size and mass of these devices while reducing their natural frequencies, a variety of techniques have been investigated. Varying the cross sections along the beam length (Dietl & Garcia, 2010; Reissman et al., 2007; Roundy et al., 2005) and the ratio of tip mass to beam mass (Dietl & Garcia, 2010; Wickenheiser, 2011) have been shown to improve the electromechanical coupling (a factor in the energy conversion rate) over a uniform cantilever beam design. Multi-beam structures can reduce the overall dimensions of the design by folding it in on itself while retaining a similar natural frequency to the original, straight configuration (Karami & Inman, 2011; Erturk et al., 2009); however, this requires a more complex analysis of the natural frequencies and mode shapes (Wickenheiser, 2012).

In resonant designs, minimizing the mechanical damping in the system enhances the power harvesting performance (Lefeuvre et al., 2005; Shu and Lien, 2006; Wickenheiser & Garcia, 2010c). Unfortunately, lightly damped systems are the most sensitive to discrepancies between the resonance and the driving frequencies. Several methods have been analyzed for tuning the stiffness of the vibrating beam in order to match a slowly varying base excitation frequency. (Challa et al., 2008) and (Reissman et al., 2009) have considered placing one or more magnets to either side of the tip mass to create either an attractive or repulsive force that changes the effective stiffness of the beam, thus allowing the natural frequencies to be adjusted to match the base excitation frequency. Similarly, (Mann & Sims, 2009) harvest energy from a magnet levitating in a cavity between two magnets; varying the spacing of the magnets changes the natural frequency of the levitation. (Leland & Wright, 2006) have proposed tuning the natural frequencies of the beam by applying an axial load; however, this technique has been found to increase the apparent mechanical damping in the structure. A similar concept has been developed for adjusting the pre-tension in extensional mode resonators (Morris et al., 2008). These methods can be considered "quasi-static" because the rate at which the natural frequencies can be tuned is often much slower than the vibration frequency. Thus, these methods are ideal if the base excitation is an approximately stationary process with frequencies concentrated in a narrow band.

The off-resonant response of these systems can be enhanced by destabilizing the relaxed state of the beam. A bi-stable cantilever beam can be created by adding a repelling magnet beyond a magnetic tip mass or by adding attracting magnets on either side. In this situation, the beam can be induced to jump from one well of attraction to the other either periodically, quasi-periodically, or chaotically, depending on the amplitude and frequency of the base excitation. Bi-stability can be realized with a "snap-through" mechanism, in which the mass moves perpendicularly to the elastic axis (Ramlan et al., 2010), using the aforementioned beam and magnetic set-up first analyzed by (Moon, 1978), and using an inherently bi-stable composite plate (Arrieta, 2010). This technique is suited for strong excitations that are able to drive the beam between the two potential wells; however, for low excitation levels the performance converges towards the linear system unless a perturbation is added to "kick" the system into the other well.

In this chapter, a technique known as frequency up-conversion is employed to generate strong off-resonant responses. This technique is based off a repetition of the bi-stable system to create a sequence of potential wells; the transition between them induces a "pluck" followed by a free response at the fundamental frequency. A similar concept has been pursued by Tieck et al. (2006), consisting of a rack placed transversely near the tip of the beam that would periodically pluck the beam as it vibrated. Other concepts utilizing mechanical rectification have been proposed for harvesting energy from buoy motion (Murray and Rastegar, 2009) and low-frequency, rotating machinery (Rastegar and Murray, 2008).

In the following sections, the equations of motion (EOMs) are derived for a uniform beam with magnetic tip mass under periodic base excitation. The eigenvalue problem for this design is then solved for the natural frequencies and mode shapes. The modal expansion is reduced to a single mode (the fundamental) in order to derive an approximate model for low frequencies well below the fundamental frequency. A simplification is derived based on neglecting the base excitation; this simplification leads to a model of the beam's excitation in terms of a sequence of plucks followed by free vibrations. A few simple case studies are presented to highlight the accuracy of this approximate model.

2. Derivation of electromechanical EOMs

The layout of the piezoelectric, vibration-based energy harvester and the nearby magnetized structures used for mechanical rectification is presented in Fig. 1. For this study, a bimorph configuration is considered, in which piezoelectric layers are bonded to both sides of an inactive substructure. Other configurations, such as the unimorph, can be modeled with few modifications, as pointed out below. Electrodes are assumed to cover the upper and lower surfaces of each layer, and they are wired together in the "parallel" configuration, as depicted. In this configuration, the voltage drop across each layer is assumed to be the same, and the charge displaced by each layer is additive, much like capacitors in parallel. Because the piezoelectric layers are on opposite sides of the neutral axis, each layer experiences opposite strains; hence, they must be poled in the same direction to avoid charge cancellation. It is assumed that the electrodes and connecting wires have negligible resistance and that the resistivity of the piezoelectric material is significantly higher than that of the external circuitry; thus, the transducer impedance is assumed to be purely reactive.

A tip mass is connected to the free end of the beam, and its center of mass is displaced axially from the connection point by a distance d_t. Tip masses are traditionally added to decrease the natural frequency of the beam and to increase the strain due to base excitation. In this situation, the tip mass is considered to be a permanent magnet and is attracted to ferromagnetic structures placed in a line parallel to the y-axis with spacing d_m between them. These structures are not magnets themselves; rather, they become magnetized due to the proximity of the magnetic tip mass. Thus, in this device, the tip mass is an active component of the excitation while fulfilling its passive role as just described.

In the following section, the EOMs for the electromechanical system presented in Fig. 1 are derived through force, moment, and charge balances while adopting the Euler-Bernoulli beam assumptions and linearized material constitutive equations. The approach taken herein is based on force and moment balances and is a generalization of the treatments by

(Erturk & Inman, 2008; Söderkvist, 1990; Wickenheiser & Garcia, 2010c). It is assumed that each beam segment is uniform in cross section and material properties. Furthermore, the standard Euler-Bernoulli beam assumptions are adopted, including negligible rotary inertia and shear deformation (Inman, 2007). Subsequently, a solution consisting of a series of assumed modes is presented, and the EOMs are decoupled into modal dynamics equations. As will be demonstrated, only the first bending mode is excited significantly by the plucking of the magnetic force. Although higher modes can be excited by higher frequency base excitation, this study focuses primarily on base excitation frequencies well below the fundamental resonant frequency.

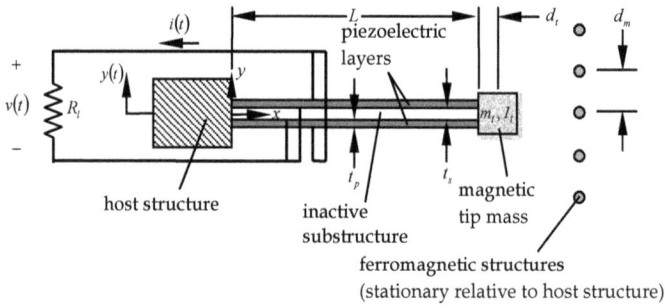

Figure 1. Layout and geometric parameters of cantilevered vibration energy harvester in parallel bimorph configuration with magnetic tip mass

2.1. Electromechanical EOMs

In this derivation, the states of the electromechanical system are the following: $w(x,t)$ is the relative transverse deflection of the beam with respect to its base, $v(t)$ is the voltage across the energy harvester as seen by the external circuit, and $i(t)$ is the net current flowing into the external circuit. The input to the system is $y(t)$, the absolute transverse displacement of the base; therefore, $w(x,t) + y(t)$ is the absolute transverse deflection of the beam.

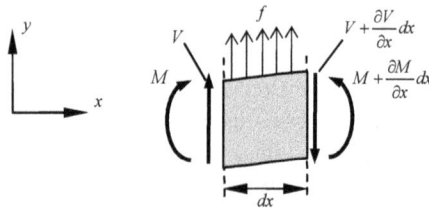

Figure 2. Free-body diagram of Euler-Bernoulli beam segment

Consider the free-body diagram shown in Fig. 2. Dropping higher order terms, balances of forces in the y-direction and moments yield

$$\frac{\partial V(x,t)}{\partial x} + f(x,t) = (\rho A)_{eff} \frac{\partial^2 w(x,t)}{\partial t^2} \qquad (a)$$

$$\frac{\partial M(x,t)}{\partial x} = -V(x,t) \qquad (b)$$

(1)

where $V(x,t)$ is the shear force, $M(x,t)$ is the internal moment generated by mechanical and electrical strain, $f(x,t)$ is the externally applied force per unit length, and $(\rho A)_{eff}$ is the mass per unit length (Inman, 2007). For the case of a bimorph beam segment, this term is given by

$$(\rho A)_{eff} = \frac{m}{L} = \frac{\rho_s t_s bL + 2\rho_p t_p bL}{L} = b(\rho_s t_s + 2\rho_p t_p) \qquad (2)$$

where m is the mass of the beam (not counting the tip mass), L is its length, b is its width, ρ_s and t_s are the density and thickness of the substrate, and ρ_p and t_p are the density and thickness of one of the piezoelectric layers, respectively. As can be seen in Eq. (2), if the segment is monolithic, $(\rho A)_{eff}$ is simply the product of the density of the material and the cross-sectional area. The externally applied force per unit length can be written as the sum of the distributed inertial force along the beam and the inertial force of the tip mass – which arise because the non-inertial frame of the base is taken as the reference – and the magnetic force applied at the center of the tip mass:

$$f(x,t) = -(\rho A)_{eff} \frac{d^2 y(t)}{dt} - m_t \frac{d^2 y(t)}{dt} \delta(x-L) - f_{mag}(t)\delta(x-L) \qquad (3)$$

where m_t is the mass of the tip mass, $f_{mag}(t)$ is the magnetic force, and $\delta(\cdot)$ is the Dirac delta function. In this study, the magnetic force is assumed to be purely in the y-direction. Although there is a stiffening effect due to the axial attractive force, it is considered negligible. The negative sign on the magnetic force indicates that it is an attractive force.

The internal bending moment is the net contribution of the stresses in the axial direction in the beam. The stress within the piezoelectric layers is found from the linearized constitutive equations

$$T_1 = c_{11}^E S_1 - e_{31} E_3$$

$$D_3 = e_{31} S_1 + \varepsilon_{33}^S E_3$$

(4)

where T is stress, S is strain, E is electric field, D is electric displacement, c is Young's Modulus, e is piezoelectric constant, and ε is dielectric constant. The subscripts indicate the direction of perturbation; in the cantilever configuration shown in Fig. 1, 1 corresponds to axial and 3 corresponds to transverse. The superscript $(\cdot)^E$ indicates a linearization at constant electric field, and the superscript $(\cdot)^S$ indicates a linearization at constant strain (IEEE, 1987). The use of Eq. (4) assumes the hypothesis of plane stress, which is reasonable since the beams are not directly loaded in the other directions, and small deflections. The stress within the substrate layer is given simply by the linear stress-strain relationship $T_1 = c_{11,s}S_1$, where $c_{11,s}$ is Young's Modulus of the substrate material in the axial direction. Since deformations are assumed small, the axial strain is the same as the case of pure bending, which is given by $S_1 = -y\partial^2 w(x,t)/\partial x^2$ (Beer & Johnson, 1992), and the transverse electric field is assumed constant and equal to $E_3 = \pm v(t)/t_p$, where $v(t)$ is the voltage across the electrodes, and the top and bottom layer have opposite signs due to the parallel configuration wiring. (This approximation is reasonable given the thinness of the layers.)

Consider the case of a bimorph beam. The bending moment along the length of the beam is

$$M(x,t) = \int_{-t_s/2-t_p}^{-t_s/2} T_1 bydy + \int_{-t_s/2}^{t_s/2} T_1 bydy + \int_{t_s/2}^{t_s/2+t_p} T_1 bydy$$

$$-m_t \frac{d^2 y(t)}{dt} d_t H(x-L) - f_{mag}(t) d_t H(x-L)$$

$$= -\left[\int_{-t_s/2-t_p}^{-t_s/2} c_{11}^E by^2 dy + \int_{-t_s/2}^{t_s/2} c_{11,s} by^2 dy + \int_{t_s/2}^{t_s/2+t_p} c_{11}^E by^2 dy \right] \frac{\partial^2 w(x,t)}{\partial x^2}$$

$$-\left[\int_{-t_s/2-t_p}^{-t_s/2} \frac{e_{31}}{t_p} bydy - \int_{t_s/2}^{t_s/2+t_p} \frac{e_{31}}{t_p} bydy \right] v(t) \left[H(x) - H(x-L) \right]$$

$$-m_t \frac{d^2 y(t)}{dt} d_t H(x-L) - f_{mag}(t) d_t H(x-L) \qquad (5)$$

$$= \underbrace{\left\{ c_{11,s} b \frac{t_s^3}{12} + 2 c_{11}^E b \left[\frac{t_p^3}{12} + t_p \left(\frac{t_p + t_s}{2} \right)^2 \right] \right\}}_{(EI)_{eff}} \frac{\partial^2 w(x,t)}{\partial x^2}$$

$$+ \underbrace{-e_{31} b (t_s + t_p) v(t)}_{\vartheta} \left[H(x) - H(x-L) \right]$$

$$-m_t \frac{d^2 y(t)}{dt} d_t H(x-L) - f_{mag}(t) d_t H(x-L)$$

where $H(\cdot)$ is the Heaviside step function. In Eq. (5), the constant multiplying the $\partial^2 w(x,t)/\partial x^2$ term is defined as $(EI)_{eff}$, the effective bending stiffness. (Note that if the beam segment is monolithic, this constant is simply the product of the Young's Modulus and the moment of inertia.) The constant multiplying the $v(t)$ term is defined as ϑ, the electromechanical coupling coefficient. Substituting Eq. (5) into Eq. (1) yields

$$(\rho A)_{eff} \frac{\partial^2 w(x,t)}{\partial t^2} + (EI)_{eff} \frac{\partial^4 w(x,t)}{\partial x^4} + \vartheta \left[\frac{d\delta(x)}{dx} - \frac{d\delta(x-L)}{dx} \right] v(t) =$$

$$-(\rho A)_{eff} \frac{d^2 y(t)}{dt} - \left[m_t \frac{d^2 y(t)}{dt} + f_{mag}(t) \right] \left[\delta(x-L) + d_t \frac{d\delta(x-L)}{dx} \right] \qquad (6)$$

which is the transverse mechanical EOM for the beam.

The electrical EOM can be found by integrating the electric displacement over the surface of the electrodes, yielding the net charge $q(t)$ (IEEE, 1987):

$$q(t) = \underset{\substack{\text{upper} \\ \text{layer}}}{\iint D_3 dA} - \underset{\substack{\text{lower} \\ \text{layer}}}{\iint D_3 dA}$$

$$= b \int_0^L \left[\frac{1}{t_p} \int_{t_s/2}^{t_s/2+t_p} -e_{31} y \frac{\partial^2 w(x,t)}{\partial x^2} dy - \frac{\varepsilon_{33}^S}{t_p} v(t) \right] dx$$

$$- b \int_0^L \left[\frac{1}{t_p} \int_{-t_s/2-t_p}^{-t_s/2} -e_{31} y \frac{\partial^2 w(x,t)}{\partial x^2} dy + \frac{\varepsilon_{33}^S}{t_p} v(t) \right] dx \qquad (7)$$

$$= \underbrace{-e_{31} b \left(t_s + t_p \right) \frac{\partial w(x,t)}{\partial x} \bigg|_{x=L}}_{\vartheta} - \underbrace{\frac{2\varepsilon_{33}^S bL}{t_p} v(t)}_{C_0}$$

where the constant multiplying the $v(t)$ term is defined as C_0, the net clamped capacitance of the segment. Eqs. (6–7) provide a coupled system of equations; these can be solved by relating the voltage $v(t)$ to the charge $q(t)$ through the external electronic interface.

2.2. Modal decoupling

The system of coupled equations (6–7) can be solved by assuming that the transverse deflection of the beam can be written as a convergent series expansion of eigenfunctions, i.e.

$$w(x,t) = \sum_{i=1}^{\infty} \phi_i(x) \eta_i(t) \qquad (8)$$

where $\phi_i(x)$ is the ith transverse mode shape function, and $\eta_i(t)$ is the ith modal displacement. Given the configuration in Fig. 1 with a tip mass having a nontrivial mass m_t and moment of inertia I_t, the eigenvalues λ_i corresponding to the mode shapes must satisfy

$$F_{cf} - \frac{m_t}{(\rho A)_{eff} L} \lambda F_{cp} - \frac{I_t + m_t d_t^2}{(\rho A)_{eff} L^3} \lambda^3 F_{cr} + \frac{I_t m_t}{(\rho A)_{eff}^2 L^4} \lambda^4 F_{cc} - \frac{2 m_t d_t}{(\rho A)_{eff} L^2} \lambda^2 \sin \lambda \sinh \lambda = 0 \qquad (9)$$

where $F_{cf} = 1 + \cos \lambda \cosh \lambda$ are the clamped-free, $F_{cp} = \sin \lambda \cosh \lambda - \cos \lambda \sinh \lambda$ are the clamped-pinned, $F_{cr} = \sin \lambda \cosh \lambda + \cos \lambda \sinh \lambda$ are the clamped-rolling, and $F_{cc} = 1 - \cos \lambda \cosh \lambda$ are the clamped-clamped eigenvalue terms, respectively (Oguamanam, 2003). The mode shape functions are given by

$$\phi_i(x) = \cos\left(\lambda_i \frac{x}{L}\right) - \cosh\left(\lambda_i \frac{x}{L}\right)$$

$$+ \frac{\sin\lambda_i - \sinh\lambda_i - \dfrac{m_t}{(\rho A)_{eff} L}\left[\lambda_i^2 \dfrac{d_t}{L}\left(\sin\lambda_i + \sinh\lambda_i\right) - \lambda_i\left(\cos\lambda_i - \cosh\lambda_i\right)\right]}{\cos\lambda_i + \cosh\lambda_i - \dfrac{m_t}{(\rho A)_{eff} L}\left[\lambda_i^2 \dfrac{d_t}{L}\left(\cos\lambda_i - \cosh\lambda_i\right) + \lambda_i\left(\sin\lambda_i - \sinh\lambda_i\right)\right]} \tag{10}$$

$$\times\left[\sin\left(\lambda_i \frac{x}{L}\right) - \sinh\left(\lambda_i \frac{x}{L}\right)\right]$$

These functions may be scaled arbitrarily and still be admissible, and in the present case are done to satisfy the following orthogonality condition:

$$\int_0^L (\rho A)_{eff}\, \phi_i(x)\phi_j(x)dx + m_t\phi_i(L)\phi_j(L) + m_t d_t\left[\frac{d\phi_i(x)}{dx}\phi_j(x) + \phi_i(x)\frac{d\phi_j(x)}{dx}\right]_{x=L}$$

$$+ \left(I_t + m_t d_t^2\right)\left[\frac{d\phi_i(x)}{dx}\frac{d\phi_j(x)}{dx}\right]_{x=L} = \delta_{ij} \tag{11}$$

where δ_{ij} is the Kronecker delta.

Substituting (8) into (6) and applying the orthogonality condition (11) results in

$$\frac{d^2\eta_k(t)}{dt^2} + 2\zeta_k\omega_k\frac{d\eta_k(t)}{dt} + \omega_k^2\eta_k(t) + \Theta_k v(t) =$$

$$-(\rho A)_{eff}\,\gamma_k\frac{d^2 y(t)}{dt^2} - \beta_k\left[m_t\frac{d^2 y(t)}{dt^2} + f_{mag}(t)\right] \tag{12}$$

at which point a modal damping term has been inserted. The kth modal short-circuit (i.e. $v(t) = 0$) natural frequency ω_k is given by

$$\omega_k = \sqrt{\frac{\lambda_k^4 (EI)_{eff}}{(\rho A)_{eff} L^4}} \tag{13}$$

for Euler-Bernoulli beams. Eq. (12) constitutes the EOM for the kth transverse vibrational mode. The modal influence coefficients appearing in Eq. (12) are given by

$$\Theta_k = \vartheta \frac{d\phi_k(x)}{dx}\bigg|_{x=L}, \quad \gamma_k = \int_0^L \phi_k(x)dx, \quad \beta_k = \phi_k(L) + d_t\frac{d\phi_k(x)}{dx}\bigg|_{x=L} \tag{14}$$

Θ_k is the modal electromechanical coupling coefficient, γ_k is the modal influence coefficient of the distributed inertial force along the beam, and β_k is the modal influence coefficient of the concentrated force at the tip. A similar decoupling of the electrical EOM (7) yields

$$q(t) = \sum_{i=1}^{\infty} \Theta_i r_i(t) - C_0 v(t) \tag{15}$$

It remains to write the applied magnetic force $f_{mag}(t)$ in terms of the modal coordinates. Due to the assumption of a symmetrical tip mass, this force is applied at its centroid, as shown in Fig. 3. It is further assumed that the ferromagnetic structures are placed uniformly with spacing d_m and that distant structures do not influence the magnetic force (a reasonable assumption given the $1/r^3$ dependency). Additionally, the rotation of the tip mass is assumed small compared to its absolute translation (base motion + relative deflection), and so its effect on the magnitude of the magnetic force is ignored. Thus, the magnetic force is approximately sinusoidal with wavelength d_m, and so it can be written in the form

$$f_{mag}(t) = F_{mag} \sin\left[\frac{2\pi}{d_m}\left(w(x_m,t) + y(t)\right)\right] \tag{16}$$

where x_m is the x-coordinate of the tip mass centroid. The magnitude of this force F_{mag} is a complicated function of the material properties and geometry of the tip mass and the ferromagnetic structures that is beyond the scope of this work (see Moon, 1978; Stanton et al., 2010). F_{mag} is normalized by the maximum static tip load the beam can support without failing. In this study, a maximum strain of 0.1% is chosen, resulting in a maximum static tip load of

$$F_{max} = \frac{(EI)_{eff}}{\left(\dfrac{t_s}{2} + t_p\right)L}(0.001) \tag{17}$$

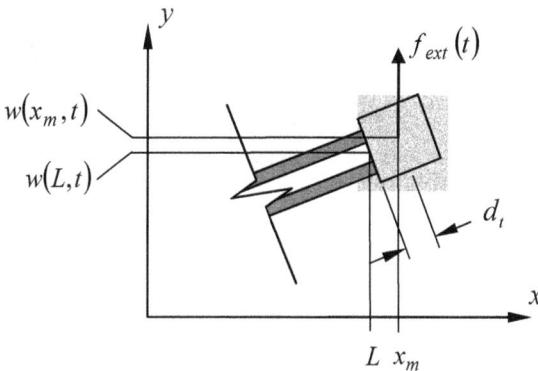

Figure 3. Tip mass coordinates used for locating the centroid in terms of modal coordinates.

The position of the tip mass centroid, shown in Fig. 3, can be written in terms of the modal coordinates:

$$w(x_m,t) \approx w(L,t) + d_t \left.\frac{\partial w(x,t)}{\partial x}\right|_{x=L} = \sum_{i=1}^{\infty} \beta_i \eta_i(t) \tag{18}$$

3. Estimates of expected power harvested

3.1. Linear case, frequency domain

In order to establish a baseline against which the effects of the magnetic force can be compared, in this section the magnetic interactions are not considered, i.e. $F_{mag}=0$. The most prevalent (e.g. duToit et al., 2005; Lefeuvre et al., 2005; Liao & Sodano, 2008; Shu & Lien, 2006) assumption of constant-amplitude, sinusoidal base excitation forms the basis for analysis of more complex periodic forcing. In this study, it is assumed that the base acceleration is a weakly stationary random process. This general framework includes the special cases of harmonic (single or multiple frequencies), white noise, band-limited noise, and periodic in mean square processes (Anderson & Wickenheiser, 2012; Lin, 1967). The average power dissipated by the load after transients have died out is given by

$$E[P(t\to\infty)] = \frac{E[v^2(t\to\infty)]}{R_l} = \frac{R_{vv}(0)}{R_l} = \frac{1}{R_l}\int_{-\infty}^{\infty}|H(\omega)|^2 \Phi_{AA}(\omega)d\omega \tag{19}$$

where $E[\cdot]$ is the expectation operator, $R_{vv}(\tau)$ is the autocorrelation function of the voltage, $H(\omega)$ is the frequency transfer function of the energy harvester between acceleration and voltage, and $\Phi_{AA}(\omega)$ is the spectral density of the base acceleration (Lin, 1967). Since the system is assumed to be stable, the power output is seen to approach a weakly stationary process as $t\to\infty$.

In order to calculate the frequency transfer function, it is assumed that the electrical load can be represented by a resistor with value R_l. Eq. (15) can then be rewritten as

$$\frac{v(t)}{R_l} = i(t) = \frac{dq(t)}{dt} = \sum_{i=1}^{\infty} \Theta_i \frac{d\eta_i(t)}{dt} - C_0 \frac{dv(t)}{dt} \tag{20}$$

Since the system of equations (12,19) is linear, the modal responses $\eta_k(t)$ and output voltage $v(t)$ are sinusoidal at the driving frequency of the base excitation. The frequency transfer function between base displacement and voltage can be derived from the EOMs, yielding

$$\frac{V(\omega)}{Y(\omega)} = \frac{R_l\sum_{j=1}^{\infty}\Theta_j^2 \dfrac{i\left[(\rho A)_{eff}\gamma_j + m_t\beta_j\right]\omega^3}{\omega_j^2-\omega^2+i2\zeta_j\omega_j\omega}}{iR_lC_0\omega+1+R_l\sum_{j=1}^{\infty}\Theta_j^2 \dfrac{i\omega}{\omega_j^2-\omega^2+i2\zeta_j\omega_j\omega}} \tag{21}$$

where $i = \sqrt{-1}$ and ω is the base excitation frequency (Wickenheiser & Garcia, 2010b). The frequency transfer function between base acceleration and displacement is simply $Y(\omega)/A(\omega) = -1/\omega^2$. In this study, however, only the fundamental mode is assumed to be excited; hence, the j subscript is dropped and the fundamental natural frequency is written as ω_n.

In order to use Eq. (18), an estimate of the spectral density of the base acceleration $\Phi_{AA}(\omega)$ is required. An overview of spectral density estimation methods can be found in (Porat, 1994), any of which can provide an approximation of the base excitation signal of the form

$$\frac{d^2 y(t)}{dt^2} = a(t) \approx \sum_{k=1}^{N} A_k \cos(\omega_k t + \varphi_k) \tag{22}$$

where the component amplitudes A_k, frequencies ω_k, and phase angles φ_k are obtained from the spectral density estimate. The number of terms needed N is often determined by a user-defined error tolerance used to capture the "quality" of the signal approximation in some optimal manner. The spectral density is then given by

$$\Phi_{AA}(\omega) = \frac{1}{2\pi} \int_{-\infty}^{\infty} R_{AA}(t) e^{-i\omega t} dt = \frac{1}{2\pi} \int_{-\infty}^{\infty} E\left[a(t_0 + t)\overline{a(t_0)} \right] e^{-i\omega t} dt$$
$$= \frac{1}{2\pi} \int_{-\infty}^{\infty} \lim_{T \to \infty} \frac{1}{T} \int_0^T \left[a(t_0 + t)\overline{a(t_0)} \right] dt_0 e^{-i\omega t} dt \tag{23}$$

Consider first the case $N = 2$, where the base excitation is composed of the sum of two sinusoids. Then, without loss of generality,

$$a(t) = \underbrace{A_1 \cos(\omega_1 t)}_{a_1(t)} + \underbrace{A_2 \cos(\omega_2 t + \varphi)}_{a_2(t)} = \frac{A_1}{2}\left(e^{i\omega_1 t} + e^{-i\omega_1 t} \right) + \frac{A_2}{2}\left(e^{i(\omega_2 t + \varphi)} + e^{-i(\omega_2 t + \varphi)} \right) \tag{24}$$

Then

$$R_{AA}(t) = R_{A_1 A_1}(t) + R_{A_2 A_2}(t) + \frac{A_1 A_2}{4} \lim_{T \to \infty} \frac{1}{T} \int_0^T \left(e^{i\left[\omega_2(t_0+t) + \varphi \right]} + e^{-i\left[\omega_2(t_0+t) + \varphi \right]} \right)\left(e^{-i\omega_1 t_0} + e^{i\omega_1 t_0} \right) dt_0$$
$$+ \frac{A_1 A_2}{4} \lim_{T \to \infty} \frac{1}{T} \int_0^T \left(e^{i\omega_1(t_0+t)} + e^{-i\omega_1(t_0+t)} \right)\left(e^{-i(\omega_2 t_0 + \varphi)} + e^{i(\omega_2 t_0 + \varphi)} \right) dt_0 \tag{25}$$

Integrating the first term in the integrand and taking the limit yields

$$\lim_{T \to \infty} \frac{1}{T} \int_0^T e^{i\left[\omega_2(t_0+t) + \varphi \right]} e^{-i\omega_1 t_0} dt_0 = \lim_{T \to \infty} \frac{1}{T} \frac{1}{i(\omega_2 - \omega_1)}\left(e^{i\left[(\omega_2 - \omega_1)T + \omega_2 t + \varphi \right]} - e^{i(\omega_2 t + \varphi)} \right)$$
$$\leq \lim_{T \to \infty} \frac{1}{T} \frac{1}{|\omega_2 - \omega_1|}(1+1) = 0 \tag{26}$$

Each of the other integrated terms also averages out to 0 in the long run; hence,

$$R_{AA}(t) = R_{A_1 A_1}(t) + R_{A_2 A_2}(t) \tag{27}$$

Then, by mathematical induction,

$$R_{AA}(t) = \sum_{k=1}^{N} R_{A_k A_k}(t) \tag{28}$$

Using this result in Eq. (23) gives

$$\begin{aligned}
\Phi_{AA}(\omega) &= \frac{1}{2\pi} \int_{-\infty}^{\infty} R_{AA}(t) e^{-i\omega t} dt = \frac{1}{2\pi} \int_{-\infty}^{\infty} e^{-i\omega t} \sum_{k=1}^{N} \frac{A_k^2}{4} \left(e^{i\omega_k t} + e^{-i\omega_k t} \right) dt \\
&= \sum_{k=1}^{N} \frac{A_k^2}{4} \left[\delta(\omega - \omega_k) + \delta(\omega + \omega_k) \right]
\end{aligned} \tag{29}$$

Using Eqs. (19,29), the average power harvested can be simplified:

$$E\left[P(t \to \infty) \right] = \frac{1}{2R_l} \sum_{k=1}^{N} \frac{A_k^2}{\omega_k^4} \left| H(\omega_k) \right|^2 \tag{30}$$

Eq. (30) indicates that the frequency transfer function $H(\omega)$ need only be evaluated at the component frequencies of the base acceleration. This equation can be rewritten in the form

$$E\left[P(t \to \infty) \right] = \sum_{k=1}^{N} A_k^2 C_k \tag{31}$$

where C_k can be interpreted as the gain of the harmonic of frequency ω_k. This gain is given by the formula

$$C_k = \frac{\left[(\rho A)_{eff} \gamma + m_t \beta \right]^2 \alpha k_e^2}{2\omega_n} \frac{\Omega_k^2}{\Lambda_k(i\Omega_k)\Lambda_k(i\Omega_k)} \tag{32}$$

where

$$\Lambda_k(i\Omega_k) = \alpha(i\Omega_k)^3 + (2\zeta\alpha + 1)(i\Omega_k)^2 + (\alpha + 2\zeta + \alpha k_e^2)(i\Omega_k) + 1 \tag{33}$$

The following non-dimensional parameters are employed in Eqs. (32-33):

$$\Omega_k = \frac{\omega_k}{\omega_n}, \quad k_e^2 = \frac{\Theta^2}{C_0 \omega_n^2}, \quad \alpha = R_l C_0 \omega_n \tag{34}$$

where Ω_k is the ratio of the frequency of the acceleration component to the fundamental natural frequency, k_e^2 is the modal electromechanical coupling coefficient, and α is the ratio of the load resistance to the modal impedance.

3.2. Nonlinear case, time domain

In the presence of the magnetic field, the EOMs become nonlinear, and the analysis based off of the frequency transfer function detailed in the previous section is no longer valid. Instead, the vibrations induced by the spatially periodic magnetic field are interpreted as a series of plucks that occur each time the tip mass crosses an unstable equilibrium point between the ferrous structures. Each pluck is followed by a free response – underdamped in this case – superposed on the relatively slow base motion. An example response showing these two superposed motions is depicted in Fig. 4. The free response at the fundamental frequency of the beam, as opposed to the frequency of the base motion, drives the majority of the energy harvested.

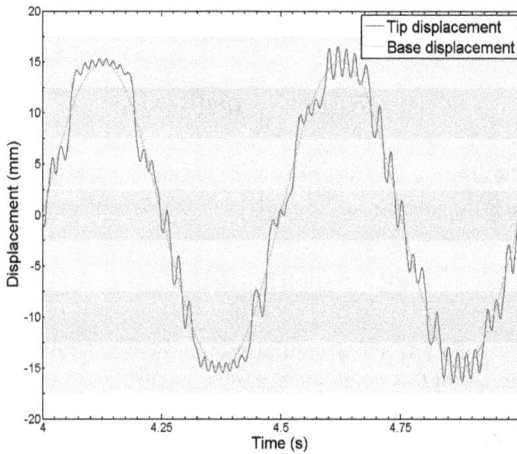

Figure 4. Absolute base and tip displacements: $F_{mag} = 0.75F_{max}$, $d_m = 5$ mm, $y(t) = Y\sin(\omega t)$, $Y = 15$ mm, $\omega = 2$ Hz. The shaded areas are the basins of attraction of the stable equilibria (Wickenheiser & Garcia, 2010b).

To analyze the energy harvested from a pluck, first the effect of the magnetic field strength on the free response is considered. To simplify the analysis, the inertial force due to base excitation is assumed to be negligible, and the effect of the energy dissipated by the resistor is approximated by an additional damping term. Hence, the total effective modal damping ratio is written as $\zeta_{eff} = \zeta + \zeta_e$, the sum of the mechanical and electrical damping. The electrical damping term can be accurately approximated as

$$\zeta_e = \frac{k_e^2}{2\sqrt{1-k_e^2}} \frac{\alpha}{1+\alpha^2} \qquad (35)$$

in the case of steady-state oscillations (Davis & Lesieutre, 1995); this formula is validated for free oscillations in the sequel. Using this damping model, the modal EOM, Eq. (12), can be written as

$$\frac{d^2\eta(t)}{dt^2} + 2\zeta_{eff}\omega_n \frac{d\eta(t)}{dt} + \omega_n^2\eta(t) = -F_{mag}\beta\sin\left[\frac{2\pi}{d_m}\left(\beta\eta(t)+y(t)\right)\right] \tag{36}$$

Linearizing Eq. (36) about the point $y(t)=kd_m$, $\eta(t)=0$, where k is an integer, gives

$$\frac{d^2\eta(t)}{dt^2} + 2\zeta_{eff}\omega_n \frac{d\eta(t)}{dt} + \left(\omega_n^2 + \frac{2\pi\beta^2}{d_m}F_{mag}\right)\eta(t) = 0 \tag{36a}$$

Thus, the term in parentheses is the square of the effective natural frequency, $\omega_{n,eff}^2$.

The amplitude of each pluck, and hence, the initial condition of the free response, is determined by the location of the unstable equilibrium between each pair of ferrous structures. In equilibrium,

$$\omega_n^2\eta(t) = -F_{mag}\beta\sin\left[\frac{2\pi}{d_m}\left(\beta\eta(t)+y(t)\right)\right] \tag{37}$$

again assuming that the effect of the electromechanical coupling is negligible. First, consider the case when the base is moving upward, i.e. $\dot{y}(t)>0$. In this case, the pluck occurs when

$$\sin\left[\frac{2\pi}{d_m}\left(\beta\eta(t)+y(t)\right)\right] = 1 \tag{38}$$

When this condition occurs, any more vertical motion of the tip results in a decreased downward magnetic force. At this point, the beam has passed over a local maximum in the magnetic potential, and it begins accelerating towards the next stable equilibrium. The response after cresting the potential hill is approximated by Eq. (36). By plugging Eq. (38) into Eq. (37), the amplitude of the pluck can be found:

$$\eta_0 = \frac{-F_{mag}\beta}{\omega_n^2} \tag{39}$$

If the times of the plucks are denoted t_k, then solving Eq. (38) for t_k, and using the fact that $y(t_k)=Y\sin(\omega t_k)$, yields

$$t_k = \frac{1}{\omega}\sin^{-1}\left(\frac{4k-3}{4}\frac{d_m}{Y}+\frac{F_{mag}\beta^2}{\omega_n^2 Y}\right), \quad k=1,...,N \text{ where } (N-1)d_m<Y\leq Nd_m \tag{40}$$

A similar formula can be derived for the pluck times when $\dot{y}(t)<0$:

$$t_{k+N} = \frac{\pi}{\omega}-\frac{1}{\omega}\sin^{-1}\left(\frac{4(N-k+1)-1}{4}\frac{d_m}{Y}-\frac{F_{mag}\beta^2}{\omega_n^2 Y}\right), \quad k=1,...,N \tag{41}$$

By examining Eq. (36), the (in this case) underdamped free response can be found to be

$$\eta(t) = \eta_0 e^{-\zeta_{eff}\omega_{n,eff}t} \cos(\omega_{d,eff}t) \tag{42}$$

where $\omega_{d,eff} = \omega_{n,eff}\sqrt{1-\zeta_{eff}^2}$ is the effective damped natural frequency, and the initial amplitude η_0 is given by Eq. (39). This solution can now be plugged into Eq. (20) to find the voltage response $v(t)$ after the pluck. Assuming that the voltage is 0 at the time of the pluck, the solution is given by

$$v(t) = -X_1 e^{-t/R_l C_0} + X_1 e^{-\zeta_{eff}\omega_{n,eff}t}\cos(\omega_{d,eff}t) + X_2 e^{-\zeta_{eff}\omega_{n,eff}t}\sin(\omega_{d,eff}t) \tag{43}$$

where

$$X_1 = \frac{\left(\Omega_{eff}\alpha\right)^2 - \zeta_{eff}\Omega_{eff}\alpha}{1 - 2\zeta_{eff}\Omega_{eff}\alpha + \left(\Omega_{eff}\alpha\right)^2}\frac{\Theta}{C_0}\eta_0, \quad X_2 = \frac{\zeta_{eff}\left(\Omega_{eff}\alpha\right)^2 - \left(\Omega_{eff}\alpha\right)^2 - \Omega_{eff}\alpha}{1 - 2\zeta_{eff}\Omega_{eff}\alpha + \left(\Omega_{eff}\alpha\right)^2}\sqrt{1 - \zeta_{eff}^2}\frac{\Theta}{C_0}\eta_0,$$

and $\Omega_{eff} = \omega_{n,eff}/\omega_n$.

The energy harvested during the free vibrations can be adequately approximated by the following formula:

$$E(t) = \frac{1}{R_l}\int_0^t v^2(\tau)d\tau \approx \frac{1}{2R_l}\int_0^t \frac{\Theta^2 R_l^2 \omega_{n,eff}^2}{1 - 2\zeta_{eff}\Omega_{eff}\alpha + \left(\Omega_{eff}\alpha\right)^2}\eta_0^2 e^{-2\zeta_{eff}\omega_{n,eff}\tau}d\tau$$

$$= \frac{\Theta^2 R_l \omega_{n,eff}}{4\zeta_{eff} - 8\zeta_{eff}^2\Omega_{eff}\alpha + 4\zeta_{eff}\left(\Omega_{eff}\alpha\right)^2}\eta_0^2\left(1 - e^{-2\zeta_{eff}\omega_{n,eff}\tau}\right) \tag{44}$$

The accuracy of this approximate formula can be seen in Fig. 5. The results of the simulation of the original EOMs, Eqs. (12,20), are plotted using a solid line, whereas Eq. (44) is plotted using a dashed line. To arrive at Eq. (44), it is assumed that the initial transients and the oscillating terms in $v^2(t)$ integrate out to 0; hence, the result is a smooth exponential curve. Although instantaneously the approximate curve may not be accurate, it matches the overall growth of the exact solution. Hence, Eq. (44) is an accurate representation of the energy harvested from a free vibration with a non-zero initial deflection and a zero initial velocity.

For a sequence of plucks, which is what occurs with the frequency up-conversion technique, it is assumed that the plucks are instantaneous and that the deflection is "reset" to the value given by Eq. (39) after each pluck. Hence, the total energy harvested during a half cycle of the base excitation is

$$E_{total} = \sum_{k=1}^N E(t_{k+1} - t_k) \tag{45}$$

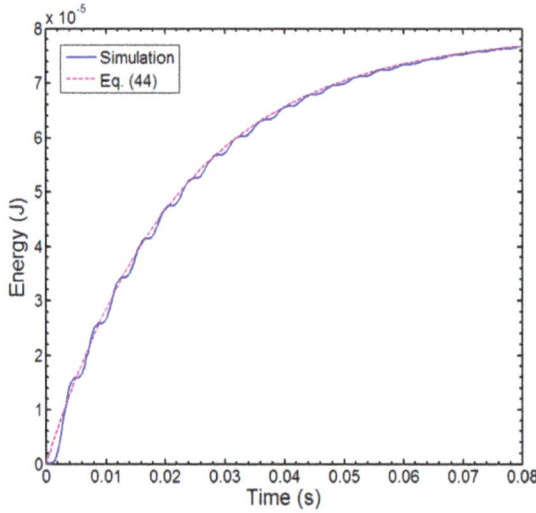

Figure 5. Energy harvested during one free vibration after an initial pluck, comparison between simulation using exact EOMs and approximation using Eq. (44). Parameters used are listed in Table 1.

4. Simulated response to sinusoidal base excitation

In this section, the response of the system to sinusoidal base excitation, $y(t) = Y \sin(\omega t)$, is presented in both the time and frequency domains. The geometry and material properties used in the following simulations are listed in Table 1. The tip mass m_t is approximately one-third of the overall beam mass, and its moment of inertia I_t is calculated assuming the mass is roughly cube-shaped. The resistor value chosen for this study is the optimal value for energy harvesting at the fundamental frequency in the limit of small electromechanical coupling, i.e. $R = 1/(C_0 \omega_{SC,1})$ (Wickenheiser & Garcia, 2010c).

The first three natural frequencies of the beam are $\omega_{SC,1} = 34.1$ Hz, $\omega_{SC,2} = 271.6$ Hz, and $\omega_{SC,3} = 806.8$ Hz. Since frequencies around and below the fundamental frequency are of interest in this study, a three-mode expansion of the beam displacement is deemed sufficient. Furthermore, since the magnetic force is applied at the tip of the beam, only the fundamental mode is significantly excited by the plucking.

The transfer functions for power harvested (normalized by $Y^2 \omega^3$) are plotted in Fig. 6. Five different values for the magnetic force strength F_{mag} are plotted alongside the baseline case of an inactive tip. For the cases with a nonzero magnetic force, the transfer functions are derived numerically. The system is simulated for 50 cycles of base motion, and the relative tip deflection is averaged over the last 20 cycles of each run in order to minimize the effects of initial transients. This process is completed 10 times at every frequency, and the results are averaged.

Beam properties:		
L	length	100 mm
b	width	20 mm
t_s	thickness of substructure	0.5 mm
t_p	thickness of PZT layer	0.4 mm
ρ_s	density of substructure	7800 kg/m³
ρ_p	density of PZT	7800 kg/m³
$c_{11,s}$	Young's modulus of substructure	102 GPa
c_{11}^E	Young's modulus of PZT	66 GPa
e_{31}	piezoelectric constant	-12.54 C/m²
ε_{33}^S	permittivity	15.93 nF/m
ζ	modal damping ratio	6.4%
Tip mass properties:		
m_t	mass	10 g
d_t	centroid displacement	5 mm
I_t	moment of inertia	1.7x10⁻⁷ kg–m²
Derived properties:		
$(\rho A)_{eff}$	mass per length	0.20 kg/m
$(EI)_{eff}$	bending stiffness	0.25 N–m²
C_0	net clamped capacitance	160 nF
k_e^2	electromechanical coupling coefficient	0.049
R	resistance	29.3 kΩ

Table 1. Geometry and material properties.

The overall trend of the responses indicates that the magnet has an increasing effect as the base excitation frequency decreases to 0. As is discussed in the sequel, at low frequencies relative to the fundamental resonance, the response converges to a sequence of free responses. In this regime, the inertial forces are negligible, and so the disturbances due to the magnetic force are relatively large. The normalized power approaches half an order of magnitude below its resonance value as $\omega \to 0$ due to the energy harvested from the plucks. As the driving frequency increases, the beam tip has less time to oscillate in each potential well, and, thus, its motion tends to converge towards the motion of the baseline case. As $\omega \to \omega_{SC,1}$, the frequency of the base motion approaches the frequency of the impulse response of the beam from the magnetic force. Hence, the time in which the beam is in free response decays to 0, and so all of the frequency response functions converge towards the baseline function, as shown in Fig. 5. A discussion of the variation in frequency response with respect to the magnet parameters F_{mag} and d_m can be found in (Wickenheiser & Garcia, 2010b).

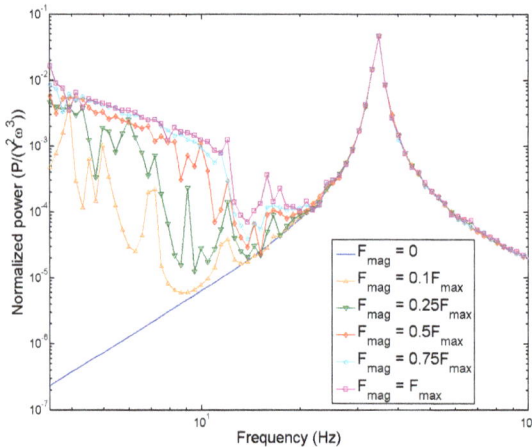

Figure 6. Normalized power harvested transfer function: $d_m = 5$ mm, $Y = 15$ mm (Wickenheiser & Garcia, 2010b).

Fig. 7 compares two methods of simulating the tip deflection response: using the original EOMs and assuming a series of undamped free responses give by Eq. (42). The most notable difference that can be seen from the figure is that the free response assumption does not take into account the base motion, which causes the solution to the EOMs to drift upward or downward depending on the sign of the base velocity. Another difference between the two models is that the assumed time of the plucks, indicated by the vertical lines and given by Eqs. (40,41), generally occur before the plucks in the actual solution. This happens because the beam has enough inertia to resist the pull of the next magnet, i.e. to overcome the potential well barrier between magnets. Only after the beam's velocity decays sufficiently does it become trapped in the next potential well in the sequence.

Fig. 8 depicts the voltage response during the same simulation that has been plotted in Fig. 7. A comparison of the two curves plotted shows an excellent agreement between the simulation results and the predicted voltage given by Eq. (43). There is a slight asymmetry in the curve representing the simulation results due to the base excitation. This discrepancy is much less pronounced than in Fig. 7 since the voltage is dominated by the velocity of the tip deflection, which is not affected directly by the base motion, unlike the tip position. The primary difference between the two curves in Fig. 8 is the error in predicting when the plucks occur, as discussed in the previous paragraph. This error causes an over-prediction in the number of cycles of free oscillation, but this error is small compared to the duration of the free response between plucks. The initial magnitude of the voltage free response is well predicted, however; this prediction is much more significant in the estimation of voltage given by Eq. (43).

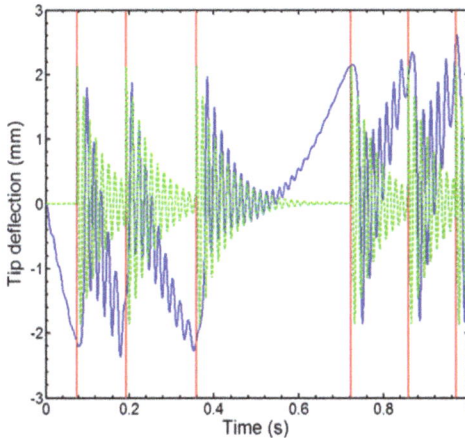

Figure 7. Comparison of simulated tip deflection from EOMs (solid) and a series of plucks, Eq. (42) (dashed). Vertical lines indicate the estimated times of plucking according to Eqs. (40,41).

Fig. 9 shows the comparison between the power frequency transfer functions of the simulation, the approximation by a series of plucks, and the linear system (without magnets). This plot is generated using the same procedure as the one used to produce Fig. 6. The most striking feature of this plot is that the simulation results are seen to converge to the approximation by a series of plucks for low frequencies and converge to the linear system approximation at frequencies approaching the fundamental resonance. As previously mentioned, at frequencies around the fundamental resonance, the beam is not allowed to vibrate freely because the pluck frequency exceeds its natural frequency. Hence, there is no exponential decay in the amplitude of the tip deflection between plucks. In this case, the forced response (i.e. particular solution) dominates the motion, and so the frequency transfer function of the nonlinear system approaches that of the linear system. At low frequencies, the base excitation term becomes negligible, and so the mechanical EOM reduces to Eq. (36), the basis for the series of plucks approximation. In this scenario, the magnetic force drives the excitation of the beam, whereas the inertial force due to the base excitation is negligible. This is manifested in the decrease in the linear response at low frequencies.

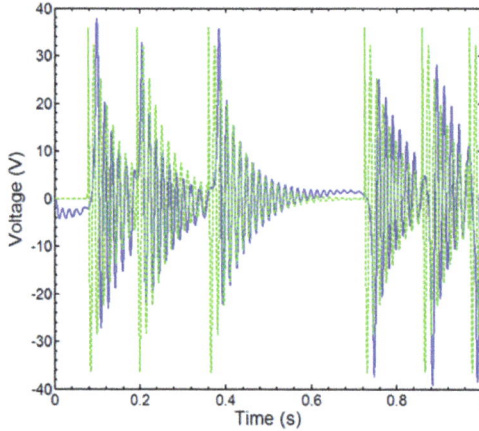

Figure 8. Comparison of simulated voltage from EOMs (solid) and a series of plucks, Eq. (43) (dashed).

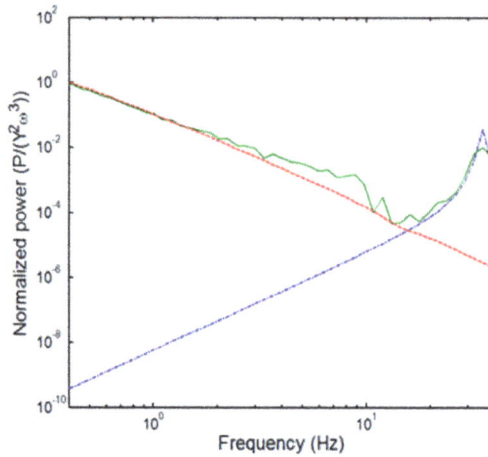

Figure 9. Comparisons of normalized power frequency transfer functions between EOMs (solid), a series of plucks (dashed), and the linear system (dash-dot).

5. Conclusions

This chapter presents an accurate means of approximating the non-linear response of the frequency up-conversion technique as a series of free responses. This simplification is based on the assumption that the base excitation is negligible, and so it only holds at low frequencies compared to the fundamental resonance of the beam. This approximation, however, is useful in the design of energy harvesters utilizing this technique as it enables power to be generated at very low frequencies. This means that the device can be designed for a fundamental frequency much higher than the nominal base excitation frequency, which tends to result in smaller and lighter transducers. At low frequencies, the approximation derived herein is shown to agree well with the simulation results of the full non-linear equations of motion in terms of displacement, voltage, and power harvested. It is confirmed through analysis of the frequency transfer function that the non-linear system converges to the approximation by a series of free responses at low frequencies and to the linear system response at frequencies around the fundamental. Hence, a combination of analytical solutions can be used to predict the energy harvesting performance of this non-linear device in lieu of simulation of the full dynamics equations.

Author details

Adam Wickenheiser
George Washington University, United States

6. References

(1987). IEEE Standard on Piezoelectricity, IEEE, New York, NY.

Anderson, B. & Wickenheiser, A. (2012). Performance analysis of frequency up-converting energy harvesters for human locomotion. Proceedings of SPIE, ISSN 0277-786X.

Arrieta, A. F.; Hagedorn, P.; Erturk, A. & Inman, D. J. (2010). A piezoelectric bistable plate for nonlinear broadband energy harvesting. Applied Physics Letters, Vol. 97, 104102, ISSN 0003-6951.

Challa, V. R.; Prasad, M. G.; Shi, Y. & Fisher, F. T. (2008). A vibration energy harvesting device with bidirectional resonance frequency tenability. *Smart Materials and Structures*, Vol. 17, 015035, ISSN 0964-1726.

Davis, C. L. & Lesieutre, G. A. (1995). A modal strain energy approach to the prediction of resistively shunted piezoceramic damping. Journal of Sound and Vibration, Vol. 184, No. 1, pp. 129–139, ISSN 0022-460X.

Dietl, J. M. & Garcia, E. (2010). Beam shape optimization for power harvesting. Journal of Intelligent Material Systems and Structures, Vol. 21, pp. 633–646, ISSN 1045-389X.

duToit N. E., Wardle, B. L. & Kim, S. (2005). Design Considerations for MEMS-Scale Piezoelectric Mechanical Vibrations Energy Harvesting. Integrated Ferroelectrics, Vol.71, pp. 121–160, ISSN 1058-4587.

Erturk, A. & Inman, D. J. (2008). A Distributed Parameter Electromechanical Model for Cantilevered Piezoelectric Energy Harvesters. *Journal of Vibrations and Acoustics*, Vol. 130, No. 4, 041002, ISSN 1048-9002.

Erturk, A.; Renno, J. M. & Inman, D. J. (2009). Modeling of piezoelectric energy harvesting from an L-shaped beam-mass structure with an application to UAVs. Journal of Intelligent Materials Systems and Structures, Vol. 20, pp. 529–544, ISSN 1045-389X.

Inman, D. J. (2007). Engineering Vibration (3rd), Pearson, ISBN 0-13-228173-2, Upper Saddle River, NJ.

Karami, M. A. & Inman, D. J. (2011). Electromechanical Modeling of the Low-Frequency Zigzag Micro-Energy Harvester. Journal of Intelligent Material Systems and Structures, Vol. 22, No. 3, pp. 271–282, ISSN 1045-389X.

Lefeuvre, E.; Badel, A.; Benayad, A.; Lebrun, L.; Richard, C. & Guyomar, D. (2005). A comparison between several approaches of piezoelectric energy harvesting. Journal De Physique IV, Vol. 128, pp. 177–186, ISSN 1155-4339.

Leland, E. S. & Wright, P. K. (2006) Resonance tuning of piezoelectric vibration energy scavenging generators using compressive axial preload. Smart Materials and Structures, Vol. 15, No. 5, pp. 1413–20, ISSN 0964-1726.

Liao, Y. & Sodano, H. A. (2008). Model of a single mode energy harvester and properties for optimal power generation. Smart Materials and Structures, Vol. 17, No. 6, 065026, ISSN 0964-1726.

Lin, Y. K. (1967). Probabilistic Theory of Structural Dynamics, McGraw-Hill, ISBN 0-88-275377-0, New York, NY.

Mann, B. P. & Sims, N. D. (2009). Energy harvesting from the nonlinear oscillations of magnetic levitation, Journal of Sound and Vibration, Vol. 319, pp. 515-530, ISSN 0022-460X.

Moon, F. C. (1978). Problems in magneto-solid mechanics, Mechanics Today, Vol. 4, pp. 307-390.

Morris, D. J.; Youngsman, J. M.; Anderson, M. J. & Bahr, D. F. (2008). A resonant frequency tunable, extensional mode piezoelectric vibration harvesting mechanism, Smart Materials and Structures, Vol. 17, No. 6, 065021, ISSN 0964-1726.

Murray, R. & Rastegar, J. (2009). Novel Two-Stage Piezoelectric-Based Ocean Wave Energy Harvesters for Moored or Unmoored Buoys. Proceedings of SPIE, ISSN 0277-786X, San Diego, CA, March, 2009.

Oguamanam, D. C. D. (2003). Free vibration of beams with finite mass rigid tip load and flexural-torsional coupling. International Journal of Mechanical Sciences, Vol. 45, pp. 963–979, ISSN 0020-7403.

Porat, B. (1994). Digital Processing of Random Signals, Prentice-Hall, ISBN 0-48-646298-6, Englewood Cliffs, NJ.

Ramlan, R.; Brennan, M. J.; Mace, B. R. & Kovacic, I. (2010). Potential benefits of a non-linear stiffness in an energy harvesting device. Nonlinear Dynamics, Vol. 59, pp. 545–558, ISSN 0924-090X.

Rastegar, J. & Murray, R. (2008). Novel Two-Stage Electrical Energy Generators for Highly Variable and Low-Speed Linear or Rotary Input Motion. Proceedings of ASME, ISBN 9780791843260.

Reissman, T.; Dietl, J. M. & Garcia, E. (2007). Modeling and Experimental Verification of Geometry Effects on Piezoelectric Energy Harvesters. Proceedings of 3rd Annual Energy Harvesting Workshop, Santa Fe, NM, February, 2007.

Reissman T.; Wolff, E. M. & Garcia, E. (2009). Piezoelectric Resonance Shifting Using Tunable Nonlinear Stiffness. Proceedings of SPIE, Vol. 7288, 72880G–1, ISSN 0277-786X.

Roundy, S.; Leland, E. S.; Baker, J.; Carleton, E.; Reilly, E.; Lai, E.; Otis, B.; Rabaey, J.M.; Wright, P.K. & Sundararajan, V. (2005). Improving power output for vibration-based energy scavengers. Pervasive Computing, Vol. 4, No. 1, pp. 28–36, ISSN 1526-1268.

Roundy, S.; Wright, P. K. & Rabaey, J. (2003). A study of low level vibrations as a power source for wireless sensor nodes. Computer Communications, Vol. 26, No. 11, pp.1131–1144, ISSN 0140-3664.

Shu, Y. C. & Lien, I. C. (2006). Analysis of power output for piezoelectric energy harvesting systems. Smart Materials and Structures, Vol. 15, No. 6, pp. 1499–1512, ISSN 0964-1726.

Sodano, H. A., Park, G. & Inman, D. J. (2004). Estimation of Electric Charge Output for Piezoelectric Energy Harvesting. Strain, Vol. 40, No. 2, pp. 49–58, ISSN 0039-2103.

Söderkvist, J. (1990). Electric Equivalent Circuit for Flexural Vibrations in Piezoelectric Materials. IEEE Transactions on Ultrasonics, Ferroelectrics, and Frequency Control, Vol. 37, No. 6, pp. 577–586, ISSN 0885-3010.

Stanton, S. C., McGehee, C. C., & Mann, B. P. (2010). Nonlinear Dynamics for Broadband Energy Harvesting: Investigation of A Bistable Piezoelectric Inertial Generator. Physica D, Vol. 239, pp. 640–653, ISSN 0167-2789.

Tieck, R. M.; Carman, G. P. & Lee, D. G. E. (2006). Electrical Energy Harvesting Using a Mechanical Rectification Approach. Proceedings of IMECE, pp. 547–553, ISBN 0-79-183790-4.

Wickenheiser, A. M. (2011). Design Optimization of Linear and Nonlinear Cantilevered Energy Harvesters for Broadband Vibrations. Journal of Intelligent Material Systems and Structures, Vol. 22, pp. 1213–1225, ISSN 1045-389X.

Wickenheiser, A. M. (2012). Eigensolution of piezoelectric energy harvesters with geometric discontinuities: Analytical modeling and validation. Journal of Intelligent Material Systems and Structures, published online, ISSN 1045-389X.

Wickenheiser, A. & Garcia, E. (2010a). Design of energy harvesting systems for harnessing vibrational motion from human and vehicular motion. Proceedings of SPIE, ISSN 0277-786X, San Diego, CA, March, 2010.

Wickenheiser, A. M. & Garcia, E., (2010b). Broadband vibration-based energy harvesting improvement through frequency up-conversion by magnetic excitation. Smart Materials and Structures, Vol. 19, No. 6, 065020, ISSN 0964-1726.

Wickenheiser, A. M. & Garcia, E. (2010c). Power Optimization of Vibration Energy Harvesters Utilizing Passive and Active Circuits. Journal of Intelligent Materials Systems and Structures, Vol. 21, No. 13, pp. 1343–1361, ISSN 1045-389X.

Chapter 11

Microscale Energy Harvesters with Nonlinearities Due to Internal Impacts

Cuong Phu Le and Einar Halvorsen

Additional information is available at the end of the chapter

1. Introduction

Energy harvesting from mechanical vibrations is one among several alternatives currently considered as power sources for wireless devices. The main motivation is the prospect of eliminating maintenance or increasing maintenance intervals by providing a means for recharging batteries or replacing batteries altogether. Energy harvesting can also be an enabling technology for applications where operating conditions, e.g. temperature, inhibit use of batteries. The prospect of reducing system size can also be a factor of interest. Vibration energy harvesting is therefore a topic of great interest in the scientific community [1-6], especially regarding miniaturized devices. For macro scale devices, commercial products have already emerged [7].

A vibration energy harvester is usually a spring-mass-damper system with a transducer that is continuously driven by the relative motion of the mass with respect to a device frame. The transducers are typically one of the three main types: electromagnetic, piezoelectric or electrostatic [1-2, 8-14]. For small scale systems, vibration energy harvesters face at least two fundamental obstacles. Reduced size necessarily means reduced mass, meaning reduced output power in an inertially driven device. Furthermore, the smaller harvesters have smaller space available for proof mass motion which again limits the distance over which work can be done.

In practical generators, mechanical end-stops are intentionally designed in order to confine the displacement of the inertial mass to the finite die dimension and to avoid spring fracture or degradation of material properties. When the acceleration amplitude is sufficient for proof-mass impacts on end-stops to occur, non-linear effects such as the jump phenomenon in the displacement vs. frequency response appear. Even though this behavior can be exploited to extend device bandwidth, operating a conventional harvester in this regime has the considerable disadvantage that the output power saturates at high excitation levels and therefore the effectiveness of the device decreases. This saturation is quite generic and has been reported for a variety of devices [15-20].

This chapter is concerned with the extent to which the internal impacts on these end-stops can be exploited by making transducing end-stops. Several prototypes utilizing impact principles in macroscale piezoelectric devices have been presented [21-28]. Here we consider microscale electrostatic energy harvesters with two types of transducers, main transducer and secondary end-stop transducers. At sufficiently strong excitations, the impact of the proof-mass onto the end-stops actuates the secondary transducer and thereby harvests the excess kinetic energy of the proof mass. Therefore, the device provides power through two states as excitation strength increases, a first stage with only primary transducer output, and a second stage with output from all transducers.

For a velocity damped generator, the end-stop limit is reached when $AQ=Z_L\omega_0^2$ where A is the package acceleration, Q is the total (loaded) quality factor, ω_0 is the angular resonant frequency and Z_L is the maximum displacement. In a previous work [29], we demonstrated the concept on an open device with a relatively high mechanical Q of about 200. Here, we demonstrate the concept on an encapsulated device with a rather low Q of about 4. As in [29], we compare to a reference device of the same die dimensions.

Section 2 of this chapter details the motivation and working principle for the impact-based electrostatic device. The MEMS-implementation of the concept, made in the Tronics MPW foundry process [31], is described in detail in Section 3, modelled in Section 4 and characterized in Section 5.

2. Device principles

A schematic model of a traditional harvester is shown in Figure 1. With such a design, the typical behavior in frequency sweeps is a clipping of the resonance peak and the occurrence of a jump phenomenon on the high frequency side of the clipped peak. With increasing amplitude, the output power eventually saturates, at least approximately. These effects have been observed in several devices, e.g. in a mesoscale electromagnetic harvester by [16], a mesoscale piezoelectric harvester [18-19] and a microscale electrostatic harvester [17]. Some examples of measured and simulated characteristics of a microscale electrostatic energy harvester from [30] are shown in Figure 2 which displays "clipping" of the response and extended up-sweep bandwidth, and Figure 3 which displays saturation.

The clipping of the response in Figure 2 and the saturation in Figure 3 are direct negative consequences of the displacement limit. Whether the end-stop impacts are elastic or give loss of kinetic energy, is not significant for the output power when the vibrations are sinusoidal and at the resonant frequency [19]. Loss at end-stop impacts mainly affects the phase relationship between the driving force and the displacement. This has consequences for the value of the jump-down frequency in the up-sweep (at about 1450 Hz in Figure 2) and the details of displacement waveform. The displacement waveform may even show period doubling or chaotic-like behaviour without significant deviation from the saturation characteristic in Figure 3, see [32]. If we are mainly concerned with vibrations at the resonant frequency, we are then free to design the end-stop with any degree of loss that we deem suitable without compromising the output power performance.

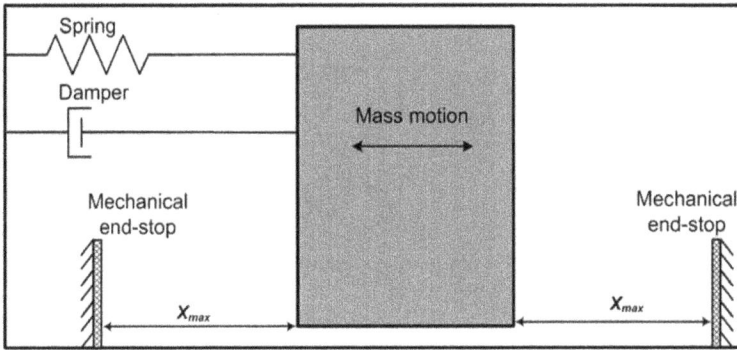

Figure 1. A schematic illustration of typical energy harvesters including a spring-dashpot mass system with use of mechanical end-stops to limit mass motion

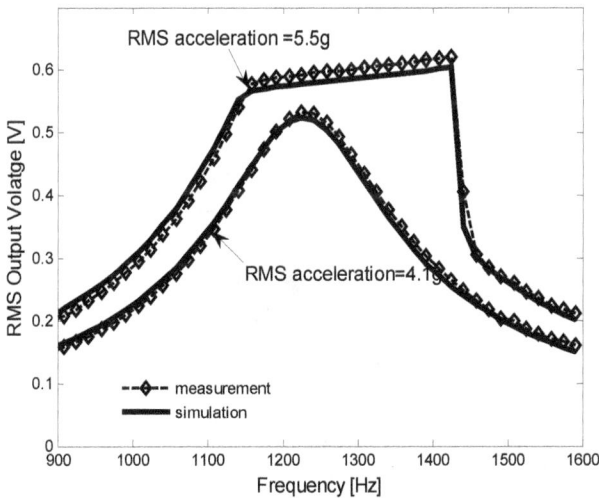

Figure 2. Up-sweep frequency response of RMS output voltage for different acceleration levels at bias voltage V_b=30V. From [30].

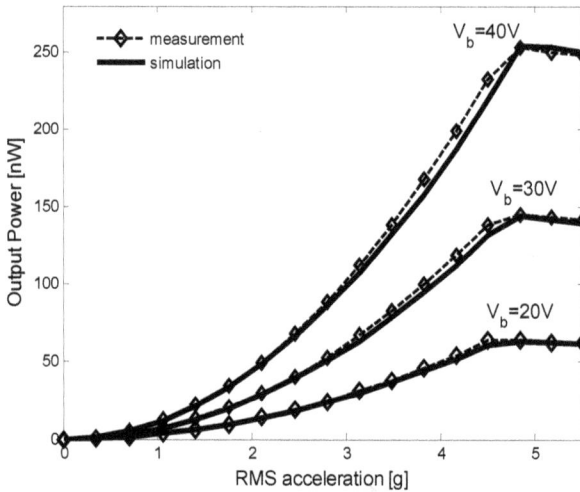

Figure 3. Output power versus acceleration amplitude for different bias voltages. From [30].

The observation that end-stop loss is not important suggests that it can be beneficial to design end-stops that are also transducers. The concept is illustrated by velocity damped end-stops in Figure 4. If these secondary transducers can scavenge significant proof-mass kinetic energy at each impact and convert it to electrical energy, we obtain power in addition to that already available from a primary transducer that will be present and associated with the proof-mass motion anyway. The questions are then how these transducing end-stops can be made and, since some chip real estate must be allocated for them, if this approach has any advantages over using the entire area for a conventional device.

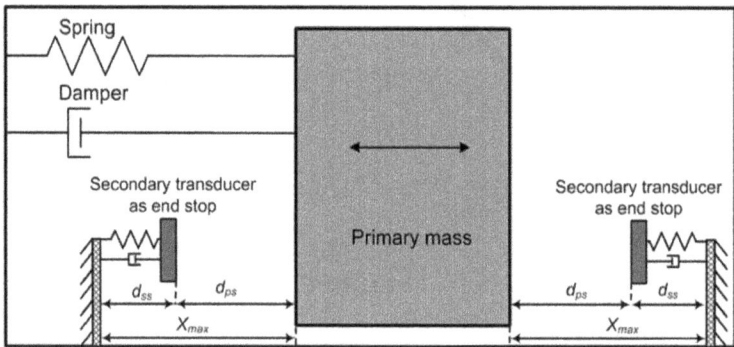

Figure 4. A schematic illustration of device concept with use of end-stops as additional transducers

3. MEMS devices

Here we consider a MEMS realisation of a device based on internal impacts as motivated in the previous section.

3.1. Impact device

Figure 5 shows an impact-device design. There are three independently suspended structures that constitute the device: one primary mass with its electrostatic transducer (ET1) and two secondary electrostatic structures with their own transducers (ET2) acting as end-stops to prevent the ET1 proof mass motion. The ET1 is an ordinary comb-drive structure driven by a movable proof mass m_p attached to four linear-springs with total stiffness k_p and a corresponding damping coefficient b_p. For in-plane motion of the primary mass, the output power is scavenged by the ET1 transducer which has two overlap-varying capacitances with opposite phase. The ET1 transducer is characterised by a capacitor finger length l_p, a capacitor finger width w_p, gap between fingers g_p, an nominal finger overlap x_{0p} and N_p fingers on each side.

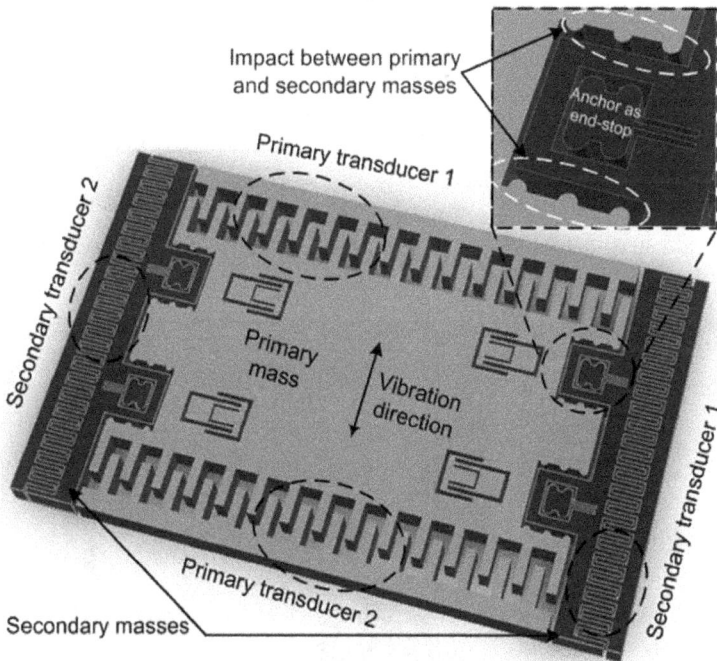

Figure 5. Schematic layout of the impact device with additional secondary electrostatic transducers functioning as end-stops for the primary mass

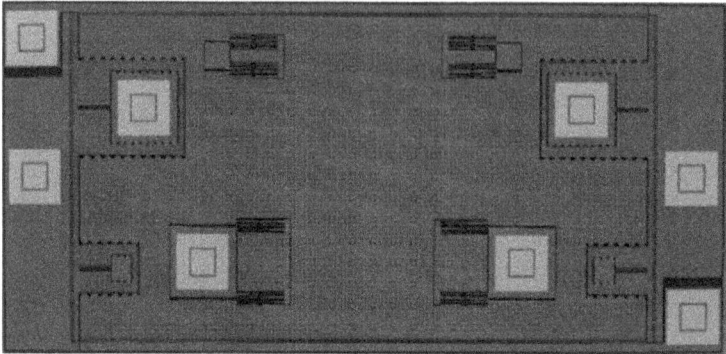

Figure 6. A total view of the impact device with primary and secondary electrostatic transducers (Photograph: Tronics Microsystems S.A.)

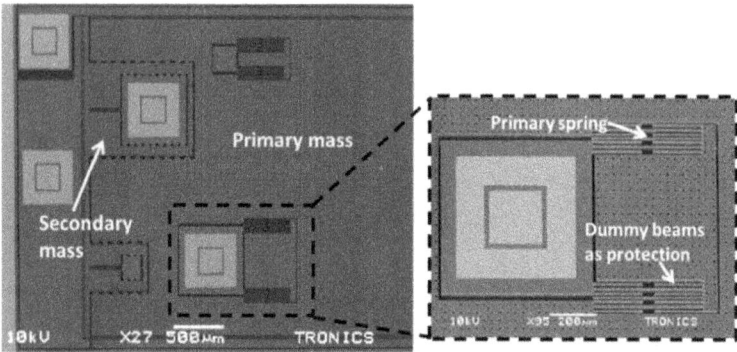

Figure 7. A close-up view of the primary and secondary transducers of the impact device (Photograph: Tronics Microsystems S.A.)

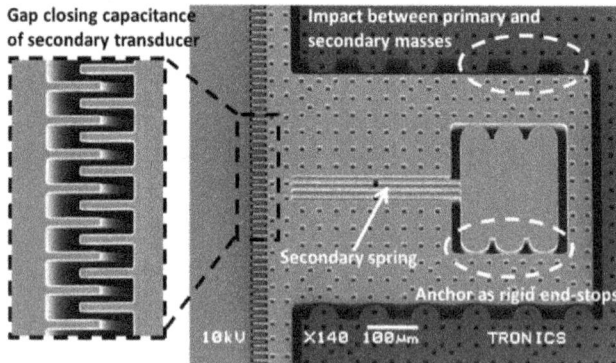

Figure 8. Secondary spring and gap-closing transducer of the secondary structure (Photograph: Tronics Microsystems S.A.)

The ET2 transducer has a mass m_s and a mechanical damping b_s. The ET2 suspension is designed as two single beams with a total spring stiffness of k_s, giving much larger resonant frequency than that of the ET1. The ET2 uses a gap varying capacitance with capacitor finger length l_s, finger width w_s, finger overlap x_{0s} and a nominal capacitor gap g_s. The number of fingers is N_s for each ET2 electrode. Rigid end-stops are used to limit the ET2 motion under overload conditions. The gap varying transducer was chosen for ET2 in order to obtain a large capacitance variation for a small displacement.

The ET1 and ET2 are accelerated in the same direction. Assuming negligible inertial actuation of the secondary structure, the impact between the primary and secondary masses occurs when the displacement amplitude of the primary mass reaches the limit d_{ps}, exciting the secondary transducers to generate extra output power. The maximum displacement amplitude of the secondary mass is d_{ss}. The impact is a contact of bumps on flat surface designed on the mass shape. The cylindrical bumps have semi-cylindrical geometry with radius R. The modelling of this kind of structures was investigated in [30].

The die is 8×4mm^2 and is fabricated in the Tronics MPW (multi-project wafer) service with high aspect ratio micromachining of the 60μm thick device layer of Silicon-on-Insulator (SOI) wafers [31]. Figure 6 shows the full view of the device. The ET2 mass m_s is significantly smaller than the ET1 mass m_p, $m_p=14m_s$. Effort is made to utilize the available area. Placing the supports within the area of the proof mass, makes sure a minimum area is wasted so the proof mass can be as large as possible while leaving the entire length of proof mass available for the comb-drives.

Figure 7 shows a close-up view of the primary and secondary masses. The ET1 proof mass is attached to four springs. The springs in this device are designed as folded flexures with released stress in the axial direction, resulting in the linear beams for transverse motion. The ET2 spring design makes use of two single beams separated by a distance of 2.5mm, giving linear behavior within the ET2-structure travel length. In order to secure predictable beam widths, protection beams oriented in parallel with the spring beams are included to reduce over-etching of the spring beam during fabrication. With this counter measure, we expect the beam cross-section to be closer to the ideal rectangular shape, and therefore its stiffness to be close to the design value. The measured resonance frequency deviates approximately 1.5% from the design value.

There are four metal pads on anchors: two placed on the ET1's anchors and two deposited on the ET2's anchors. They connect to voltage sources used for external biasing in the experiments. Four remaining metal pads are placed on the fixed electrodes to connect the external load resistances.

Figure 8 presents details of the ET2. We see the gap-closing transducer and the bump geometry for the contact regions between the ET1 and ET2 structures. The spring anchors of the ET2 structure also function as rigid end-stops that restrict maximum displacement of both ET1 and ET2 structures to avoid contact between fixed and counter electrodes. All of the device parameters for the ET1 and the ET2 are listed in Table 1.

Parameters	Primary structure	Secondary structure
Die dimensions	8×4mm²	
Device thickness, t	60µm	
Length of capacitor fingers, l_p, l_s	25µm	30 µm
Width of capacitor fingers, w	4µm	4µm
Gap between capacitor fingers, g_p, g_s	3.0µm	3.5µm
Number of capacitor fingers on each electrode, N_p, N_s	416	225
Nominal capacitor finger overlap, x_{0p}, x_{0s}	10µm	25 µm
Length of spring	500µm	350µm
Width of spring	6.5µm	5.2µm
Distance between primary and secondary masses, d_{ps}	4.0µm	
Distance between secondary mass and rigid end stops, d_{ss}	3.0µm	
Bump radius, R	30µm	

Table 1. Parameters of the impact device: primary and secondary electrostatic transducers

3.2. Reference device

Figure 8 shows a view of the reference device with its in-plane overlap-varying transducer. The transducer is similar to the primary transducer of the impact device. Both device prototypes have the same chip dimension. The reference transducer has a larger area for the proof mass and a slightly higher transducer capacitance than the ET1. This is due to more space being available within the same chip real-estate when there are no transducing end-stops. The reference proof mass is suspended by four folded flexure beams. The beams are connected to fours anchors acting as rigid end-stops to confine maximum displacement of the proof mass. This device was described in detail in [30] where end-stop modeling was studied.

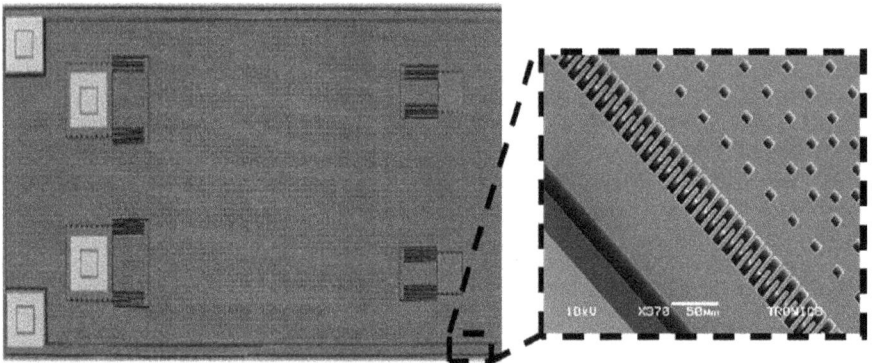

Figure 9. A view of the reference device with the same die dimension [30] (Photograph: Tronics Microsystems S.A.)

The reference transducer is also biased externally. The output voltage is simply connected to load resistance via the metal pads deposited on the fixed electrodes. Further details of the reference-device geometry can be found in [30].

4. Modelling

As a check that the device operates according to our understanding, we will compare measurements to simulation. At the lumped-model level, the device dynamics is governed by a few nonlinear differential equations that can be solved by a variety of numerical tools. We prefer to use a circuit simulator as a solver, i.e. LT-SPICE, and therefore need to formulate the dynamics as an equivalent circuit. The overall scheme of the modelling is the same as we previously used for our previous MEMS devices [30]. Special features here are that there are 3 mechanical degrees of freedom in the impact device, the proof mass position and the position of each of the two secondary structures, and that the impacts are very crucial for the operation.

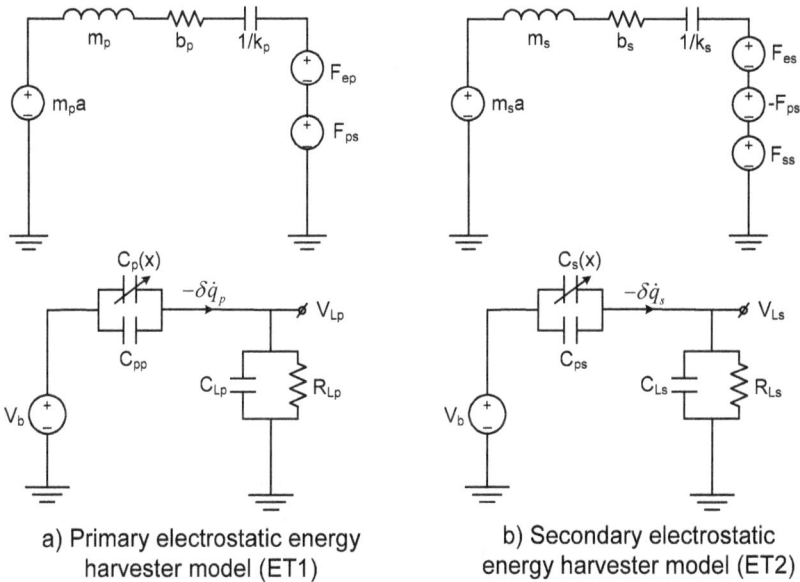

a) Primary electrostatic energy harvester model (ET1)

b) Secondary electrostatic energy harvester model (ET2)

Figure 10. Lumped model of the impact device: a) primary electrostatic transducer (ET1) and b) secondary electrostatic transducer (ET2) in one port

The device is modeled as showed in Figure 10 which gives equivalent circuits for the mechanical and electrical parts of the primary structure and of a secondary structure. The proof mass displacement of the ET1 and ET2 are characterized by two variables x_p and x_s giving the displacement from the nominal position. An impact between the primary and secondary masses takes place for the relative displacement beyond the limit of d_{ps}. Similar to the reference device, the ET2 design also has its own rigid end-stops to prevent its proof mass motion from extremely high acceleration which probably causes collapsing effects. The rigid end-stops are engaged for secondary displacements larger than a maximum distance d_{ss}. The force F_{ss} between the ET2 and the rigid end-stops are modeled using behavioral voltage sources as described in [30]. All model parameters of the impact device are listed in Table 2. Both ET1 and ET2 have the same bias voltage V_b. Due to the similar design of the reference device and the primary ET1, their optimal load resistances are almost equal. For simplicity the ET2s are also connected to the same load resistance R_L as the reference device and ET1, giving a straightforward later comparison between the outputs of the two devices.

Parameters	Primary ET1	Secondary ET2
Inertial proof mass, m_p, m_s	2.1mg	0.15mg
Spring stiffness, k_p, k_s	115.1N/m	29.5N/m
Damping coefficient, b_p, b_s	4.0e-3Ns/m	0.2e-3Ns/m
Nominal variable capacitance, C_{0p},C_{0s}	1.3pF	1.6pF
Parasitic capacitance, C_{pp}, C_{ps}	17.9pF	6.0pF
Load resistance, R_{Lp}, R_{Ls}	4.9MΩ	4.9MΩ
Load parasitic capacitance, C_{Lp}, C_{Ls}	4.2pF	2.0pF

Table 2. Model parameters of the impact device: primary and secondary electrostatic transducers

5. Measurements

Figure 11 shows the frequency response of the impact device compared with the reference device response for an RMS acceleration of 0.71g and a bias voltage V_b=7V in linear regime. At the small acceleration level, the primary mass motion is below limit and then there is no impact between the masses. The output power of the impact device is mainly from the ET1 transducers. The simulation results fit well to the measured results. The ET1 behaves similarly to the reference device. In design, the resonance frequency of the ET1 is the same as that of the reference device, but due to the over-etching effects, the ET1 resonance frequency is slightly smaller, about 1168Hz. The output power of the reference device is bigger than that of the ET1 at the same frequency. For example at the resonance frequency, the ET1 output power is 0.06nW, about three times less than the value of the reference device. The lower output power of the primary transducers originates from the smaller primary mass m_p<m and the smaller transducer capacitance C_{0p}<C_0. This is due to the area sacrificed for the secondary transducers in the impact design.

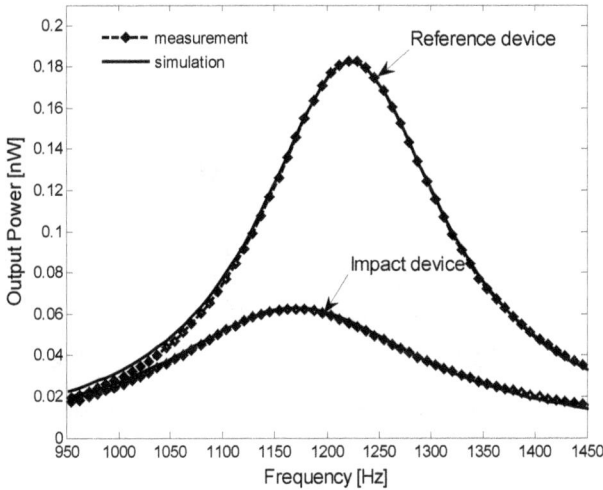

Figure 11. Frequency response of the impact device compared to the reference device for bias voltage V_b=7V and RMS acceleration of 0.71g

Fig. 12 shows the output power for frequency up-sweeps for each transducer in the reference and impact devices at A_{rms}=5.5g, which is sufficient to cause impacts between the primary and secondary masses. Compared with the reference output power, the primary output power is still smaller around the resonant frequency. The output power of the secondary transducer is significant in a frequency range from 1.15kHz to 1.30kHz. For example, at a frequency f=1.22kHz, the secondary output power is 108nW, but the output power of the reference and primary transducers is only 3.0nW and 1.4nW respectively. The energy from the impact is effectively utilized by the high transduction of the secondary transducers in the impact device. The nonlinear effect of the rigid end-stops is evident in the reference output response as saturation of the output power and occurrence of the jump phenomenon. The impacts have no performance benefit beyond up-sweep bandwidth enhancement in the reference device. Performance of the impact device is also modeled and simulated. The simulation results capture the main features of the measurements and thereby confirm that the essential mechanisms in the device have been identified.

A wider response bandwidth is obtained in simulation for the secondary transducer, while the primary transducer response behaves qualitatively like the measured result. The main differences between the measured and simulated results can be explained from the modeling of the impacts. We have seen in the simulations that when varying the loss in the impact model, the bandwidth is affected so inaccuracy in the loss representation can be at least partly responsible for the discrepancy. In addition, the design values for the device geometry have been used in the model and therefore small deviations in the distance of travel before impact could influence the impact events and thereby the bandwidth.

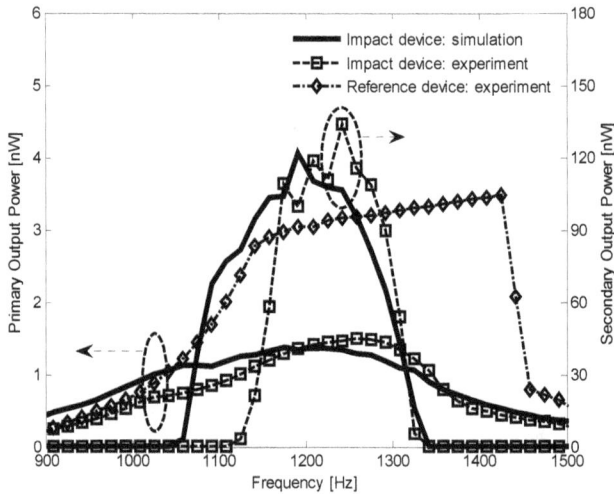

Figure 12. Output power frequency responses of the primary and secondary transducers in the impact device compared to the output from the reference device at RMS acceleration of 5.5g and bias voltage V_b=7V

Figure 13 compares the output powers of the reference and impact devices under bias voltage V_b=7V and at their resonance frequencies. For the impact device, the total output power is the sum of the primary and secondary output powers. For small accelerations, the primary mass does not impact on the secondary mass. The total output power is only contributed from the primary transducer. As a result, the total output power of the impact device is less than that of the reference device. For RMS accelerations larger than 3.5g, the primary mass begins to impact on the secondary structure. The secondary output voltage is dominant for RMS accelerations greater than 4.5g, giving a total output power significantly higher than that of the reference device. For example, the total output power of the impact device is approximately 200nW at an RMS acceleration of 5.5g, 33 times greater than RMS output power of the reference device which is 6.1nW. Further increase of the acceleration amplitude causes the secondary structure's motion to be limited by its own rigid end-stops. In this case, the impact device saturates at much higher output power level than the reference device does. Since the proof-mass displacement constraint is the same for both devices and the masses are about the same, this means that the secondary transducers have provided a dramatic increase in the harvester effectiveness.

Figure 14 illustrates the frequency response of the total output power of the impact device and the reference device in frequency up- and down sweeps at an acceleration amplitude of 5.5g and bias voltage V_b=7V. The total output power of the impact device is considerably higher than that of the reference device in the frequency range of secondary-transducer activation. There is no jump phenomenon or hysteresis in the frequency response of the impact device at such accelerations. This differs from the reference device.

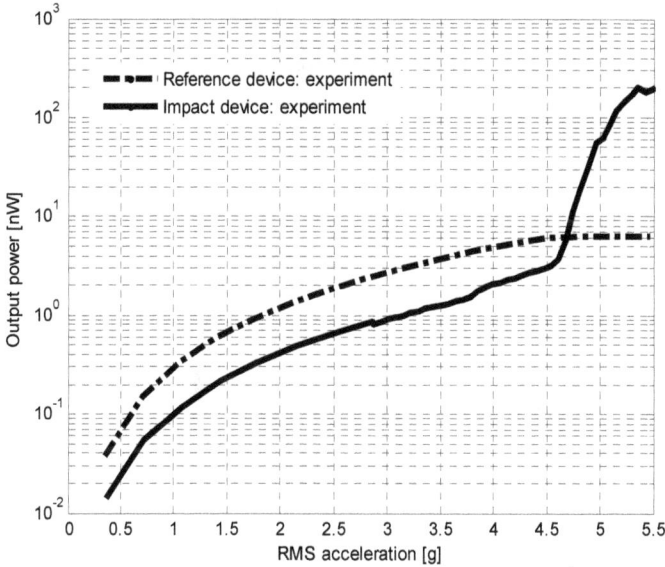

Figure 13. Comparison between measured output power of the reference and impact devices for bias voltage V_b=7V at resonance frequencies

One notable difference between this encapsulated low-Q device and the unpackaged high-Q device presented earlier [29] is the large difference in the RMS acceleration that is required for end-stop engagement. It is rather obvious that the difference in Q is responsible for this, but for both devices a subsequent additional increase in RMS acceleration is necessary before the output of the secondary transducers become appreciable, in the present device from 3.5g to 4.5g. For future designs, measures should be taken to ensure that the end-stop transducers are effective already when actuated a small distance so as to narrow down this range of RMS accelerations.

The merits of the impact-device concept can be quantified through the figure of merit *energy harvester effectiveness* as defined in [1]. Some example values are given for the present packaged device and a previous unpackaged device in Table 3. For conventional energy harvesters operating in the linear regime with displacement less than the limit X_{max}, the effectiveness is proportional to the acceleration amplitude. Then, it degrades as the acceleration amplitude increases beyond the value needed to reach the maximum amplitude X_{max}. This behaviour is displayed by the reference-device values in the table. With the active end-stops, the extra power improves the harvester effectiveness under displacement-limited operation. For the packaged devices presented in this book chapter, the effectiveness of the impact device is 4.25%, while this value is only of 0.11% for the reference device in the impact regime. The high mechanical quality factor Q in the previous unpackaged devices gives an even larger effectiveness up to 23.12%. Microscale energy harvesters have typical effectiveness in the range from 1% to 10% and it is lower for smaller displacement limits [1]. The impact devices therefore achieve effectiveness values under displacement-limited operation that are comparable, or even favourable, in comparison with other device prototypes in [9, 15, 17-18, 33-36] with the same scale of the displacement limit. Together these examples show that there is much to be gained from transducing end-stops.

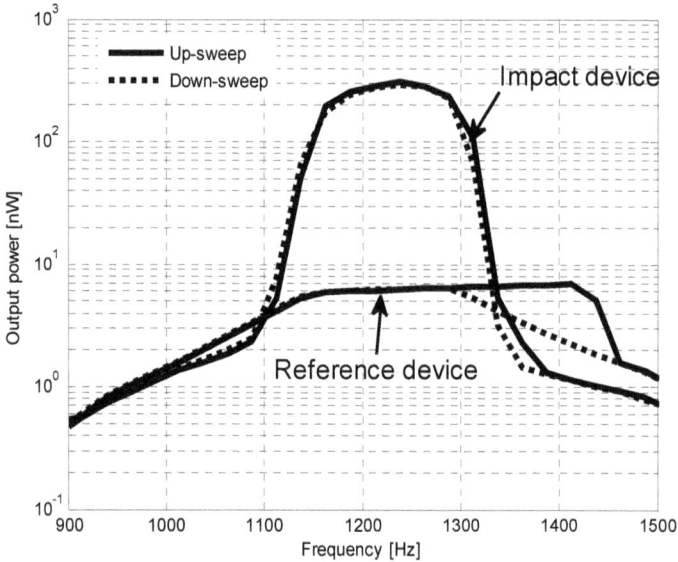

Figure 14. Frequency response of the measured output power of the reference and impact devices for a RMS acceleration of 5.5g and bias voltage V_b=7V

Packaged prototype, X_{max}=7μm			Unpackaged prototype, X_{max}=10μm [29]			
RMS acceleration amplitude [g]		Effectiveness [%]		RMS acceleration amplitude [g]		Effectiveness [%]
		Reference device	Impact device			Reference device	Impact device
Linear regime	2.10	0.074	0.037	Linear regime	0.04	11.18	9.56
	4.19	0.145	0.078		0.06	17.42	14.81
Impact regime	5.15	0.139	3.017	Impact regime	1.76	5.40	14.44
	5.50	0.106	4.249		1.87	5.09	23.12

Table 3. Comparison of harvester effectiveness between the reference and impact devices

6. Conclusion

An electrostatic energy harvester with in-plane overlap-varying transducers on the primary mass and with secondary gap varying transducers as end-stops has been designed, modeled and characterized. The simulations are consistent with the measurement results. The performance was compared with that of a standard in-plane- overlap-varying type device. With the transducing end-stops we have seen that a considerable performance boost is obtained, with output power up to a factor 33 over the reference device, even though the reference device performed a factor of 3.4 better at low acceleration levels. None of the typical jump phenomena were observed in up and down frequency sweeps for this device. The frequency response of the impact device had approximately the same bandwidth as the reference device had on down sweeps.

Author details

Cuong Phu Le and Einar Halvorsen
Department of Micro and Nano Systems Technology,
Faculty of Technology and Maritime Sciences, Vestfold University College, Tønsberg, Norway

Acknowledgement

This work was financially supported by the Research Council of Norway under grant 191282. We thank Prof. Eric Yeatman and Prof. Oddvar Søråsen for useful discussions and suggestions.

7. References

[1] Mitcheson P D, Yeatman E M, Rao G K, Holmes A S, Green T C. Energy harvesting from human and machine motion for wireless electronic devices. Proceedings of the IEEE 2008;96(9) 1457-1486.

[2] Beeby S P, Tudor M J, White N M. Energy harvesting vibration sources for microsystems applications. Measurement Science and Technology 2006;17 175-195.

[3] Roundy S, Wright P K, Rabaey J. A study of low level vibrations as a power source for wireless sensor nodes. Computer Communications 2003;26 1131-1144.

[4] Roundy S, Steingart D, Frechette L, Wright P, Rabaey J. Power sources for wireless sensor networks. Computer Science 2004;2920:1-17.

[5] Starner T, Paradiso J A. Human generated power for mobile electronics. In: Piguet C. (ed.) Low-Power Electronics. Boca Raton FL: CRC Press; 2004. p1-35.

[6] Cantatore E, Ouwerkerk M. Energy scavenging and power management in networks of autonomous microsensors. Microelectronics Journal 2006; 37(12) 1584-1590.

[7] http://www.perpetuum.com/fsh.asp/ (accessed 19 June 2012).

[8] Anton S R, Sodano H A. A review of power harvesting using piezoelectric materials (2003–2006). Journal of Smart Materials and Structures 2007;16 1–21.

[9] Williams C B, Yates R B. Analysis of a micro-electric generator for microsystems. Sensors and Actuators A: Physical 1996;52 8-11.

[10] Mitcheson P D, Miao P , Stark B H, Yeatman E M, Holmes A S, Green T C. MEMS electrostatic micropower generator for low frequency operation. Sensors and Actuators A: Physical 2004;115 523–529.

[11] Naruse Y, Matsubara N, Mabuchi K, Izumi M, Suzuki S. Electrostatic micro power generator from low frequency vibration such as human motion. Journal of Micromechanics and Microengineering 2009;19 094002.

[12] Ferrari M, Ferrari V, Guizzetti M, Marioli D, Taroni A. Piezoelectric multifrequency energy converter for power harvesting in autonomous microsystems. Sensors and Actuators A: Physical 2008;142 329–335.

[13] Koukharenko E, Beeby S P, Tudor M J, White N M, O'Donnell T, Saha C, Kulkarni S, Roy S. Microelectromechanical systems vibration powered electromagnetic generator for wireless sensor applications. Journal of Microsystems Technology 2006;12(10) 1071–1077.

[14] Beeby S P, Torah R N, Tudor M J, Glynne-Jones P, O'Donnell T, Saha C, Roy S. A micro electromagnetic generator for vibration energy harvesting. Journal of Micromechanics and Microengineering 2007;17 1257–1265.

[15] Tvedt L, Blystad L-C J, Halvorsen E. Simulation of an electrostatic energy harvester at large amplitude narrow and wide band vibrations. In: Proceedings of Symposium on Design, Test, Integration, Packaging of MEMS/MOEMS, 9-11 April 2008, Nice, France; 2008.

[16] Soliman M S M, Abdel-Rahman E M, El-Saadany E F, Mansour R R. A wideband vibration-based energy harvester. Journal of Micromechanics and Microengineering 2008;18 115021.

[17] Hoffmann D, Folkmer B, Manoli Y. Fabrication, characterization and modeling of electrostatic micro-generators. Journal of Micromechanics and Microengineering 2009;19 094001.

[18] Blystad L-C J, Halvorsen E. A piezoelectric energy harvester with a mechanical end stop on one side. Journal of Microsystems Technology 2011;17 505-511.

[19] Blystad L-C J, Halvorsen E, Husa S. Piezoelectric MEMS energy harvesting driven by harmonic and random vibrations. IEEE Transactions on Ultrasonics, Ferroelectrics, and Frequency Control 2010;57(4) 908-919.

[20] Liu H, Tay C J, Quan C, Kobayashi T, Lee C. Piezoelectric MEMS energy harvester for low-frequency vibrations with wideband operation range and steadily increased output power. Journal of Microelectromechanical systems 2011;20(5) 1131-1142.

[21] Moss S, Barry A, Powlesland I, Galea S, Carman G P. A broadband vibro-impacting power harvester with symmetrical piezoelectric bimorph-stops. Journal of Smart Materials and Structures 2011;20 045013.

[22] Umeda M, Nakamura K, Ueha S. Energy storage characteristics of a piezo-generator using impact induced vibration. Japanese Journal of Applied Physics 1997, 36:3146–3151.

[23] Xu C-N, Akiyama M, Nonaka K, Watanabe T. Electrical power generation characteristics of PZT piezoelectric ceramics. IEEE Transactions on Ultrasonics, Ferroelectrics, and Frequency Control 1998;45 1065–1070.

[24] Funasaka T, Furuhata M, Hashimoto Y, Nakamura K. Piezoelectric generator using a LiNbO3 plate with an inverted domain. In: Proceeding of Symposium on IEEE Ultrasonics, 5–8 October, Sendai, Japan; 1998.

[25] Yoon S H, Lee Y H, Lee S W, Lee C. Energy-harvesting characteristics of PZT-5A under gunfire shock. Materials Letters 2008;62 3632-3635.

[26] Djugum R, Trivailo P, Graves K. A study of energy harvesting from piezoelectrics using impact forces. The European Physical Journal Applied Physics 2009;48(01) 11101.

[27] Gu L, Livermore C. Impact-driven, frequency up-converting coupled vibration energy harvesting device for low frequency operation. Journal of Smart Materials and Structures 2011;20 045004.

[28] Le C P, Halvorsen E, Søråsen O, Yeatman E M. An electrostatic energy harvester with power-extracting end stops driven by wideband vibrations. In: Proceedings of the PowerMEMS2011 Workshop, 5-18 November, Seoul, Korea; 2011.

[29] Le C P, Halvorsen E, Søråsen O, Yeatman E M. Microscale electrostatic energy harvester using internal impacts. Journal of Intelligent Material Systems and Structures 2012; DOI: 10.1177/1045389X12436739.

[30] Le C P, Halvorsen E. MEMS electrostatic energy harvesters with end-stop effects. Journal of Micromechanics and Microengineering 2012;22 074013.

[31] http://www.tronicsgroup.com/full-service-MEMS-foundry/ (accessed 19 June 2012).

[32] Kaur S, Halvorsen E, Søråsen O, Yeatman E M. Numerical analysis of nonlinearities due to rigid end-stops in energy harvesters. In: Proceedings of the PowerMEMS2010 Workshop, 30 Nov-3 December, Leuven, Belgium; 2010.

[33] Mizuno M, Chetwynd D G. Investigation of a resonance microgenerator. Journal of Micromechanics and Microengineering 2003;13 209–216.

[34] Miyazaki M, Tanaka H, Ono G, Nagano T, Ohkubo N, Kawahara T, Yano K. Electric-energy generation using variable-capacitive resonator for power-free LSI: efficiency analysis and fundamental experiment. In: Proceeding of the 2003 International Symposium on Low Power Electronics and Design, 25-27 August, Seoul, Korea; 2003.

[35] Despesse G, Chaillout J, Jager T, Leger J M, Vassilev A, Basrour S, Charlot B. High damping electrostatic system for vibration energy scavenging. In: Proceeding of the Joint Conference on Smart Objects and Ambient Intelligence, 12-14 October, Grenoble, France; 2005.

[36] Miao P, Mitcheson P D, Holmes A S, Yeatman E M, Green T C, Stark B H. MEMS inertial power generators for biomedical applications. Journal of Microsystem Technology 2006;12 1079–1083.

Chapter 12

Non-Linear Energy Harvesting with Random Noise and Multiple Harmonics

Ji-Tzuoh Lin, Barclay Lee and Bruce William Alphenaar

Additional information is available at the end of the chapter

1. Introduction

Harvesting energy from background mechanical vibrations in the environment has been proposed as a possible method to provide power in situations where battery usage is impractical or inconvenient. The most commonly used method for energy harvesting is to generate power from the vibrations of a piezoelectric material [1-3]; other methods include electromagnetic inductive coupling [4-6] and charge pumping across vibrating capacitive plates [7-10]. It has been shown that a piezoelectric cantilever attached to a vibrating structure can be used to power wireless transmission nodes for sensing applications [9]. In order to generate sufficient power, the frequency of the vibration source must match the resonant frequency of the piezoelectric cantilever. If the source vibrates at a fixed, known frequency, the dimensions of the cantilever, and the proof mass can be adjusted to ensure frequency matching. Many naturally occurring vibration sources do not have a fixed frequency spectrum, however, and vibrate over a broad range of frequencies. Lack of coupling of the piezoelectric cantilever to the off-resonance vibrations means that only a small amount of the available power can be harvested.

Recent reports have shown that the resonant frequency of a simply supported beam [11] or a piezoelectric cantilever [12] can be tuned by applying an axial force. Research also show that the resonant frequency of a cantilever can also be manipulated by applying a transverse force on the cantilever [13,14]. (In all these cases, the cantilevers response remained within the linear regime.) In principle, this effect could be developed into an active tuning scheme which matches the cantilever resonance to the maximum vibrational output of the environment at any particular time. Calculations indicate, however, that the power consumed by active tuning completely offsets any improvement obtained in the scavenging efficiency [15]. More promising are passive tuning schemes in which a fixed force modifies the frequency response of the cantilever beam, without requiring additional power input.

For example, an attractive magnetic force acting above the cantilever beam reduces the spring constant of the cantilever and lowers the resonance frequency [13,14], while an attractive force acting along the axis of the cantilever applies axial tension, and increases the resonance frequency [12]. While this can be used to tune the resonant frequency, there is no increase in output power, and the cantilever motion can even be dampened by the magnetic force and the resulting power output reduced [12,13].

The use of a magnetic force to introduce non-linear oscillation in cantilever motion has recently been reported [16-18]. A pendulum made with piezoelectric material [16] was used to study the energy output under different strengths of random Gaussian noise. An improvement of between 400% and 600% was observed compared to a standard linear oscillator. A piezomagnetoelastic structure [17] with two external magnets was studied, in which chaotic motion was observed outside the resonance frequency. It was further reported [18] that the softening response of a cantilever due to a magnetic attractor expands the response bandwidth and also increases the off resonant amplitude significantly.

Stochastic motions have been long observed with a pendulum in a repulsive magnetic field [19-20] In a generalization effort, the optimal relationship among the physical parameters for a coupling enhancement was provided in [16] [Cottone et al., 2009] using Duffing oscillator. Improvements for the non-linear system have been attributed to an advantage in the amplification of the vibration response from energy harvesters in the stochastic regime [17-18].

Here, we will first demonstrate how this capability can be used to improve power output from a broadband vibration source, having a 1/f frequency dependence (pink noise) [21]. Note that a 1/f vibration spectrum describes a vibration source in which the power spectral density of the vibration is inversely proportional to frequency. Since many naturally occurring vibration sources display a 1/f dependence, this provides evidence that the magnetic coupling could be used for more efficient energy harvesting in practical settings.

The second part of this chapter provides an in-depth study of the response of a magnetically coupled cantilever at different frequencies [22-23]. It is our observation that amplification of the cantilever output occurs not only under stochastic motion but also due to subharmonic and ultraharmonic resonance in the vicinity of the main resonant frequency. The partial solutions of subharmonic and ultraharmonic are intrinsically embedded in the magnetic coupled equation as derived in forced oscillations of weakly nonlinear systems [24]. For a particular weakly coupled cantilever experimented in this paper, maximum output is maintained at the resonant frequency through combination of ultra-harmonic components. In a singly parametric excited scan of voltage production with non-linear piezoelectric cantilever, four distinct types of efficiency improvements are observed, in which the signal is amplified above the linear cantilever operation: (1) ultraharmonic amplification below resonance; (2) stochastic amplifications in multi-frequency and multi-amplitude oscillations; (3) ultra-sub-harmonic amplification at multiple quarter frequencies; (4) sub-harmonic amplification at one-third frequencies. For data analysis, a 1-D non-linear system coupled with piezoelectric charge production is modeled to illustrate the dynamic functions.

2. Non-linear dynamics in Pink noise background

2.1. Experimental setup and vibration backgroud

Figure 1 shows the set-up for the magnetically coupled piezoelectric cantilever measurements. The cantilever is manufactured using commercially available unimorph piezoelectric discs composed of a 0.9 mm thick PZT layer deposited on a 1 mm thick brass shim (APC International, MFT-50T-1.9A1). The disc is cut into a 13 mm wide by 50 mm long strip, and clamped at one end to produce a 44 mm long cantilever. The PZT layer extends 25 mm along the length of the cantilever, and the remainder is brass only. The proof mass (including the magnet and an additional fixture that holds the magnet) weighs 2.4 gm, while the cantilever itself weighs 0.8 gm. The electrical leads are carefully soldered with thin lead wires (134 AWP, Vishay) to the top side of the PZT and the bottom side of the shim [21].

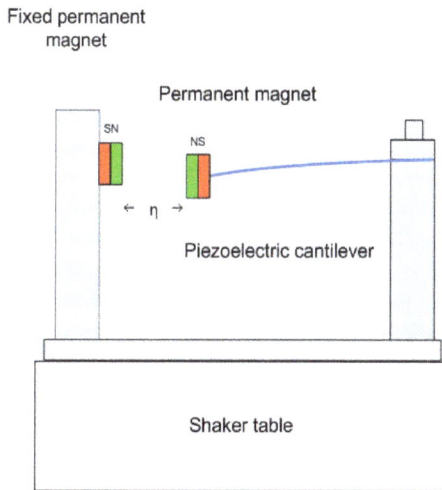

Figure 1. The experimental set-up for the magnetically coupled (non-linear) piezoelectric cantilever. The magnetic force is repulsive and bi-directional.

Vibration is generated by a shaker table (Labwork ET-126) driven by an amplified pink noise source (Labwork Pa-13 amplifier). The pink noise is generated numerically, with amplitude and crest factor set to -4dB and 1.41, respectively. The average shaker table acceleration is 7.5 m/s2, independent of the magnetic coupling. A custom Labview data acquisition program measures output voltage from the cantilever beam and the acceleration from the shaker table, once every second. The voltage peak to peak (Vpp) is measured by an oscilloscope (Agilent 54624A), and the dc voltage is detected with a digital multi-meter (YOGOGAWA 7561). A 5mm diameter round rare earth magnet (Radio Shack model 64-1895) is attached to the vibrating tip of the cantilever beam, while a similar opposing magnet is attached directly to the shaker table frame, with repulsive force. The distance between the magnets is adjusted to 5.5 mm, to make the magnetic force comparable to the spring force of the cantilever.

2.2. Experiment results

The voltage generated by the cantilever in response to the pink noise source is measured using three different circuits, (shown in Figures 2(a), 3(a), and 4(a)). In each case, the output from the coupled cantilever is compared with the output from the same cantilever in the uncoupled situation (with the opposing magnet removed). In Figure 2, the piezoelectric cantilever beam is wired directly to an oscilloscope with a 1 M Ohm input impedance and the peak-to-peak output voltage, Vpp is measured. As shown in Figure 2 (b) the cantilever output is seen to fluctuate as a function of time, reflecting the random nature of the vibrations. For much of the time, the output from the coupled and uncoupled cantilevers is similar. However, occasionally, very large voltage spikes are observed in the output from the coupled cantilever, that are not observed for the uncoupled case. The voltage peak to peak spans to 5.7 V (min. 0.7 V and max. 6.4 V) with the coupled setup and only 2.2 V (min. 0.9 V to max. 3 V) volts with the uncoupled cantilever. The overall RMS powers for the uncoupled cantilever are 3.95 μW and 4.85 μW for the coupled case. The ratio of the maximal voltage output from the coupled to the uncoupled is 2.1.

In Figure 3, the voltage generated by the piezoelectric cantilever beam is rectified, using 0.4 V forward biased diodes, and detected across a 22 μF capacitor and a 1 M Ohm resistor in parallel. As shown in Figure 3(b), the amplitude of the voltage output with this measurement circuit is most of the time higher in the coupled case than in the uncoupled case. This is because the RC decay time of the circuit is larger than the time between the large amplitude deflections of the cantilever. The average voltage measured across the capacitor or the voltage integration over time is approximately 50% higher in the coupled case.

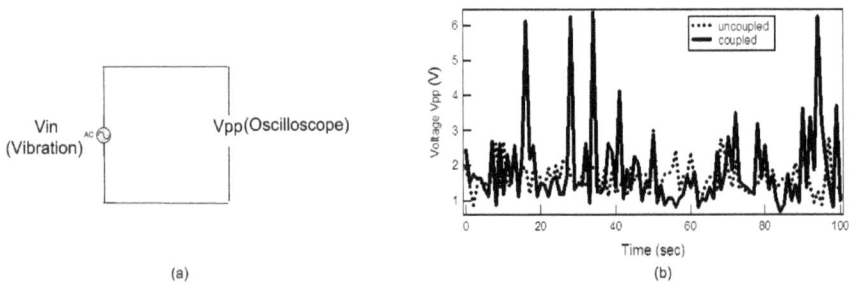

(a)

(b)

Figure 2. (a) The open circuit measurement on V_{PP} directly from the piezoelectric cantilever, and (b) the higher swing voltage reflects the voltage generated by coupling setup with larger cantilever motions.

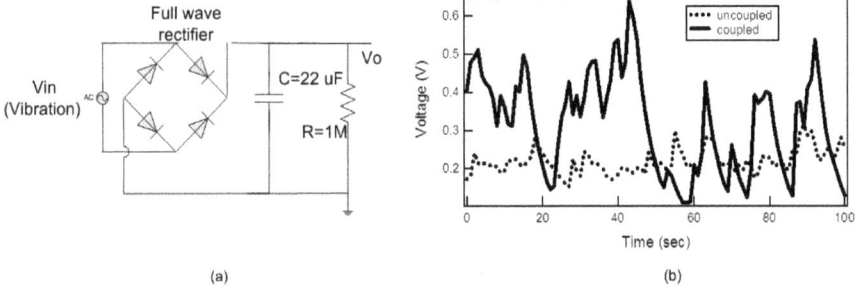

(a)   (b)

Figure 3. (a) The schematic of a rectified circuit with a 1 M Ohm resister, and (b) the fluctuations of the voltage indicate that more power being generated by the magnetic coupled cantilever.

In Figure 4, the rectified voltage is measured directly across the 22 µF capacitor without the 1 M Ohm resistor. As shown in Fig. 4(b), the voltage across the capacitor increases with time, until a maximum charging voltage is achieved. The maximum voltage measured across the capacitor is approximately 50% higher in the coupled case than in the uncoupled case. Note that there is a time delay for the coupled cantilever to achieve a higher voltage than the uncoupled cantilever. This is due to the time passing before the first large amplitude deflection occurs. The random nature of the motion means that this time will vary from run to run, however, on average the coupled cantilever output will be consistently higher than the uncoupled output. Note that in addition to producing more power, the higher voltage output enables circuit operation without a step-up transformer, eliminating the power loss in the transformer.

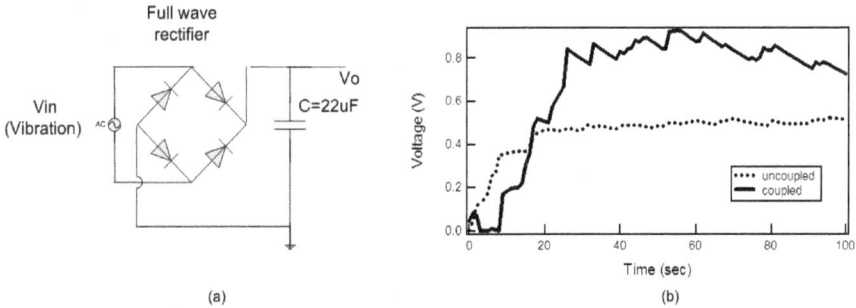

(a)   (b)

Figure 4. (a) The schematic of the storage circuit, and (b) DC voltage output measured on the storage capacitor indicating more charge is stored with the magnetic coupling setup.

2.3. Discussion

It is instructive to compare the force exerted on the cantilever in the coupled and uncoupled cases. To do this, an empirical measure of the magnetic force is obtained using the experimental set-up shown in Figure 5.

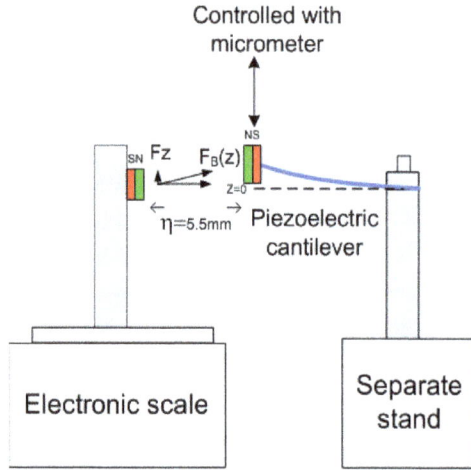

Figure 5. The magnetic force component function, F_z, is determined by the electronic scale versus the manual deflection of the cantilever.

The opposing magnet is mounted onto a measurement scale, and the position of the magnetized cantilever is manipulated by pushing up and down at the end of a cantilever beam, simulating flexure movement. The deflection z is measured using a micrometer, while the reading on the scale provides the force between the two magnets. The details of the force measurments were shown in [22]. Only the magnetic force in the z direction, F_z, contributes to the resultant spring force. At z=0, the force is zero in the z direction because the two magnetic forces only repel each other in the longitudinal direction. F_z increases as the angles between the two magnets increase until the overlap between the two magnets is zero. At this point, F_z decreases with increasing distance because the force is inversely proportional to the distance squared.

The spring force, the magnetic force and the resultant force (spring plus magnetic) are plotted in Figure 6,

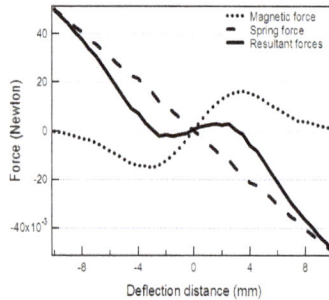

Figure 6. The plot shows the magnitude of the magnetic forces exerted on the cantilever beam, the spring forces and the resultant forces.

The resultant force is significantly reduced compared to the bare spring force near z=0. The coupled system has three equilibrium points where the resultant force is zero, compared to the single equilibrium point of the bare spring force. Because the resultant force in the region of the three equilibrium points is relatively small, transitions between the three points occurs relatively easily. Note that the middle equilibrium is unstable, therefore when the piezoelectric cantilever is set up for the coupling experiment, the cantilever is off the equilibrium point toward ground in static state as shown in Figure 1. In Figure 7 the potential energy is plotted for both the uncoupled and coupled systems. The potential energy is calculated by direct integration of the force with respect to the displacement, z. This gives for the uncoupled case, and for the coupled case. For the coupled case, the resultant potential is raised, with two local minima symmetric to z=0. This double well structure allows easy movement of the cantilever beam even when excited by non resonant forces. Once it passes the local high potential, it drifts to the other side of the balance, resulting in an increased total deflection distance. This can be seen by considering the possible motion of the cantilever beam having a kinetic energy, h, which is large enough to surmount the potential barrier at z=0. With the same random acceleration background the coupled cantilever can travel further distance than the uncoupled one. The voltage output, which depends on the movement of the cantilever, therefore, increases. The ratio of the maximum displacement in the coupled and uncoupled systems determined from Figure 7 is 2.4. This is comparable to the ratio of maximum voltage output in the coupled and uncoupled systems, which was seen in Figure 2 (b), at 2.1.

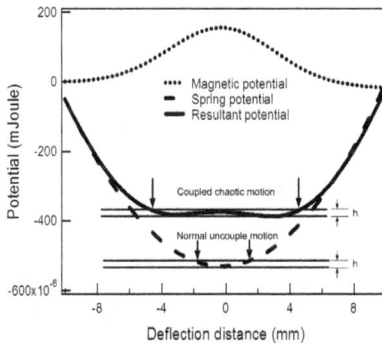

Figure 7. The direct integration from the measured forces function in Fig. 6 leads to the magnetic potential, spring potential and the resultant spring potential. The responding range in the coupled and the uncoupled cantilever is defined by the same potential height, h.

The magnetic coupling (although a passive force requiring no energy) introduces a symmetric force which acts in the opposite direction to the spring force around z=0. Being comparable in magnitude to the spring force, the magnetic force compensates the spring potential, and introduces a double valley in the potential energy profile. Under the influence of the modified spring potential, the magnetically coupled cantilever responds to a random vibration source (like the pink noise) by moving chaotically between the two minima in the potential energy profile. As compared with the non-chaotic motion of the uncoupled cantilever around the single z=0 potential minimum, this produces larger cantilever deflection and more voltage output from the piezoelectric cantilever. The oscillations around the resonance frequency are unstable and chaotic, but persistent. The modified spring potential is higher, and flatter than the bare spring potential, making the magnetic coupled cantilever easier to excite in the random frequency region. The experiments show that the ratio of the open circuit peak to peak voltage output and the potential well are closely related. Future work includes the design and implementation of modified potential wells and further analysis of the gain due to the modified potential wells.

3. Resonance broadening in broad band spectrum

3.1. Experiment setup

The experiment set up is the same as Figure 1. In all measurements, the shaker table acceleration is set to approximately 4.2 m/s² at resonant freqeuncy, and the frequency swept from 0 to 30 Hz in 0.5 Hz steps. The opposing magnet fitted at the free end of the cantilever supplies a symmetrical, repulsive force about the balance of the cantilever during vibration. The horizontal separation between the magnets (designated by η) is adjusted to be approximately between 6 to 6.5 mm. This separation is found to provide the best compensation for the spring force, and makes the effective restoring force as small as possible near the equilibrium point.

3.2. Experiment result with open circuit

Figure 8 shows both the output of the piezoelectric cantilever as a function of shaker table vibration frequency for the linear and non-linear case. The voltage generated by the piezoelectric cantilever beam is directed measured by oscilloscope treated as an open circuit. At the resonance frequency (measured to be 9.5 Hz) the output of the cantilever was 53 V, and the peak height, resonance frequency and line width are all approximately the same for the linear and non-linear states (here linear refers to the non-coupled state, while non-linear refers to the magnetically coupled state). On either side of the main resonance, however, there is additional output observed for the non-linear cantilever, which is not observed in the linear state. As can be seen from a comparison of the linear and the non-linear runs, the overall amplitude profile of the non-linear run is much larger in the sense of a broadband distribution, although there are gaps between peaks in the overall pattern of the non-linear output.

Figure 9 shows the output of both the linear and non-liner cantilever measured as a function of time at selected frequency to illustrate the comparison of the linear and non-linear dynamics. The voltage output of the non-linear cantilever evolves with frequency, while being amplified close to the resonance frequency. The spectrum shows a variety of amplified motions and harmonics. For example, at a driving frequency as low as 6.5 Hz (between 6-7.5Hz) (Figure 9(a)) both the linear and non-linear cantilever motions follow the vibrations of the shaker table, producing periodic oscillations. The amplitude of the oscillations for the non-linear cantilever is 5 times larger than those for the linear cantilever, however. At the resonant frequency (Figure 9(b)) both linear and non-linear cantilevers oscillate at the driving frequency with equal amplitudes. At 13 Hz (Figure 9(c)) the linear cantilever motion continues to follow the vibrations of the shaker table, producing low amplitude periodic oscillations. The non-linear cantilever motion is aperiodic and has a magnitude which is on average 3 times larger than that of the linear cantilever. At 16 Hz (Figure 9 (d)) the non-linear cantilever produces a 3 times larger peak to peak amplitude than the linear cantilever, and shows multiple and periodic "half-way" vibrations. At 20Hz (Figure 9 (e)) the non-linear cantilever shows a 5 times larger amplitude at the frequency of 6.7Hz than the linear output at 20 Hz.

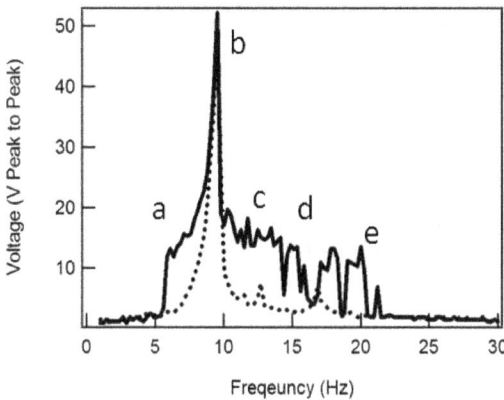

Figure 8. The voltage output (peak to peak) of the piezoelectric cantilever measured as a function of frequency (dash line for linear and solid line for non-linear state).

Note should be taken that there are two unexpected small peaks at 12.5 Hz and 17 Hz for the linear response. The peaks at 12.5 Hz and 17 Hz on the experiment data come from the torsion and standing wave oscillations. It is the result of how the piezoelectric cantilever was facilitated with magnet and its fixture as the proof mass. The cantilever is relatively thin and droops naturally due the weight of proof mass a few millimeters (as shown in Figure 1) to a curve. The L-shape fixture that holds the magnet was bolted with a screw on one side parallel to the brass shim. The magnet is then attached on the other side of the L-shape fixture, perpendicular to the brass shim in such way to make magnetic coupling. During the process, the cantilever was deformed and twisted slightly. As a result, the combined proof mass is slightly located off the center of the cantilever beam resulting in weight imbalance and torsion mode resonance. The fixture also creates an area where the free end is rigid with the fixture, which acts like a semi-fixed end, paving a way for a standing wave vibration when the cantilever is excited. Finite Element Analysis (FEA) simulating the structure and dimensions confirms that the first 3 modes of vibration include bending, torsion and standing wave oscillations.

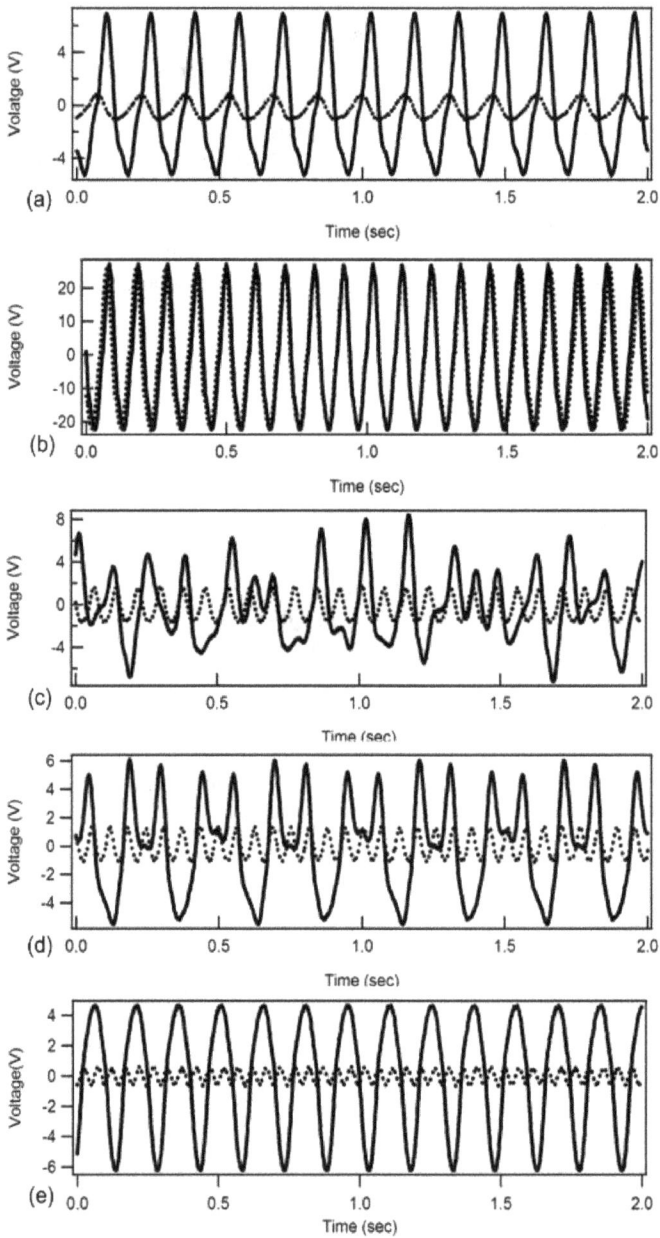

Figure 9. The output of the linear (dash line) and non-linear (solid) system in the time domain: (a) 6.5 Hz (b) 9.5Hz at resonance; (c) 13 Hz ; (d) 16Hz ;(e) 20Hz

3.3. Theoretical simulations

The dynamics of the piezoelectric cantilever is modeled by a 1-D driven spring-mass system coupled with the piezoelectric effect under the influence of a magnetic force Fm(z) [17-18]:

$$m\ddot{z} + d\dot{z} + kz + F_m(z) + \sigma V = mA(\omega)cos(\omega t),$$

(1)

with mass m=0.0024 kg, damping coefficient d=0.0075 kg/sec, spring constant k=8.55 N/m, and angular frequency ω. Here, z is the vertical deflection of the cantilever, V is the generated voltage, σ=5x10-6 N/V is the coupling coefficient, and A is the acceleration of the shaker table (A=4.2 m/sec2 measured at resonance frequency). The voltage output is related to the deflection of the piezoelectric cantilever through:

$$\dot{V} + \frac{1}{R_l C_l}V + \theta\dot{z} = 0$$

(2)

where R_l is the equivalent resistance, C_l is the equivalent capacitance and 1/ ($R_l C_l$)= 0.01 , and θ=1250 is the piezoelectric coupling coefficient in the measured circuit. The transverse magnetic force (in the z direction) is determined from the force between two magnetic dipoles (Kraftmakher, 2007):

$$F_m(z,\eta) = \frac{-3u_0 M^2 a(bz)(4\eta^2 - (bz)^2)}{4\pi((bz)^2 + \eta^2)^{\frac{7}{2}}}$$

(3)

where M is the dipole magnetization, u0 is the permeability in air, and η is the horizontal separation between the magnets at z=0. The correction factors a and b are included to compensate for the flexure motion of cantilever and the magnetic force along the cantilever axis [16]. The magnetization M is determined by direct measurement of the axial force between the cantilever and a fixed magnet using a reference scale [22].

The solution to the coupled differential equations (1) and (2) is determined using Maple software to give the voltage output versus time for a given driving frequency, magnetic force function, and separation η. In order to fit our experiment data, the magnetic force Fm(z) was modified by a and b parameters and used for our calculation, where M = 0.011Am2, η = 6.5 mm, a = 1.04 and b = 1.21. As in the experiment, the output is calculated for t = 0 to 10 seconds, and the maximum peak-to-peak output over the last 2 seconds obtained. The result of the frequency domain is showed in Figure 10, which resembles the experimental result as seen in Figure 8.

Both the experiment and simulation figures show broadband vibration for the non-linear configuration between 6-20Hz. The simulation in Figures 11(a)-(e) reproduces many of the features observed in the experiment in Figures 9(a)-(e). The rest of Figures 11-15 reveals more about the complexity of the multiple harmonics in the non-linear systems. The simulations of the time domain with the corresponding frequency selected from experiment are shown in Figures 11(a)-15(a). Figures 11(b)-15(b) illustrate the velocity vs. voltage output of the piezoelectric cantilever in both the linear and non-linear cases. Figures 11(c)-15(c) are the Fourier transform of the coupled cantilever cases in Figures 11(a)-15(a), respectively, showing the compositions of frequency components for the non-linear states. The following section will discuss the multiple harmonic components directly derived from the non-linear dynamics simulations.

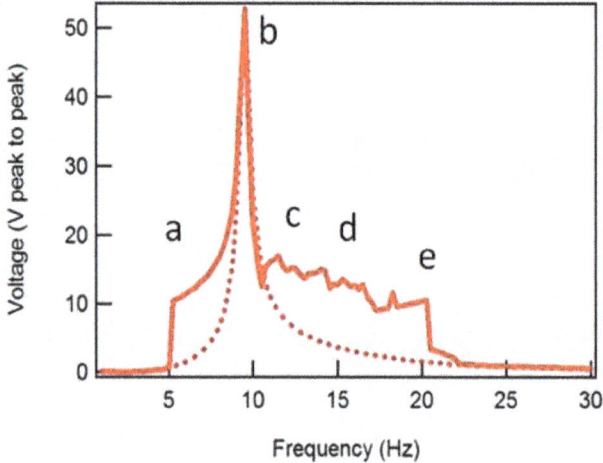

Figure 10. The simulated voltage output (peak to peak) of the piezoelectric cantilever is plotted in the frequency domain (dash line for linear and solid line non-linear).

3.4. Multiple harmonics analysis

At a driving frequency of 6.5 Hz, as seen in Figure 11(a), both the linear and non-linear cantilever motion follow the vibrations of the shaker table, producing periodic oscillations. The amplitude of the oscillations for the coupled cantilever, however, is approximately 5 times larger than those for the linear cantilever, as seen in the experiment in Figure 9(a). The velocity vs. voltage in Figure 11(b) shows that the coupled cantilever has non-linear component in voltage production. Further analysis through Fourier transformation indicates that the non-linear cantilever shows the combination of the excited 6.5 Hz harmonic (dominant and high amplitude) and the 20 Hz ultraharmonic (3 times the excited frequency), as seen in Figure 11(c).

At the resonant frequency of 9.5 Hz (Figure 12(a)) both non-linear and linear cantilevers oscillate at the driving frequency with equal amplitude of voltage output. The responses for both the coupled and uncoupled cantilever at resonant frequency are almost identical in the voltage output. The velocity vs. voltage in Figure 12(b) shows a little non-linearity at 90° and -90° of the vibration cycles. Through Fourier transformation as seen in Figure 12(c), the non-linear cantilever shows some components of vibration at the excited 9.5 Hz harmonic (dominant) and the 29 Hz ultraharmonic (3 times the excited frequency).

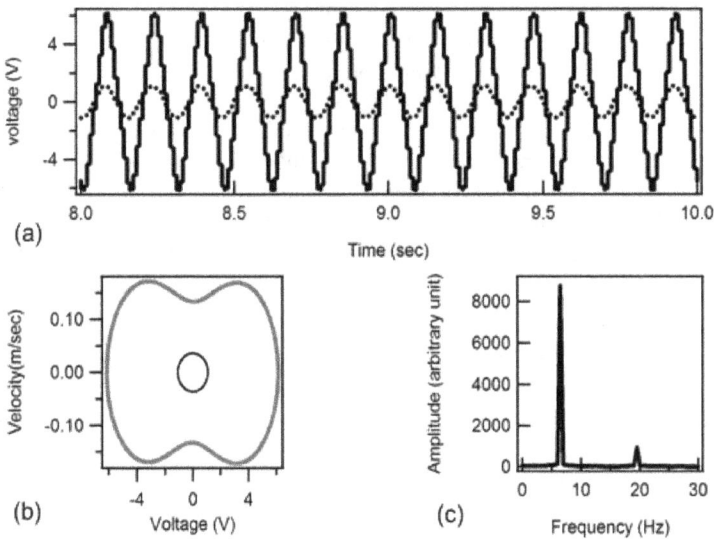

Figure 11. The theoretical analysis of excited frequency at 6.5 Hz. (a) the time domain voltage output, dash line for linear and solid line for non-linear states; (b) the velocity vs. voltage output, dark line for linear and light line for non-linear state; (c) the Fourier transform of the non-linear state from the data of Figure 5(a).

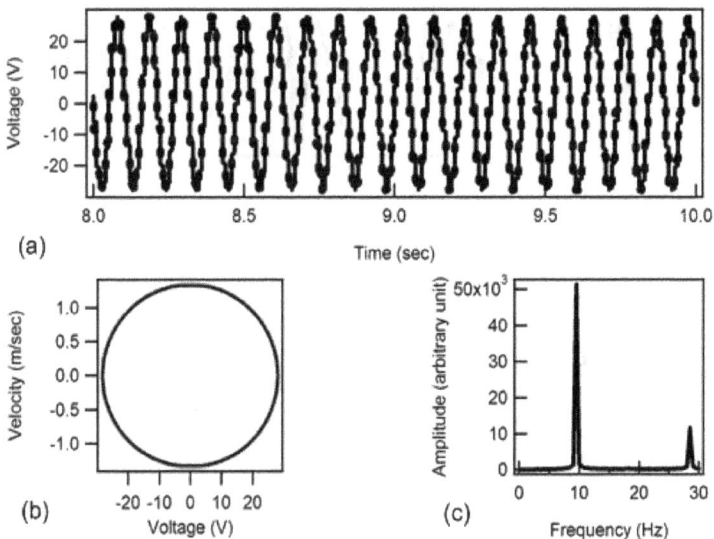

Figure 12. The theoretical analysis of excited frequency at 9.5 Hz. (a) the time domain voltage output, dash line for linear and solid line for non-linear state; (b) the velocity vs. voltage output, light line for linear and dark line for non-linear state; (c) the Fourier transform of the non-linear state from the data of Figure 12(a).

The response for the non-linear cantilever is chaotic at 13 Hz as seen in Figure 13(a), but with average 3 folds larger magnitude than the linear one. The velocity vs. voltage relation in Figure 13(b) shows chaotic motions for the coupled cantilever. Using Fourier transformation for Figure 13(a) results in Figure 13(c), the coupled cantilever shows the linear response of a small portion of 13 Hz component combined with a large amplitude distribution at lower frequency that are attributed to the chaotic motion. Note that the small peaks at 12.5 Hz and 17 Hz are not observed in the simulation as seen and discussed in the experiment section. This small torsion and standing wave bending resonance are not accounted for by the simplified 1-D model used to simulate the spring mass damping model such as an ideal cantilever.

At 16Hz, the non-linear cantilever is periodic (Figure 14(a)) and is 3 times larger (peak to peak) in magnitude than the uncoupled one, with double prone of low frequency in the upper cycle. Apparently, it is and composed of different frequency and multiple haromonic motion, with large magnitude than the uncoupled motion. The evidence is also shown in the velocity vs. voltage relationship in Figure 14(b), where 3 different cyclic loops are identifiable. Fourier transformation from time data in Figure 14(a) proves that the non-linear cantilever delivers ultra-sub-harmonic vibration at $n*(16/4)$ Hz, where, n=integer in Figure 14(c).

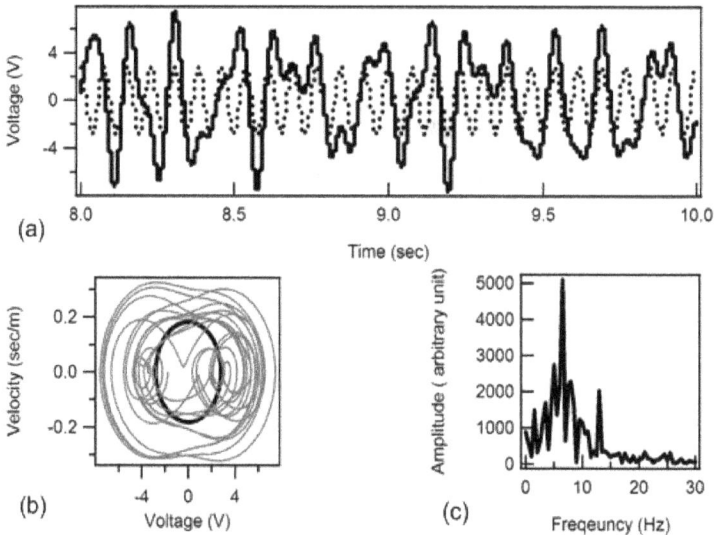

Figure 13. The theoretical analysis of excited frequency at 13 Hz. (a) the time domain voltage output, dash line for linear and solid line for non-linear states; (b) the velocity vs. voltage output, light line for linear and dark line for non-linear state; (c) the Fourier transform of the non-linear state from the data of Figure 13(a).

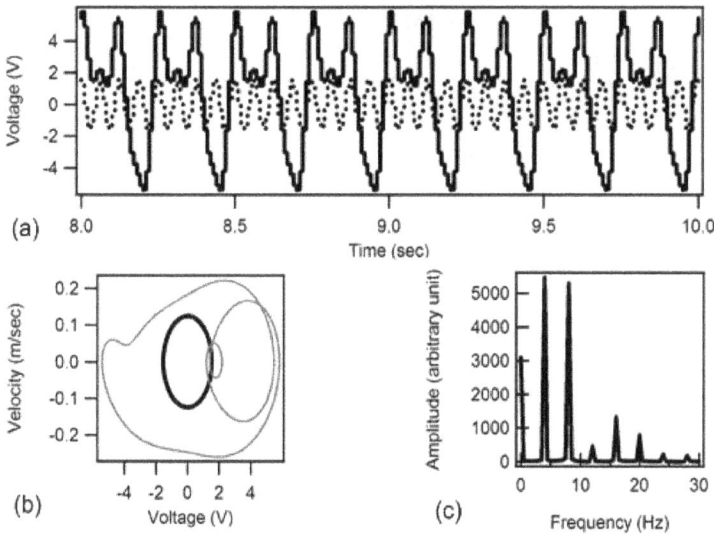

Figure 14. The theoretical analysis of excited frequency at 16 Hz. (a) the time domain voltage output, dash line for linear and solid line for non-linear states; (b) the velocity vs. voltage output, light line for linear and dark line for non-linear state; (c) the Fourier transform of the non-linear state from the data of Figure 14(a).

At 20Hz, the response for the non-linear cantilever is periodic and also 3 folds larger peak to peak magnitude than the linear one as seen in Figure 15 (a). The velocity vs. voltage in Figure 15(b) shows some combination of cyclic motions for the non-linear cantilever. Through Fourier transformation, the coupled cantilever shows subharmonic at 6.7 Hz (dominant), excite frequency/3, and 20 Hz in Figure 15(c).

The combination of the stochastic and various harmonic features have three to five folds greater voltage production than the linear standard narrow band piezoelectric cantilever. Together with the un-damped resonant response enhance the performance well beyond that of a standard energy harvester.

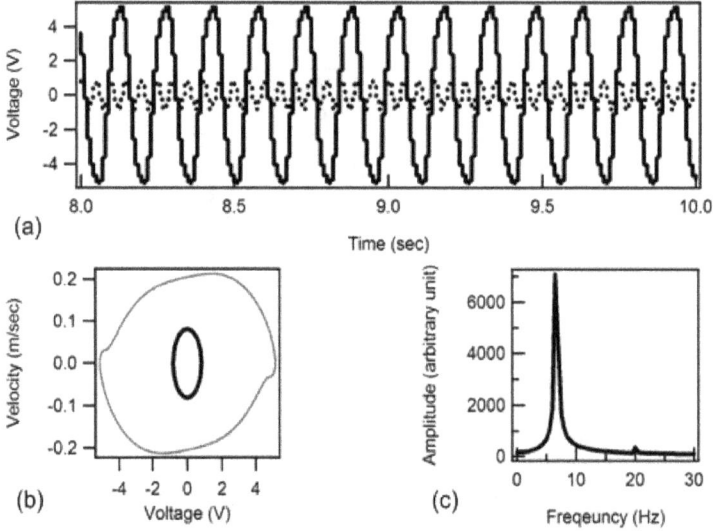

(a)

(b)

(c)

Figure 15. The theoretical analysis of excited frequency at 20 Hz. (a) the time domain voltage output, dash line for linear and solid line for non-linear states; (b) the velocity vs. voltage output, light line for linear and dark line for non-linear state; (c) the Fourier transform of the non-linear state from the data of Figure 15(a).

3.5. Experience result with storage capacitor

Figure 16 (a) shows the output of the other PZT cantilever with similar specs as a function of shaker table vibration frequency for the case where the opposing magnet is fixed to the shaker table. The voltage generated by the piezoelectric cantilever beam is rectified, and detected across a 22 μF capacitor and 1 M Ohm resistor in parallel, using the circuit shown in Figure 3 (a). The results from two measurement runs in the coupled state are shown, together with the output of the cantilever measured in the uncoupled state. (This is obtained by removing the opposing magnet.)   At the resonance frequency, (measured to be approximately 10 Hz) the output of the cantilever exceeds 16 V, and the peak height, resonance frequency and linewidth are all approximately the same for the coupled and un-coupled states. On either side of the main resonance, however, there are additional output observed for the coupled cantilever, which is not observed in the uncoupled state. As can be seen from a comparison of the two coupled runs, the frequency distribution of the peaks are the result of the multiple harmonics, as predicted in the open circuit.

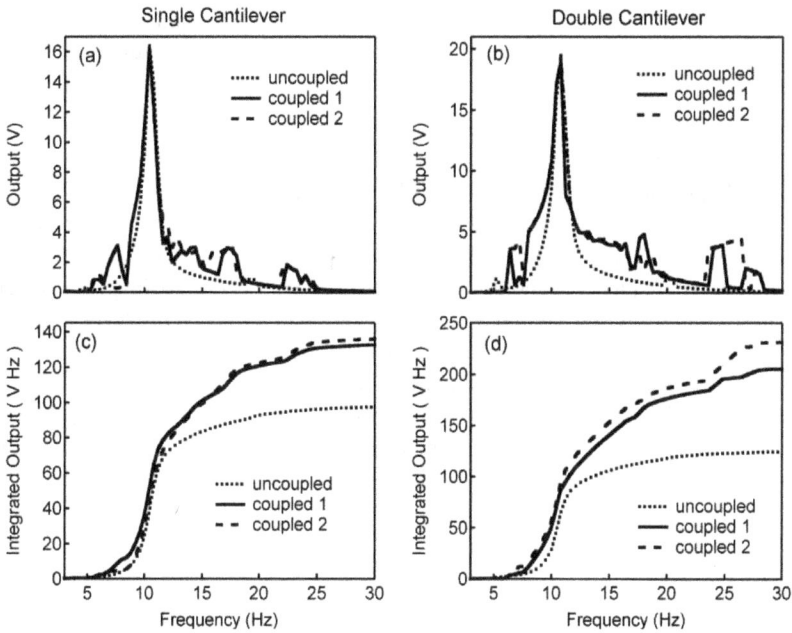

Figure 16. Voltage output of the piezoelectric cantilever as a function of shaker table frequency for **(a)** single cantilever **(b)** double cantilever. Integrated voltage output as a function of frequency for **(c)** single cantilever and **(d)** double cantilever.

Also measured was a double cantilever system, (as shown in Fig. 16(b)), in which the second magnet is connected to an opposing cantilever (having resonant frequency of around 60Hz) rather than to a fixed point. As shown in Fig. 16 (b), the results are similar to the single cantilever system, except that the double cantilever system shows a larger overall increase in off-resonance output. The overall improvement in the harvesting efficiency can be illustrated by plotting the integrated voltage output of the cantilever beam as a function of frequency. For both the single (Fig. 16 (c)) and double (Fig. 16 (d)) cantilever systems, the integrated voltage output over the 0-30 Hz bandwidth shows a substantial increase in the coupled versus the uncoupled case. The total improvement is 31%-87%, with some variation between measurement runs.

4. Conclusion

Piezoelectric cantilevers have been widely studied for energy scavenging applications, but suffer from poor output power outside of a narrow frequency range near the cantilever resonance. In this chapter, we have demonstrated how power output can be enhanced by applying a simple passive external force. When a symmetrical and repulsive magnetic force is applied to a piezoelectric cantilever beam to compensate the cantilever spring force, this lowers the spring potential and increases the output when driven by a random pink noise vibrational source. The principle may be applied to other vibration energy harvesting devices such as electromagnetic and capacitive types in random naturally pink noise environments.

In the parametrically excited piezoelectric cantilever experiments, linear and non-linear performances were compared. Overall, four distinct types of efficiency improvements appear in the non-linear configuration, in which the signal is amplified above the linear cantilever response: low frequency ultraharmonic amplification; stochastic amplifications in multi-frequency and multi-amplitude oscillations; ultra-sub-harmonic amplification at multiple quarter frequencies; subharmonic amplification at one-third frequencies. Taken together, the stochastic, sub-harmonic and ultra-harmonic response produces an average of three to five-fold increase in voltage production. For energy harvesting purposes, the combination of the four features together with the un-damped resonant response enhances the performance well beyond that of a standard energy harvester. Furthermore, an analytical model of the bi-stable dynamics produces results consistent with those observed experimentally. The simulation tool could be deployed in the future investigation for non-linear energy harvester design for broadband and beyond natural harmonic applications.

Author details

Ji-Tzuoh Lin* and Bruce William Alphenaar
Department of Electrical and Computer Engineering, University of Louisville, Louisville, KY, USA

Barclay Lee
Department of Bioengineering, California Institute of Technology, Pasadena, CA, USA

Acknowledgement

The effort was funded by the Department Of Energy DE-FC26-06NT42795 and the U.S. Navy under Contract DAAB07-03-D-B010/TO-0198. Technical program oversight under the Navy contract was provided by Naval Surface Warfare Center, Crane Division.

5. References

[1] Elvin NG, Elvin A and Choi D 2003 A self-powered damage detection sensor J. Strain Anal. Eng. Des. 38 115-24
[2] Ottman G K Hofmann H F and Lesieutre G A 2003 Optimized piezoelectric energy harvesting circuit using step-down converter in discontinuous conduction mode IEEE Trans. Power Electron. 18 696-703
[3] Roundy S 2004 A piezoelectric vibration based generator for wireless electronics, Smart Mater. Struct. 13 1131

---

* Corresponding Autor

[4] Kulah H. and Najafi K. 2004, An electromagnetic micro power generator for low-frequency environmental vibrations, 17th IEEE Int. conf. Micro Electro Mechanical System, 1004 (MEMS '04) pp 237-40

[5] von Büren T and Tröster G, Design and optimization of a linear vibration-driven electromagnetic micro-power generator, Sens. Actuators A 135 (2007) 765-775

[6] S P Beeby, R N Torah, M J Tudor, P Glynne-Jones, T O'Donnell, C R Saha and S Roy, A micro electromagnetic generator for vibration energy harvesting, J. Micromech. Microeng. 17 (2007) 1257–1265

[7] Hsi-wen Lo and Yu-Chong Tai, Parylene-based electret power generators, J. Micromech. Microeng. 18 (2008) 104006-104014

[8] Sterken T, Fiorini P, Baert K, Puers R, and Borghs G, An electret-base electrostatic micro-generator transducers, The 12th International Conference on Solid State Sensors, Actuators and Microsystems. Boston, June 6-12, 2003 pp 1291-1294

[9] Shad Roundy, Paul Kenneth Wright and Jan M. Rabaey, Energy scavenging for Wireless Sensor Networks with Special Focus on Vibrations, Kluwer Academic Publishers, 2004.

[10] Y Chiu and V F G Tseng, A capacitive vibration to electricity energy converter with integrated mechanical switches, J. Micromech. Microeng. 18 (2008) No 10 104004-10412

[11] Eli S Leland and Paul K Wright, Resonance tuning of piezoelectric vibration energy scavenging generators using compressive axial preload. *Smart Mater. Struct.*15 (2006) 1413-1420

[12] Dibin Zhu, Stephen Roberts, Michael J. Tudor, and Stephen P. Beeby, Closed Loop Frequency Tuning of a Vibration-Based Microgenerator, Proceedings of PowerMEMS 2008/microEMS2008, November 2008.

[13] Vinod R Challa, M G Prasad, Yong Shi and Frank T Fisher, A vibration energy harvesting device with bidirectional resonance frequency tunnablity. *Smart Mater. Struct.*17 No 1 (2008) 015035

[14] Ji-Tzuoh Lin, Walter Jones, Bruce Alphenaar, Yang Xu and Deirdre Alphenaar. Passive magnetic coupling for enhanced piezoelectric cantilever response for energy scavenging applications. 17th International Symposium on Applications of Ferroelectrics (ISAF 2008 EH017) Feb 24-27 2008

[15] Shad Roundy and Yang Zhang, Toward self-tuning adaptive vibration-based microgenerator, Proceedings of SPIE Volume 5649 Smart Structure, Devices and System II February 2005 pp.373-384

[16] F. Cottone, H. Vocca, L. Gammaiton, Non-linear Energy Harvesting, Physical Review Letter,102, 080601 2009

[17] A. Erturk, J. Hoffmann, D. J. Inman, A piezomagnetoleastic structure for broadband vibration energy harvesting, Applied Physics Letter, 94 254102 2009

[18] Samuel C. Staton, Clark C. Mcgehee, Brian P. Mann, Reversible hysteresis for broadband magnetopiezoelastic energy harvesting, Applied Physics Letter, 95, 174103 2009

[19] Duchesne, B., C.W. Fischer, C.G. Gray, and K. R. Jeffrey. "Chaos In The Motion of An Inverted Pendulum: An Undergraduate Laboratory Experiment," Am. J. Phys. 1991; 59 (11)

[20] A. Siahmakoun, V. A. French, and J. Patterson. "Nonlinear Dynamics of A Sinusoidally Driven Pendulum In a Propulsive Magnetic Field," Am. J. Phys. 1997; 65 (5)

[21] Ji-Tzuoh Lin and Bruce Alphenaar. "Enhancement of Energy Harvested from a Random Vibration Source by Magnetic Coupling of a Piezoelectric Cantilever," Journal of Intelligent Material Systems and Structures, 2010; Vol. 21 Issue 13, 1337-1341

[22] Ji-Tzuoh Lin, Barclay Lee, and Bruce Alphenaar. "Magnetic Coupling of Piezoelectric Cantilever for Enhanced Energy Harvesting Efficiency" Smart Mater. Struct. 2010; 19 045012

[23] Ji-Tzuoh Lin, Kevin Walsh and Bruce Alphenaar. Enhanced Stochastic, Subharmonic and Ultraharmonic Energy Harvesting, Journal of Intelligent Material and Structure Systems,(accepted May 4th on line) 2012

[24] A. Prosperetti "Subharmonics and Ultraharmonics in The Forced Oscillations of Weakly Nonlinear Systems." American Journal of Physics Vol. 44 No. 6 1976

[25] Kraftmakher, Y. "magnetic Field of A Dipole And the Dipole-Dipole Interaction," Eur. J. Phys. 2007; 28, 409

Chapter 13

Self-Powered Electronics
for Piezoelectric Energy Harvesting Devices

Yuan-Ping Liu and Dejan Vasic

Additional information is available at the end of the chapter

1. Introduction

According to the piezoelectric direct effect, piezoelectric material can generate electrical energy from a mechanical vibration. To collect the energy, a piezoelectric element is attached on a vibrating structure, and an interface circuit is connected to the piezoelectric element to transfer the generated energy to the load. If the collected energy is dissipated by Joule effect in a resistive load, the vibration of the structure will be reduced. This is called structural passive shunt damping [1]. If the collected energy is stored in a capacitor or a battery, an energy recovery system is obtained. Piezoelectric energy harvesters (PEH) have emerged as a prominent theme of researchers whose interest is growing. The advances in low-power components or design methodology have reduced the power consumption of mobile devices, and therefore allow the feasibility of self-powered autonomous electronic devices. This opens the possibility for completely self-powered devices, and the notion of small generators producing enough power for low-consumption devices, like wireless sensor network (WSN). Ambient vibrations are presented in many different environments such as automobiles, buildings, structures (bridges, railways), industrial machinery, etc. Since 2002, numerous studies have been published on the topic of the energy harvesting. Tang et al. [2] and Khaligh et al. [3] made a long synthesis and developed a state of the art for vibration piezoelectric energy harvester. It demonstrates the interest of researchers in this topic. Moreover, piezoelectric materials have high power density [4]. The power density of the piezoelectric generator harvested the energy from vibrations is about 250 $\mu W/cm^3$. In comparison, the power density of the electrostatic generator that harvested energy from vibrations is only about 50 $\mu W/cm^3$. The vibration-to-electricity convertor can be also performed by electromagnetic transducers [5, 6], but the power density cannot be high as the piezoelectric generator. Piezoelectric technologies have received much attention, as they have high electromechanical coupling and no requirement of the external voltage source.

Figure 1 shows the configuration of piezoelectric energy harvesting system. A piezoelectric element is attached on a host structure and a charging circuit is connected to the piezoelectric elements. The piezoelectric elements convert the vibration energy of the host structure into electrical energy, and then the generated electrical energy is stored in a storage buffer. The voltage generated across the piezoelectric element is an AC voltage, so the basic charging circuit is an AC/DC converter or said a rectifier.

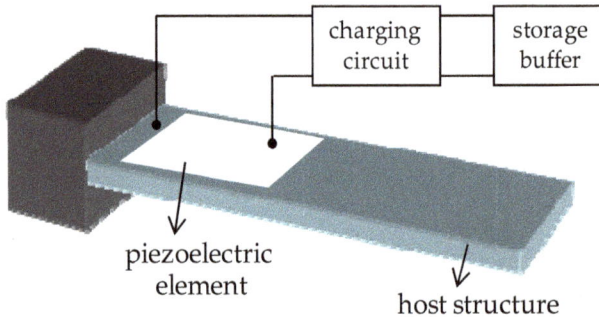

Figure 1. Configuration of piezoelectric energy harvester

Since the piezoelectric element has large clamped capacitance, an impedance matching circuit is required to maximize the generated power. It was known that an inductor can be added to compensate the contribution of the piezoelectric clamped capacitor, but it cannot be adaptive to the environmental variations and the value of the inductance is too large in a low frequency range. To overcome this drawback, switching-type charging circuits were proposed and popularly used in recent years. In the switching circuits, the switches are operated synchronously with the vibration of the host structure in order to optimize the power flow.

Several synchronized switching circuit topologies and corresponding switching laws were proposed. They can be classified into two groups according to the placement of the rectifier and the active switches, as shown in Fig. 2. The first group of the switching circuits places the switches between piezoelectric element and the rectifier, such as parallel-SSHI (Synchronized Switching Harvesting on an Inductor) and series-SSHI [7-17]. This group of techniques is used to modify the waveform of the piezoelectric voltage, *i.e.* the voltage across the piezoelectric element, in order to increase the collected power in the weakly coupled structure. The second group places the switches between the rectifier and the storage buffer, such as SSDCI [18]. This group of the techniques is used to modify the charging current flowing into the storage buffer in order to fasten the charging speed [19] and to make a load adaptation.

Although the synchronized switch techniques can increase the collected energy or charging speed, they also require energy to supply active components. Therefore, it is of interest to achieve totally self-powered interfaces. In fact, in all switching techniques, the operation principles of the active switches are similar. In most of the time, the switches are cut-off. The switches only conduct at the local extreme values of the displacement or at the zero-crossing values of the velocity. In other words, there should be a control loop to measure vibration and then control the switching action. Figure 3 shows this control loop with the self-powered electronics.

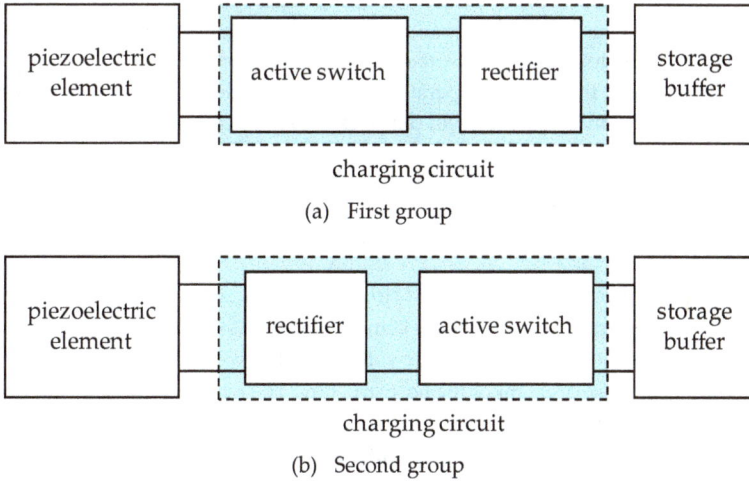

(a) First group

(b) Second group

Figure 2. Two groups of charging circuits

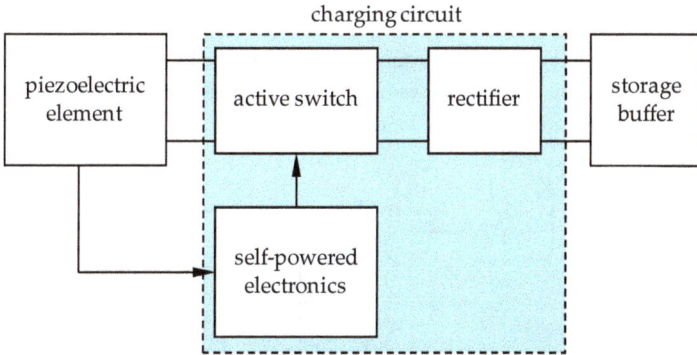

Figure 3. Piezoelectric energy harvester with self-powered synchronized switching techniques

In energy harvesting device, there are four parts that influence the performances, including 1) piezoelectric material properties, 2) configuration of the host structures, 3) charging circuit topologies and 4) storage buffers. In this chapter, we will only focus on the design of the charging circuit, especially on the self-powered electronics. The SSHI techniques will be used as a targeted circuit to add the self-powered function since it is easy to implement and owns high efficiency. We will neglect the massy theory of synchronize switching techniques, but focus on the operation principle of the self-powered charging circuit.

2. Synchronized interface circuit of piezoelectric harvester

2.1. Equivalent circuit of piezoelectric harvester

The piezoelectric energy harvester (PEH) consists of a piezoelectric elements bounded on a host structure. A mechanical model based on a spring–mass system gives a good description of the vibration behavior near the resonance of the host structure. Therefore, for simplicity, this system can be modeled as a one degree-of-freedom system of a mass M, a spring K and a damper D. Assuming the system is operated in the linear region, the differential governing equation of this electromechanical system can be expressed as equations (1) and (2):

$$M\ddot{u} + D\dot{u} + Ku + \alpha v_p = f \tag{1}$$

$$i_p = \alpha\dot{u} - C_p\dot{v}_p \tag{2}$$

where f is external force exerted on the host structure, i_P is outgoing current generated from the piezoelectric element, α is force-voltage coupling factor, C_P is the clamped capacitance of the piezoelectric element, v_P is piezoelectric voltage, i.e. the voltage across the piezoelectric element and u is the displacement of the host structure. The piezoelectric element generates AC voltage from the vibration. To store the energy in the DC form, a rectifier, i.e. AC/DC converter is required to connect the storage buffer at the piezoelectric element. Equations (1) and (2) are linear equations, but the rectifier is a non-linear circuit and it is not easy to analyze. To make the analysis more intuitive, we adopt the concept of equivalent circuit in this chapter.

According to equations (1) and (2), the equivalent circuit of the PEH can be illustrated in Fig. 4. It should be noted that force f and velocity u are equivalent to a voltage source and a flowing current in the mechanical impedance respectively.

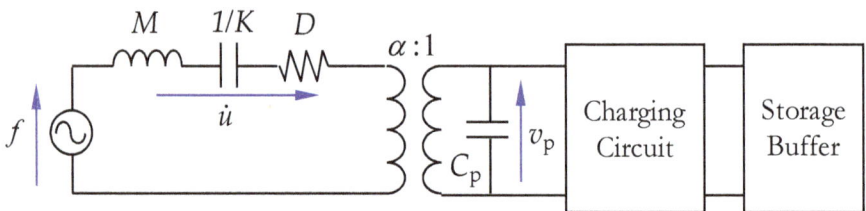

Figure 4. Equivalent circuit of the piezoelectric energy harvesting system

To calculate the generated power, the analysis is simplified by two basic assumptions:

1. The piezoelectric element and host structure are given, so the transformation ratio α is a constant;
2. The displacement u is sinusoidal at the specific frequency.

Therefore, the displacement u and velocity $\dot{u}$ can be represented as:

$$u = U_M \sin \omega t \tag{3}$$

$$\dot{u} = \omega U_M \cos \omega t \tag{4}$$

where u is the velocity of the host structure, U_M is the amplitude of the displacement, ω is the angular frequency of the host structure, and t represents time parameter. Considering the simple case when the shunt circuit is only composed of passive components first, $i.e.$ resistors, capacitors and inductors, the voltage across the piezoelectric element v_p will be also sinusoidal as shown in equation (5).

$$v_p = V_p \cos(\omega t - \theta) \tag{5}$$

where V_p is the amplitude of the piezoelectric voltage v_p and θ is the phase difference between the velocity u and the piezoelectric voltage v_p. The generated power can be obtained by multiplying piezoelectric voltage v_p and vibration velocity u. Applying the equations (4) and (5), the average power P_m generated from the piezoelectric element can be expressed as:

$$P_m = \frac{\alpha \omega U_M V_P}{2} \cos(\theta) \tag{6}$$

Equation (6) obviously shows the factor which can influence the generated power. In the case of constant velocity, $i.e.$ ωU_M is a constant, the power is determined by the amplitude of the piezoelectric voltage V_p and the phase angle θ. Actually, the amplitude of the piezoelectric voltage V_p and the phase difference θ are determined by the characteristics of the connected circuit. When the impedance of the connected circuit is zero (short circuit), the piezoelectric voltage is zero. There is no power flow out of the piezoelectric element. When the impedance of the connected circuit is infinite (open circuit), the velocity and the piezoelectric voltage have 90º phase difference due to the capacitor C_p. There is no power flow out of the piezoelectric element as well. These two critical cases both cannot output any power from the piezoelectric element but the fundamental reasons are not the same. Actually, if the phase difference is not zero, the power generated from piezoelectric material may not only flow from the piezoelectric material to the connected circuit, but may also flow in the opposite direction. To obtain the largest power from the piezoelectric element, high piezoelectric voltage V_p and zero phase difference ($\theta = 0$) are both required [19].

It should be noted that above analysis is based on the assumption that vibration velocity u is a constant value. However, the vibration velocity is not independent of the piezoelectric voltage. According to the equivalent circuit in Fig. 4, the velocity can be expressed as:

$$\dot{u} = \frac{f - \alpha v_p}{Z_L} \tag{7}$$

where Z_L is the mechanical impedance, $i.e.$

$$Z_L = \sqrt{\left(K - M\omega^2\right)^2 + \left(D\omega\right)^2} \tag{8}$$

It can be seen that the vibration velocity may decrease with increasing voltage in Eq. (7). We would like to connect an impedance matching circuit to collect more electrical energy, but the vibration energy may be suppressed. Finally, we may not collect more power from the piezoelectric element. Hagood et al. [1] used this effect to increase the structural damping, which was called piezoelectric shunt damping. However, it is not desired case in the piezoelectric energy harvester.

To avoid the influence of the piezoelectric voltage on the vibration, the force coupling factor α should be sufficient small. With this condition, equation (7) can be re-written as:

$$\dot{u} = \frac{f}{Z_L} \tag{9}$$

Accordingly, the piezoelectric voltage is independent of the vibration velocity. This means that the energy extraction from the piezoelectric element does not disturb the vibration behavior of the structure and the magnitude of the velocity can be viewed as a constant value. Therefore, with the weakly coupling assumption, the equivalent circuit can be further simplified as shown in Fig. 5, where the velocity magnitude ωU_M is a constant.

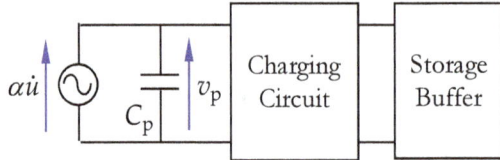

Figure 5. Equivalent circuit of the low-coupled piezoelectric energy harvester

The weakly coupling assumption is reasonable when the strain energy of the host structure is much larger than that of the piezoelectric element. Shu et al. [21] gave a more clear definition of the weakly coupled structure, that:

$$\frac{k^2}{\zeta} = \frac{\omega_0^2 - \omega_S^2}{\zeta\omega_S^2} \ll 1 \tag{10}$$

where ζ is the damping ratio of the host structure and k is the electromechanical coupling factor. ω_0 and ω_S are open-circuit resonant frequency and short-circuit resonant frequency of the host structure respectively. Piezoelectric microstructure is weakly coupled since the product of the thickness of piezoelectric element and its Young's modulus is very small compared to the base material. Actually, only a small amount of mechanical energy is taken from the structure and converted into electricity in the low-coupled structure. The synchronized switching techniques increase the piezoelectric voltage and also put piezoelectric voltage in phase with the constant velocity. This is the underlying reason that the power out of the piezoelectric element can be increased in synchronized switching techniques. In another viewpoint, synchronized switching techniques are also equivalent to increase the coupling factor of the weakly coupled structure.

2.2. Operation principle of synchronized switching techniques

The standard interface circuit of the PEH is a full-wave bridge rectifier. It converts AC voltage to DC voltage for storage buffers, such as batteries. The schematic diagram of the full-wave bridge rectifier is shown in Figure 6(a). Figure 6(b) shows the waveforms of standard DC approach. When the absolute value of the piezoelectric voltage v_p is less than load voltage V_{DC}, the diode bridge is in the open-circuit state and the clamped capacitor C_p is charged or discharged. Once the absolute value of the piezoelectric voltage v_p reaches load voltage V_{DC}, the diodes conduct and piezoelectric elements charge the storage buffer. Accordingly, the current i_p is flowing to the load discontinuously and the velocity $\dot{u}$ is not in phase with the piezoelectric voltage v_p, so the full-wave bridge rectifier is not an efficient solution for the PEHs.

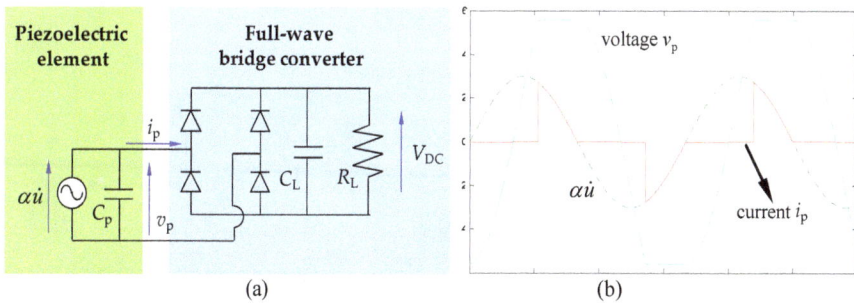

Figure 6. (a) Equivalent circuit and (b) waveforms of the piezoelectric element connected to full-wave bridge rectifier

To improve the efficiency of the energy conversion, Guyomar et al. proposed the SSHI techniques (synchronized switching harvesting on an inductor) [9, 11]. Figure 7 shows two types of SSHI techniques: series-SSHI and parallel-SSHI. Figure 8 shows the corresponding key waveforms. In SSHI techniques, a bi-directional switch Q and an inductor L are added in series or in parallel with the piezoelectric element. In most of the time, the switch Q is in open circuit state. When the local extreme displacements or zero vibration velocities occur, the switch is conducted in a very short period. In this short period, the clamped capacitor C_p makes the resonance with the inductor L and the piezoelectric voltage v_p inverses. Accordingly, the SSHI circuit increases the magnitude of the piezoelectric voltage and puts piezoelectric voltage v_p in phase with the vibration velocity, which indicates that more energy is extracted from the vibration source. The results also show that the energy stored in the clamped capacitor C_p is extracted by the LC resonance circuit and thus the piezoelectric voltage can be increased [22, 23]. The SSHI technique is equivalent to enlarge the coupling factor in the weakly coupled structure [21]. There are several synchronized switching techniques derived from SSHI techniques, such as DSSH [24]. These derived techniques still need a pulse signal which is synchronized with the vibration to drive the switches. However, obtaining this synchronized signal consumes the energy. In some cases, the energy used in the synchronized circuit may be larger than the increasing energy by the SSHI technique. So, it is of interest to design an efficiently self-powered synchronized circuit.

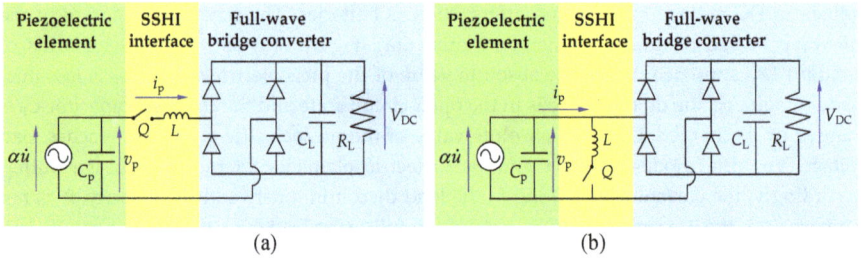

(a)   (b)

Figure 7. Equivalent circuit of (a) series-SSHI converter and (b) parallel-SSHI converter

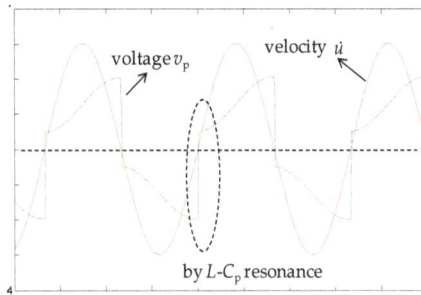

Figure 8. Key waveforms of SSHI converters

Both series-SSHI and parallel-SSHI techniques are efficient and easy to implement, but they are different from the viewpoint of the self-powering. The main difference is the magnitude of the outgoing current, *i.e.* i_P in Fig. 7. In the series-SSHI converter, the current i_P is mostly equal to zero since the switch Q is cut-off. The current i_P is only flowing when the switch Q is conducted at the local extreme displacements. On the contrary, in the parallel-SSHI converter, the switch Q is not in the current path between the piezoelectric element and the rectifier, so the outgoing current i_P flows to the load in most of the time. This basic characteristic will influence the self-powered loop design in the following section.

3. Self-powered electronics for synchronized switching

In fact, the standard full-wave bridge rectifier is a passive design, so it can be viewed as fully self-powered interface. However, the full-wave rectifier is not efficient in the weakly-coupled structure, and the synchronized switching circuits are usually used in PEH. These switching circuits require a switching control signal. In Fig. 3, it was shown that there should be an extra control loops to generate this control signal. Figure 9 shows more detail about configuration of the self-powered PEH. It can be seen that a self-powered charging circuit includes three parts: 1) a main synchronized switching converter, including active switches and a rectifier, 2) a signal conditioner to generate the control signal for the switching circuit, and 3) a DC power supply for the active electronics. The main switching converter in this section is the SSHI converter, which has already mentioned in last section. Therefore, the key point of this section is the signal conditioner and the DC power supply. The signal conditioner should be able to sense the vibration behavior and provide the corresponding real-time signal. In this section, we will start from the sensing signal and then extend it to the design rule of the practical self-powered electronics.

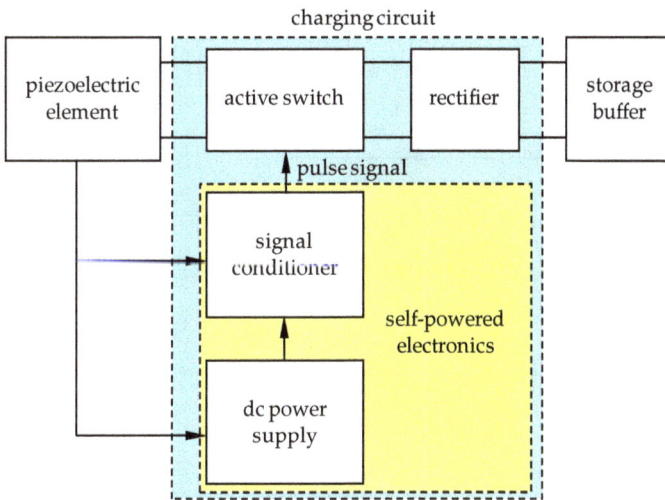

Figure 9. Configuration of self-powered electronics

3.1. Modified SSHI circuit for Low-voltage drop

Considering the micro-scale PEH, the piezoelectric voltage is typically below 5V [25]. In such a low voltage, the diode voltage drops becomes a series problem. In the original SSHI technique, it owns totally six diodes (Fig. 7) in the implemented circuit, including two diodes in the bi-directional switch and four diodes in bridge rectifier. It means that each current path includes three diode voltage drops. It is a significant voltage drops in the micro-scale PEH. Figure 10(a) shows the modified series-SSHI which owns lower voltage drops. In this design, the bi-directional switch is embedded into the bridge rectifier by using the body diode of the MOSFET. In fact, D_{a2} and D_{b2} in Fig. 10(a) are possible be further neglected as shown in Fig. 10(b) and there are only 2 diodes in the SSHI converter

[26]. The operation principles of the modified SSHI converters are similar as the original version. However, in four diodes design, the outgoing current i_p can be auto switched off by diodes D_{a2} and D_{b2} in the end of the SSHI process. It means that the control signal of the switches Q_a and Q_b can be a square signal but not a pulse signal, it is easier to generate square signal typically. However, in two diodes design of Fig. 10(b), the switches Q_a and Q_b requires the accurate pulse signal. Another modified Series-SSHI is presented Fig. 10(c). Half of the rectifier bridge can be replaced by the switching elements and another half of the rectifier can be replaced by the middle point filtering capacitance, allowing the removal of the diodes [27].

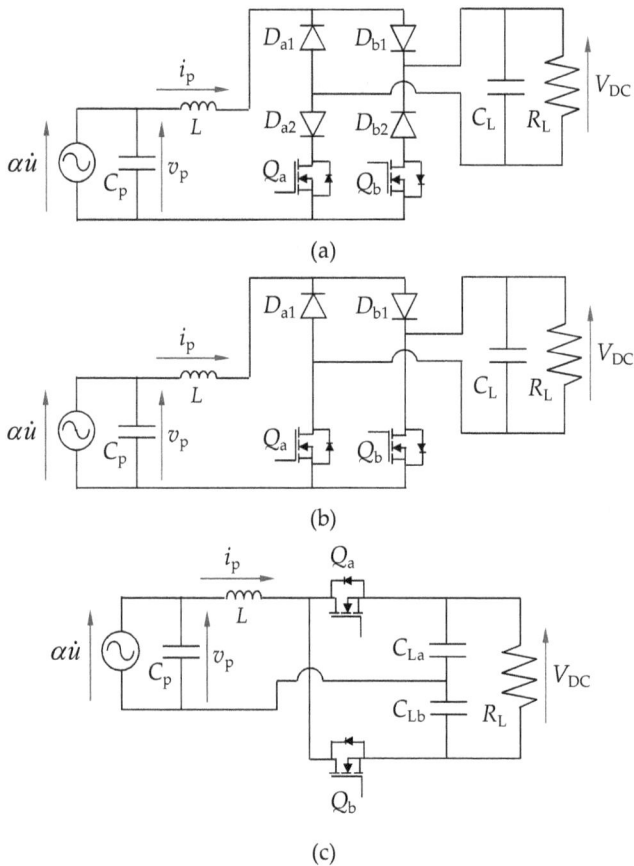

Figure 10. Modified series-SSHI converter with low voltage drop: (a) Four diodes design (b) Two diodes design (c) Diode-less series SSHI

3.2. Sensing signal

The basic concept of the self-powered issue is how to generate the switching signals synchronously with ultra-low power consumption. The SSHI techniques require the displacement or velocity signals of the host structure to control the switches. In fact, the vibration amplitude is not easily detected by displacement sensor or velocity sensor compactly. The piezoelectric element itself is the best candidate to sense the vibration. However, the piezoelectric voltage or outgoing current may not be proportional to the displacement or velocity. Equation (2) shows the relationship between the piezoelectric voltage and vibration displacement. According to Eq. (2), piezoelectric voltage v_p is only proportional to the vibration displacement u when the outgoing current i_p is equal to zero, *i.e.* the open circuit state of the piezoelectric element. Similarly, the outgoing current is only proportional to the vibration velocity $\dot{u}$ when the piezoelectric voltage v_p is equal to zero, *i.e.* the short circuit state of the piezoelectric element. Choosing the displacement or the velocity signals as the control criterion is much different in the practical electronics design. According to the operation principle of SSHI techniques, a peak detector should be used in the displacement signal and a zero-crossing detector should be used in the velocity signal.

In series-SSHI technique, piezoelectric element is mostly in the open-circuit state. The current is only flowing when the switch is conducted. Therefore, the series-SSHI converter is a possible case that we can use the piezoelectric voltage directly to represent the displacement. However, in the parallel-SSHI technique, the current is mostly continuously flowing to the load. The piezoelectric voltage is not proportional to the displacement perfectly. The outgoing current in parallel-SSHI technique is also not proportional to the velocity because piezoelectric element is not in the short circuit state. It means that we cannot use the piezoelectric voltage or the outgoing current directly to synchronize the vibration, and there should be some compensation in the signal conditioner. Since the phase of vibration velocity is 90 degrees lag than that of the displacement, piezoelectric voltage is easier to follow the vibration velocity by the phase compensation of the electrical network.

In brief summary, there are two possible ways to make the control loops of the self-powered function. Figure 11 illustrated these two control loops with different signal conditioners. As shown in Fig. 11(a), the peak detector is more suitable for series-SSHI converter because the piezoelectric voltage is synchronized with the displacement. The peak detector also works with the parallel-SSHI converter, but the phase delay is larger. The concept of the zero-crossing detector is shown in Fig. 11(b). It can be used for both series-SSHI and parallel-SSHI techniques, but a phase compensation is necessary. To avoid the phase compensation, a second piezoelectric patch should be added. Typically, the zero-crossing detector is easier to implement by electronics, but it requires a low-pass filter to avoid the high-frequency noises when the switch of the SSHI converter is active.

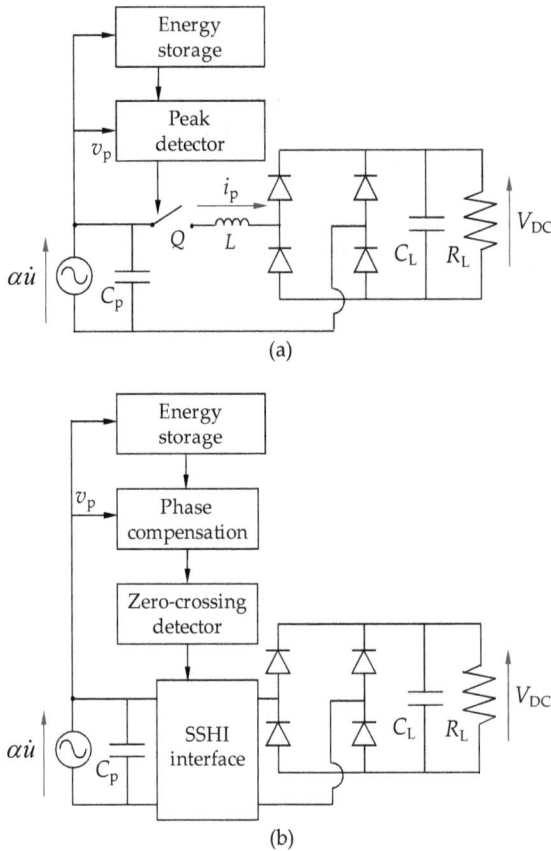

Figure 11. Control loops of the self-powered function in SSHI techniques
(a) with a peak detector (b) with a zero-crossing detector

3.3. Peak detector

Figure 12 show a simple peak detector design [26-29]. It includes three parts: a switch, a comparator and an envelope detector. To easily explain the principle of operation, we assume the diodes are ideal first. In the beginning of the positive piezoelectric voltage, the capacitor C_3 is charged by the piezoelectric voltage. The emitter-base polarization of the transistor T_2 is reverse bias, so the transistors T_2 and T_1 are both blocked. The voltage on capacitor C_3 is equal to the piezoelectric voltage v_p before the peak voltage, if the diode voltage drop is neglected. When the piezoelectric voltage reaches a peak value, it means that the piezoelectric voltage v_p will start to decrease. However, because there is no discharging current path for the capacitor C_3, the capacitor voltage v_3 will keep the same voltage value after the peak voltage of v_p. Therefore, diode D_3 is reverse bias. Once the voltage difference between the piezoelectric voltage v_p and the capacitor voltage v_3, *i.e.* $(v_3 - v_p)$, is greater than the threshold voltage of T_2, the transistor T_2 starts conducting. The capacitor C_3 discharges through the path D_2-T_2-R_1-T_1, and conduct the transistor T_1, which is the switch of the SSHI converter. Therefore, the piezoelectric voltage v_p starts to inverse and the outgoing current i_p starts to flow. When the outgoing current i_p reaches zero again by the L-C_p resonance, the harvesting process is interrupted by the diode D_1. At the end of the harvesting process, the capacitor C_3 is totally discharged. This peak detector can work both in series-SSHI converter and the parallel-SSHI converter. However, in the parallel-SSHI converter, a small voltage difference is required between the piezoelectric voltage v_p and capacitor voltage v_3 to turn T_2 on. In this case, the phase delay is larger and it depends on the load value. It should be noted that this peak detector only allows the current flowing in single direction in the SSHI switch. For bi-directional current, two peak detectors must be adopted.

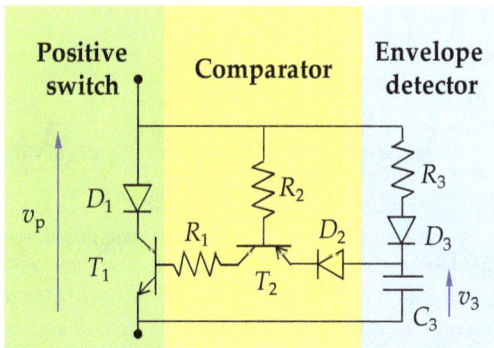

Figure 12. Positive peak detector for SSHI technique

3.4. Zero-crossing detector

Except the peak detector, another choice of the sensing circuit is the zero-crossing detector. Different from the voltage peak detector, the zero-crossing detector should detect the zero velocity or said zero equivalent current. Figure 10(b) has shown the completed system of the self-powered SSHI based on zero-crossing detector. There are three parts. First, the velocity is sensed through a conditioning circuit. Second, a comparator is required to make a zero-crossing detector in order to obtain the pulse driving signals for the switches. Third, a DC power source is required to supply the active electronics above.

Chen et al. [15, 16] applied the velocity as the switching criterion to self-powered series-SSHI and parallel-SSHI converters, which called V-SSHI. Figure 13 shows the V-SSHI design. In Chen's design, an extra piezoelectric patch was used. A current sensing resistor R_s is connected in parallel with the piezoelectric patch and a passive low-pass filter is used to reduce the high frequency noise, introduced by the SSHI switching. The high frequency noise of the velocity signal can be very large, so it is impossible to apply the sensing signal to the comparator directly. The current sensing resistor used herein must be small enough to neglect the effect of the piezoelectric capacitance, *i.e.* $i_p \approx \alpha\dot{u}$. The low-pass filter should be carefully designed to guarantee there is no phase lag. It should be mentioned that the output terminal of the comparator is a square signal but not a pulse signal, so it can only drive the SSHI-switch Q_a and Q_b in Fig. 10(a). Although the circuit topology in Fig. 10(a) owns an extra diode voltage drop compared to the circuit topology in Fig. 10(b), the peak detector in Fig. 12 also own an extra diode D_1 series with the switch T_1. Consequently, there are no difference of the voltage diode drops between the peak detector and the zero-crossing detector.

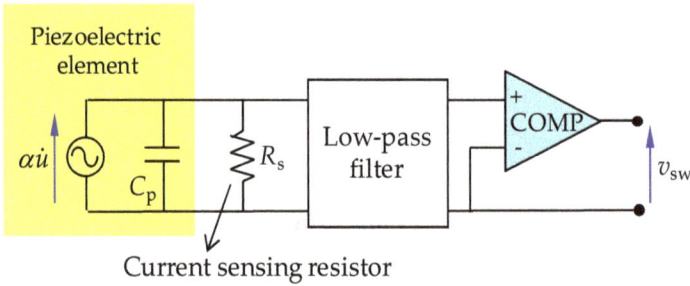

Figure 13. Equivalent current (velocity) sensing by a resistor and a passive low-pass filter

On the other hand, Ben-Yaakov and Krihely [30] adopted the zero-crossing detector in the self-powered rectifier for the output terminal of piezoelectric transformers. Their circuit is similar as the parallel-SSHI converter. Figure 14 shows their design of the signal conditioner. The signal conditioner consists of a passive differentiator (C_{dif} and R_{dif}) with a hysteresis resistor (R_{hys}) and a comparator. The passive differentiator is used to obtain the velocity signal from the piezoelectric voltage and the hysteresis resistor R_{hys} prevents undesired triggers. The output terminal of the passive differentiator and the piezoelectric reference terminal are then connected to the negative input and the positive input of the comparator respectively. Different from Chen's design, only one piezoelectric patch is used here and thus the detecting signal of the velocity has larger phase lag.

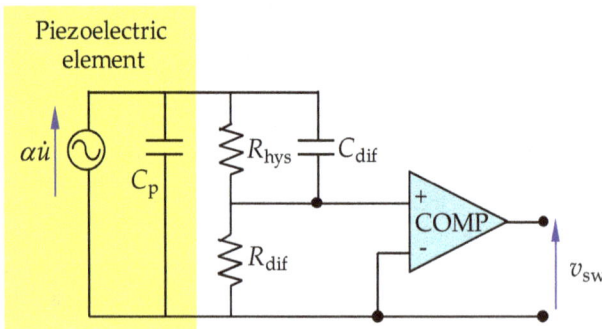

Figure 14. Equivalent current (velocity) sensing by a resistor and a passive low-pass filter

It should be noted that the comparator requires a DC power source. The auxiliary DC power source can be obtained from piezoelectric element with the diode rectifier directly. Figure 15 shows this auxiliary DC power supply.

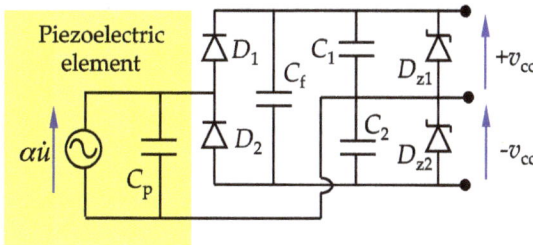

Figure 15. Auxiliary power supply for the comparator

Although the zero-crossing detector seems more complex than the peak detector, the zero-crossing detector can have less phase delay than the peak detector [15]. In addition, both the peak detector and zero-crossing designs potentially can be integrated into the ASIC design. Combing the zero-crossing detector, the auxiliary power supply and the SSHI interface, the completed self-powered series-SSHI converter for the piezoelectric energy harvester is shown in Fig. 16. It should be noted that there are three independent piezoelectric elements on the host structure. Since piezoelectric element is mostly in open-circuit state in series-SSHI converter, it is equivalently connect to high impedance. Therefore, adding the small sensing resistance influences the piezoelectric voltage in the SSHI converter. Another independent piezoelectric element is thus required to avoid phase lag of the velocity in the case of zero-crossing detector. However, an additional independent piezoelectric is not necessary in Richard's design, *i.e.* the self-powered peak detectors with the SSHI switch. The comparisons between above three self-powering techniques are summarized in Table 1.

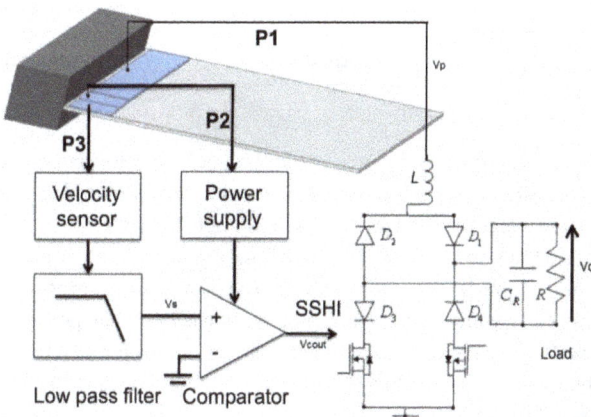

Figure 16. Self-powered SSHI converter based on a zero-crossing detector [15]

	Richard et al.	Chen et al.	Ben-Yaakov & Krihely
Signal conditioner	Peak detector	Zero-crossing detector	Zero-crossing detector
Principle			
Waveforms			
Suitable converter	Series-SSHI & parallel –SSHI	Series-SSHI & parallel -SSHI	Series-SSHI & parallel – SSHI
Phase lag	Medium (Series-SSHI), Large (Parallel-SSHI)	Small	Large
Piezoelectric element	Single piezoelectric patch	Require additional piezoelectric patches for sensing	1) Single piezoelectric patch (larger phase lag) 2) Additional patches (small phase lag)
Voltage drops	2 diode drops	2 diode drops	2 diode drops

Table 1. The comparison of each design

To summarize the comparison, we can say that the peak detector technique proposed by Richard et al. is an interesting solution as it is reliable and robust and can even use the signal from the piezoelectric transducer itself to operate (*i.e.*, no additional patch is needed). However, the command of the switching signal is not perfectly generated on maximum and minimum displacement values. The phase delay appears because of the voltage gap of discrete components (diodes and transistors) that perform the detection and the switching command, and it depends on the piezoelectric voltage level. Moreover, in the case of the parallel-SSHI converter, this phase delay also depends on the load. The zero crossing technique proposed by Chen et al. is the most accurate solution because the load impedance does not delay the sensing signal. However, the drawback of this technique is that another independent piezoelectric patch is required with a self-powered supply and that a low-pass filter must be added to avoid the high-frequency noise generated by the SSHI switching. The Ben-Yaakov & Krihely's technique simply used the signal from the single piezoelectric transducer, so their velocity detection is not accurate and is influenced by the load effect. Specificaly, in this solution, the switching commands do not perfectly fit the value of the zero velocities and owns poorer performance. In addition, this technique also requires an extra power supply for the comparator.

4. Conclusion

This chapter focused on the charging circuit of the piezoelectric energy harvester. The operation principle of the series-SSHI and parallel-SSHI converter was explained. Both two SSHI techniques can enlarge the collected power efficiently on a weakly coupled structure. Considering the application of the micro-scale energy harvester, the low voltage drop SSHI converter was developed by using the body diodes in MOSFET switches. It can neglect maximum 4 diodes drops, which is a significant value in micro-scale energy harvester. Two kinds of signal conditioners, *i.e.* the peak detector and the zero-crossing detector were demonstrated. The peak detector can have easy topology and high reliability with single piezoelectric patch, but it generates a phase delay especially in the parallel-SSHI converter. Chen's zero-crossing detector has less phase lag and thus the switching command is no difference in the series-SSHI converter and the parallel-SSHI converter, but it is more complex and requires additional piezoelectric patches. Ben-Yaakov's zero-crossing detector adopted the single piezoelectric patch, but it owns large phase lag.

Author details

Yuan-Ping Liu and Dejan Vasic
SATIE, ENS Cachan, France

Acknowledgement

The authors are grateful to ELECERAM Technology Co., Ltd. and Miézo Technology Co. Ltd. for providing us with the different types of piezoelectric elements used in this research work.

5. References

[1] Hagood NW. and Flotow A. Damping of structural vibrations with piezoelectric materials and passive electrical networks. Journal of Sound and Vibration 1991; 146(2) 243-268.

[2] Tang L, Yang Y, Soh CK. Toward Broadband Vibration-based Energy Harvesting. Journal of Intelligent Material System and Structure 2010; 21(18) 1867-1897.

[3] Khaligh A, Zeng P, Zheng C. Kinetic Energy Harvesting Using Piezoelectric and Electromagnetic Technologies-State of the Art. IEEE Transactions on Industrial Electronics 2010; 57(3) 850-860.

[4] Roundy S, Wright PK, Rabaey J. A study of low level vibrations as a power source for wireless sensor nodes. Computer Communications 2003; 26(11) 1131-1144.

[5] Inman D, Priya SJ. Energy Harvesting Technologies: Springer; 2009.

[6] Arnold DP. Review of Microscale Magnetic Power Generation. IEEE Transactions on Magnetics Magnetics 2007. 43(11) 3940-3951.

[7] Lefeuvre E, Badel A, Richard C, Guyomar D. Piezoelectric energy harvesting device optimization by synchronous electric charge extraction. Journal of Intelligent Material Systems and Structures 2005; 16(10) 865-876.

[8] Lefeuvre E, Badel A, Richard C, Guyomar D. High performance piezoelectric vibration energy reclamation. SPIE proceeding: Smart Structures and Materials & Nondestructive Evaluation and Health Monitoring 2004, San Diego (CA), USA. Vol. 5390, pp 379-387.

[9] Guyomar D, Badel A, Lefeuvre E, Richard C. Toward energy harvesting using active materials and conversion improvement by nonlinear processing. IEEE Transactions on Ultrasonics Ferroelectrics and Frequency Control 2005; 52(4) 584-595.

[10] Lefeuvre E, Badel A, Richard C, Guyomar D., Piezoelectric energy harvesting device optimization by synchronous electric charge extraction. Journal of Intelligent Material System and Structure 2005; 16(10) 865-876.

[11] Badel A, Guyomar D, Lefeuvre E, Richard C, Piezoelectric energy harvesting using a synchronized switch technique. Journal of Intelligent Material System and Structure 2006; 17(8-9) 831-839.

[12] Badel, A.; Benayad, A.; Lefeuvre, E.; Lebrun, L.; Richard, C. & Guyomar, D. (2006). Single Crystals and Nonlinear Process for Outstanding Vibration Powered Electrical Generators. IEEE Trans. on Ultrason., Ferroelect., Freq. Contr., Vol. 53, 673-684.

[13] Badel, A.; Sebald, G.; Guyomar, D.; Lallart, M.; Lefeuvre, E.; Richard, C. & Qiu, J. (2006). Piezoelectric vibration control by synchronized switching on adaptive voltage sources: Towards wideband semi-active damping. J. Acoust. Soc. Am., Vol. 119, No. 5, 2815-2825.

[14] Minazara E, Vasic D, Costa F, Poulin G. Piezoelectric diaphragm for vibration energy harvesting. Ultrasonics 2006; 44 e699-e703.

[15] Chen YY, Vasic D, Costa F, Wu WJ, Lee CK. A self-powered switching circuit for piezoelectric energy harvesting with velocity control. The European Physical Journal Applied Physics 2012, 57 30903.

[16] Chen YY, Vasic D, Costa F, Wu WJ, Lee CK. Self-powered Piezoelectric Energy Harvesting. IEEE IECON 2010, Phoenix, Arizona, USA, 7-10 november 2010

[17] Yang YW, Tang LH. Equivalent Circuit Modeling of Piezoelectric Energy Harvesters. Journal of Intelligent Material System and Structure 2009; 20(18) 2223-2235.

[18] Wu WJ, Wickenheiser AM, Reissman T, Garcia E. Modeling and experimental verification of synchronized discharging techniques for boosting power harvesting from piezoelectric transducers. Smart Material and Structures 2009, 18 055012 doi:10.1088/0964-1726/18/5/055012.

[19] Liu YP, Vasic D, Costa F. Piezoelectric Energy Harvester Circuit for Capacitive Storage Buffer. Electrimacs 2011, June 6-8, Paris, France, 2011.

[20] Liu YP, Vasic D, Costa F, Wu WJ, Lee CK. Velocity-Controlled Switching Piezoelectric Damping Based on Maximum Power Factor Tracking and Work Cycle Observation. Proceedings of 19th International Conference on Adaptive Structures and Technologies (ICAST 2008), paper no. 40, Ascona, Switzerland, October 6-9, 2008.

[21] Shu YC, Lien IC, Wu WJ. An improved analysis of the SSHI interface in piezoelectric energy harvesting, Smart Material Structures 2007, 16 2253–64 doi:10.1088/0964-1726/16/6/028.

[22] Lesieutre GA, Ottman GK, Hofmann HF. Damping as a result of piezoelectric energy harvesting. Journal of Sound and Vibration 2004; 269(3-5) 991–1001.

[23] Liang JR, Liao WH. Piezoelectric energy harvesting and dissipation on structural damping. Journal of Intelligent Material System and Structure 2009. 20(5) 515–27.

[24] Lallart M, Garbuio L, Petit L, Richard C, Guyomar D. Double synchronized switch harvesting (DSSH): a new energy harvesting scheme for efficient energy extraction Double synchronized switch harvesting (DSSH): a new energy harvesting scheme for efficient energy extraction. IEEE Transactions on Ultrasonics, Ferroelectrics and Frequency Control 2008; 55(10) 2119-2130.

[25] Lee BS, Wu WJ, Shih WP, Vasic D, Costa F. Power Harvesting using Piezoelectric MEMS Generator with Interdigital Electrodes. IEEE Ultrasonics Symposium, New-York USA, 28-31 October 2007

[26] Lallart M, Guyomar D. An optimized self-powered switching circuit for non-linear energy harvesting with low voltage output. Smart Materials and Structures 2008; 17(3) 035030. doi:10.1088/0964-1726/17/3/035030.

[27] D. Guyomar and M. Lallart, Nonlinear conversion enhancement for efficient piezoelectric electrical generators, in Ferroelectrics, Sciyo/Intech, 2010, ISBN 978-953-307-439-9.

[28] Richard C, Guyomar D, Lefeuvre E. Self-powered electronic breaker with automatic switching by detecting maxima or minima of potential difference between its power electrodes. Patent PCT/FR2005/003000 (publication number: WO/2007/063194) 2007.

[29] M. Lallart, Y-C. Wu, D. Guyomar, Switching Delay Effects on Nonlinear Piezoelectric Energy Harvesting Techniques, IEEE Trans. on Industrial Electronics, vol. 59, no. 1, January 2012

[30] Ben-Yaakov S, Krihely N. Resonant rectifier for piezoelectric sources. Applied Power Electronics Conference and Exposition, APEC March 2005. vol. I, pp. 249-253.

Chapter 14

Modeling Aspects of Nonlinear Energy Harvesting for Increased Bandwidth

Marcus Neubauer, Jens Twiefel, Henrik Westermann and Jörg Wallaschek

Additional information is available at the end of the chapter

1. Introduction

Over the last years, the field of energy harvesting has become a promising technique as power supply of autonomous electronic devices. Those use the surrounding energy, such as vibrations, temperature gradients or radiation, for conversion in electrical energy. Mechanical vibrations are an attractive source due to their high availability in technical environments, thus numerous research groups are working on this topic. The most important conversion methods for ambient vibrations are electromagnetic, electrostatic and piezoelectric. All those techniques have been successfully demonstrated in the past. [11] provides an overview of the basics in energy harvesting.

Due to the fact, that vibration energy harvester generates the most energy when the generator is excited at its resonance frequency, the converter needs to be tuned to the main external frequency of the individual environment. If the excitation frequency shifts, the performance of the generator may reduced drastically. In practical use the vibration of an environment may vary in a large spectrum. To overcome this disadvantage researchers work hard to increase the working bandwidth of an energy harvester.

This chapter is a contribution to the current state of the art for modeling broadband energy harvesting generators. In the first part the electromechanical model is derived in terms of using lumped parameters. The system is based on a piezoelectric bimorph structure. The coupled differential equations for the case of a simple electrical circuit are derived and furthermore the possibility to enhance the energy extraction is analyzed. The use of generator arrays to archive a high power outputs in a wide frequency range is discussed. In detail, the Synchronized Switch Harvesting on Inductor (SSHI) technique is studied. Further the modeling of the promising piezomagnetoelastic energy harvesting technique is covered in the last part.

2. Linear piezoelectric energy harvesting system

This section is devoted to the modeling of a linear piezoelectric bimorph for energy harvesting. This will be the basis for the proposed nonlinear techniques with enhanced bandwidth duscussed in the following sections.

2.1. Modeling of piezomechanical structures

Fundamental for the energy harvesting techniques presented in the following is the piezoelectric bimorph. In the following, the general modeling of piezomechanical structures is given, and further on applied to the case of a bending bimorph.

The calculations are based on the potential energy stored in a piezoelement,

$$U = \frac{1}{2} \int_V (T_i S_i + D_3 E_3) \, dV, \tag{1}$$

where T, S, D, E represent the mechanical stress and strain as well as the electrical displacement and field. V is the volume of the piezoelement. A one-dimensional strain distribution within the piezoelement in axis direction i is assumed. According to [6] the axis of polarization is defined as x_3. Therefore the transversal effect is represented by $i = 1$, where the mechanical strain is normal to the direction of polarization, and the longitudinal effect by $i = 3$, where the mechanical strain is in the direction of polarization. After some mathematical calculations, see [9] the energy can be written as

$$U = \frac{1}{2} \frac{Q_p^2}{C_p} + \frac{1}{2} \frac{1}{s_{ii}^E} \int_V \left[S_i^2 + \frac{k_{3i}^2}{1 - k_{3i}^2} \Delta S_{i,3}^2 \right] dV + \frac{1}{2} \frac{\alpha^2}{C_p} (\ell_i \bar{S}_i)^2 - \frac{1}{2} 2 \frac{\alpha}{C_p} \ell_i \bar{S}_i Q_p. \tag{2}$$

The electrical charge at the electrodes is termed Q_p, while the mechanical compliance in axis direction i is given as s_{ii}. Further on, the material coupling of the piezoelement is given by $k_{3i} = d_{3i}^2 / (s_{ii}^E \varepsilon_{33}^T)$. In the following, the stiffness c_p in x_i direction, the capacitance C_p of the piezoceramics and the piezoelectric force factor α are introduced as

$$c_p = \frac{1}{s_{ii}^E} \frac{V}{\ell_i^2},$$

$$C_p = \varepsilon_{33}^T \left(1 - k_{3i}^2 \right) \frac{A_{el}}{\ell_3},$$

$$\alpha = \frac{d_{3i}}{s_{ii}^E} \frac{A_{el}}{\ell_i}. \tag{3}$$

They depend on the area of electrodes A_{el}, the piezoelectric constant d_{3i}, the permittivity ε_{33} as well as the geometry of the piezoelement (length ℓ_3 between electrodes and length ℓ_i of the piezoelement in direction of mechanical strain).

The energy terms in Equation 2 can be classified as the stored electrical energy, the stored mechanical energy and the converted energy. For convenience of the following calculations,

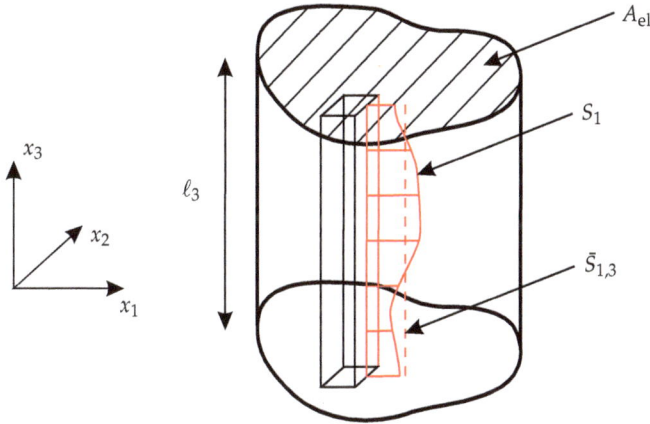

Figure 1. Piezoelement with uniaxial strain distribution.

the mechanical strain is split into the mean value $\bar{S}_i$, the mean strain $\bar{S}_{i,3}$ along the x_3 axis (between the electrodes), and the difference $\Delta S_{i,3}$ between the actual strain and $\bar{S}_{i,3}$,

$$\bar{S}_{i,3}(x_1, x_2) = \frac{\int\limits_0^{\ell_3} S_i(x_1, x_2, x_3)\mathrm{d}x_3}{\ell_3}, \quad \Delta S_{i,3} = S_i - \bar{S}_{i,3}. \tag{4}$$

The reason for this representation is that piezoelectric systems with a homogeneous strain distribution is readliy described by $\Delta S_{i,3} = 0$, which strongly simplifies the calculations. Additionally, the influence of an uneven strain distribution can be seen in the term $\Delta S_{i,3}$. See Figure 1 for an illustration of these definitions.

In case of a continuous system it is reasonable to discretize it for the further analysis. The mechanical deformation is then described by n degrees of freedom (DOF) q_i, while the charge Q_p is the electrical DOF,

$$\mathbf{q} = \begin{bmatrix} \mathbf{q}_{\text{mech}} \\ Q_p \end{bmatrix}. \tag{5}$$

Each mechanical DOF is associated with a global mode shape, which defines the mechanical strain distribution S_i within the piezoelectric volume (and the rest of the mechanical system). The overall strain distribution is then the sum of all mode shapes. In order to rewrite the energy term in Equation 2, the term $-\ell_i \bar{S}_i$, which represents the mean deformation of the piezoelement in x_i direction, will be represented by

$$- \ell_i \bar{S}_i = \sum_{k=1}^n \kappa_k q_k. \tag{6}$$

Here, a mechanical coupling vector $\boldsymbol{\kappa}$ is introduced. In that form, the energy terms in Equation 2 can be rewritten as

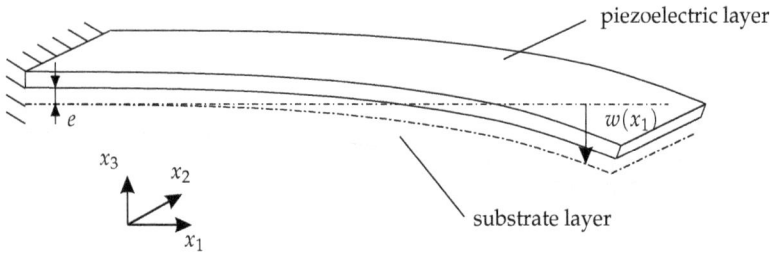

Figure 2. Piezoelectric bimorph.

$$-\frac{1}{2}2\frac{\alpha}{C_\mathrm{p}}\ell_i\bar{S}_iQ_\mathrm{p} = \frac{1}{2}\mathbf{q}^\mathrm{T}\begin{bmatrix} 0 & \frac{\alpha}{C_\mathrm{p}}\boldsymbol{\kappa} \\ \frac{\alpha}{C_\mathrm{p}}\boldsymbol{\kappa}^\mathrm{T} & 0 \end{bmatrix}\mathbf{q}$$

$$\frac{1}{2}\frac{\alpha^2}{C_\mathrm{p}}(\ell_i\bar{S}_i)^2 = \frac{1}{2}\mathbf{q}^\mathrm{T}\begin{bmatrix} \frac{\alpha^2}{C_\mathrm{p}}\boldsymbol{\kappa}\boldsymbol{\kappa}^\mathrm{T} & 0 \\ 0 & 0 \end{bmatrix}\mathbf{q}$$

$$\frac{1}{2}\frac{1}{s_{ii}^E}\int_V\left[S_i^2 + \frac{k_{3i}^2}{1-k_{3i}^2}\Delta S_{i,3}^2\right]dV = \frac{1}{2}\mathbf{q}^\mathrm{T}\begin{bmatrix} \mathbf{C}_\mathrm{mech} & 0 \\ 0 & 0 \end{bmatrix}\mathbf{q}. \tag{7}$$

The potential energy can be rewritten as

$$U = \frac{1}{2}\mathbf{q}^\mathrm{T}\mathbf{C}\mathbf{q}, \tag{8}$$

so that the stiffness matrix of the system follows as

$$\mathbf{C} = \begin{bmatrix} \mathbf{C}_\mathrm{mech} + \frac{\alpha^2}{C_\mathrm{p}}\boldsymbol{\kappa}\boldsymbol{\kappa}^\mathrm{T} & \frac{\alpha}{C_\mathrm{p}}\boldsymbol{\kappa} \\ \frac{\alpha}{C_\mathrm{p}}\boldsymbol{\kappa}^\mathrm{T} & \frac{1}{C_\mathrm{p}} \end{bmatrix}. \tag{9}$$

The term $\mathbf{C}_\mathrm{mech}$ represents the 'mechanical' stiffness matrix of the piezoelement, which can be deduced in the same way as standard mechanical systems.

2.2. Piezoelectric bimorph

Now we can apply the above obtained results for the piezoelectric bimorph. We are considering the general case of a piezoelectric layer which has a distance e to the neutral axis of the beam. The coordinate axes are defined in such a way that the origin is at contact between the piezoelectric and the substrate layers at the clamped end, see Figure 2. The x_1 axis is in beam direction and the deformations occur in x_3 direction, which is also the direction of polazization. With Euler-Bernoulli assumptions the strain is only applied in x_1 direction. That means the transversal effect of the piezoceramics is utilized. The strain terms described in the previous section then read for the case of the clamped beam:

$$\bar{S}_{1,3} = \frac{\int\limits_0^{\ell_3} S_1 dx_3}{\ell_3} = -\left(e + \frac{\ell_3}{2}\right) w''(x_1, t),$$

$$\Delta S_{1,3} = S_1 - \bar{S}_{1,3} = \left(\frac{\ell_3}{2} - x_3\right) w''(x_1, t),$$

$$\bar{S}_1 = -\left(e + \frac{\ell_3}{2}\right) \frac{\int\limits_0^{\ell_1} w''(x_1, t) dx_1}{\ell_1} = -\left(e + \frac{\ell_3}{2}\right) \frac{w'(\ell_1) - w'(0)}{\ell_1}. \tag{10}$$

The bending of the beam is described by $w(x_1, t)$. This term will be split into the part depending on coordinate x_1 and the part depending on time t,

$$w(x_1, t) = W(x_1)q(t), \quad w''(x_1, t) = W''(x_1)q(t). \tag{11}$$

In this example, only one mechanical degree of freedom is used to describe the vibrations. This is typically a reasonable approximation when the system vibrates close to one of its eigenfrequencies. In this way, the general stiffness matrix according to Equation 9 reduces to

$$\mathbf{C} = \begin{bmatrix} c_{\text{mech}} + \frac{\alpha^2}{C_p}\kappa^2 & \frac{\alpha}{C_p}\kappa \\ \frac{\alpha}{C_p}\kappa & \frac{1}{C_p} \end{bmatrix}, \tag{12}$$

and the mechanical coupling is written as

$$\kappa = \left(e + \frac{\ell_3}{2}\right)\left(W'(\ell_1) - W'(0)\right), \tag{13}$$

while the piezoelectric coupling reads

$$\frac{\alpha}{C_p} = \frac{k_{31}^2}{1 - k_{31}^2} \frac{1}{d_{31}} \frac{\ell_3}{\ell_1}, \quad \frac{\alpha^2}{C_p} = \frac{1}{s_{11}^E} \frac{k_{31}^2}{1 - k_{31}^2} \frac{\ell_2 \ell_3}{\ell_1}. \tag{14}$$

Terms of the kind $\frac{\alpha^2}{C_p}\kappa^2$, which determine the increase in eigenfrequencies between short circuit electrodes and isolated electrodes are obtained as

$$\frac{\alpha^2}{C_p}\kappa^2 = \frac{1}{s_{11}^E} \frac{k_{31}^2}{1 - k_{31}^2} \frac{\ell_2 \ell_3}{\ell_1} \left(e + \frac{\ell_3}{2}\right)^2 \left(W'(\ell_1) - W'(0)\right)^2. \tag{15}$$

For a better understanding of these terms, it is useful to introduce the area moment of inertia I^{PZT} of the piezoceramics around its own center of gravity and the moment of inertia I^{nF} around the neutral axis of the beam. Additionally, also the difference $I^{\text{nF}} - I^{\text{PZT}}$ is included in the results,

$$I^{PZT} = \frac{\ell_2 \ell_3^3}{12},$$

$$I^{nF} = \frac{\ell_2 \ell_3^3}{12} + \left(e + \frac{\ell_3}{2}\right)^2 \ell_2 \ell_3,$$

$$I^{nF} - I^{PZT} = \left(e + \frac{\ell_3}{2}\right)^2 \ell_2 \ell_3. \tag{16}$$

With these definitions, the coupling terms can be expressed as

$$\frac{\alpha^2}{C_p} \kappa^2 = \frac{1}{s_{11}^E} \frac{k_{31}^2}{1 - k_{31}^2} \frac{I^{nF} - I^{PZT}}{\ell_1} \left(W'(\ell_1) - W'(0)\right)^2. \tag{17}$$

This result can be used to discuss different geometries and types of bimorphs. Obviously a beam that consists only of piezoelectric material does not have any coupling at all, because the distance e is exactly one half of the thickness of the piezoelectric layer, $e = -\frac{\ell_3}{2}$. This means the term $I^{nF} - I^{PZT}$ vanishes. Contrary to this, a beam which is made of two identical piezoelectric layer is represented by $e = 0$ because of the symmetry, and a coupling exists. However, yet more efficient is the design of bimorphs or trimorphs with a substrate layer, which moves the neutral axis away from the surface of the piezoelectric layer. This results in a positive value $e > 0$. The best type is a symmetric trimorph with identical piezoelectric layers on both sides of the substrate layer. Here the distance equals half of the substrate layer. More details about the optimization of bimorphs can be found in [10].

In general, the piezomechanical system can be described by the following differential equations,

$$\begin{bmatrix} m_{mech} & 0 \\ 0 & 0 \end{bmatrix} \begin{bmatrix} \ddot{q} \\ \ddot{Q}_p \end{bmatrix} + \begin{bmatrix} c_{mech} + \frac{\alpha^2}{C_p}\kappa^2 & \frac{\alpha}{C_p}\kappa \\ \frac{\alpha}{C_p}\kappa & \frac{1}{C_p} \end{bmatrix} \begin{bmatrix} q \\ Q_p \end{bmatrix} = \begin{bmatrix} F(t) \\ -u_p(t) \end{bmatrix}, \tag{18}$$

with the modal mass m_{mech}, the external force $F(t)$ and the voltage $u_p(t)$ at the electrodes of the piezoelement.

2.3. Linear energy harvester

Based on these results, the simplest and linear energy harvester can be modeled. In this case only a resistor is connected as an electrical load at the electrodes of the piezoelement. Therefore the voltage u_p is dependent on the time derivative of charge $\dot{Q}_p$ and the differential equations of motions for this case including damping d_{mech} read

$$\begin{bmatrix} m_{mech} & 0 \\ 0 & 0 \end{bmatrix} \begin{bmatrix} \ddot{q} \\ \ddot{Q}_p \end{bmatrix} + \begin{bmatrix} d_{mech} & 0 \\ 0 & R \end{bmatrix} \begin{bmatrix} \dot{q} \\ \dot{Q}_p \end{bmatrix} + \begin{bmatrix} c_{mech} + \frac{\alpha^2}{C_p}\kappa^2 & \frac{\alpha}{C_p}\kappa \\ \frac{\alpha}{C_p}\kappa & \frac{1}{C_p} \end{bmatrix} \begin{bmatrix} q \\ Q_p \end{bmatrix} = \begin{bmatrix} F(t) \\ 0 \end{bmatrix} \tag{19}$$

We seek for the amplitudes of the stationary oscillations,

$$\hat{q} = \left(-\Omega^2 \mathbf{M} + j\Omega \mathbf{D} + \mathbf{C} \right)^{-1} \begin{bmatrix} \hat{F} \\ 0 \end{bmatrix}, \tag{20}$$

with the corresponding system matrices according to Equation 19. The amplitudes of the time signals are marked by a hat. With the stationary charge amplitude $\hat{Q}_p$ the instantaneous power $p(t)$ can be calculated,

$$p(t) = R i_p^2(t) = R\Omega^2 \hat{Q}_p^2 \cos^2 (\Omega t), \tag{21}$$

where i_p is the current. The energy that is dissipated in the resistor will be treated as the stationary harvested energy $E_{h,stat}$. It is then the integral of the power p,

$$E_{h,stat} = \int_0^{\frac{2\pi}{\Omega}} p(t) dt = \pi R\Omega \hat{Q}_p^2. \tag{22}$$

This result, normalized to force amplitude, is shown in Figure 3 versus the load resistance R and the excitation frequency Ω, normalized to the mechanical eigenfrequenzcy ω_0. For this study, the following system parameters are used:

$$m_{mech} = 0.005\text{kg},$$

$$d_{mech} = 0.1212\text{Ns/m},$$

$$c_{mech} = 341.2651\text{N/m},$$

$$\alpha\kappa = 0.002,$$

$$C_p = 83.676\text{nF}. \tag{23}$$

Obviously the harvested energy is highly frequency dependent. Only in a narrow frequency range around the eigenfrequency the efficiency is high. Also the resistor must be tuned for maximum harvested energy. The system is less sensitive towards changes of the resistance, but one can show that the optimal resistance which is optimal for low coupling and/or high damping is obtained as

$$R_{opt} = \frac{1}{\Omega C_p} \tag{24}$$

3. Array configuration

Utilizing an array configuration for the extension of the generator bandwidth is a commonly used approach. The idea is simple and powerfull: multiple generator elements are tuned to slightly (a few %) different eigenfrequencies. A major factor is the connection to the electrical system; if each individual element uses its own bridge rectifier, the elements are electrically uncoupled and their output power simply can be summed up. However, individual bridge

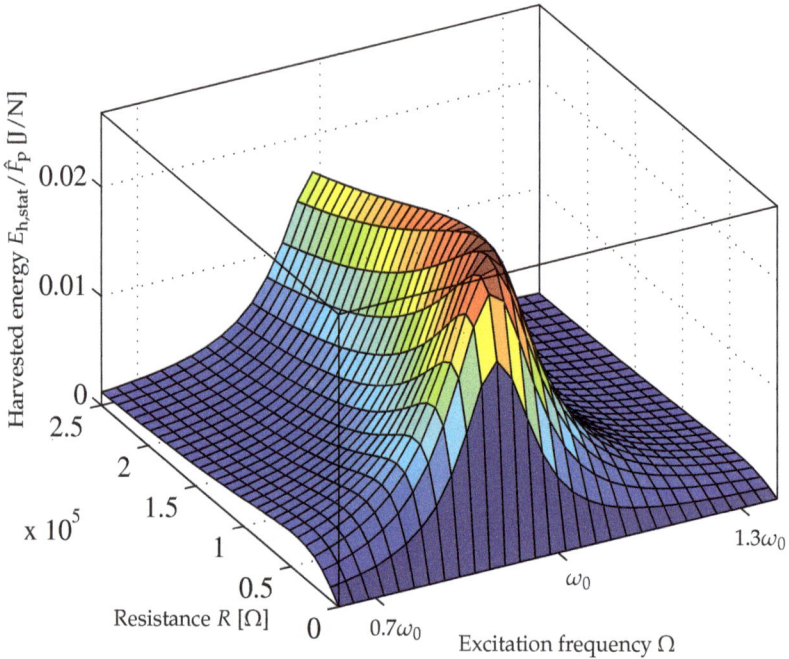

Figure 3. Harvested energy $E_{h,stat}$ versus normalized excitation frequency Ω and load resistance R.

rectifiers are a comparable big effort, a huge number of elements and cables is needed, further the voltage drop over each diode is also summed up, so that the losses increase. As soon the individual elements are made of one part the electrodes are connected naturally, which results automatically in parallel connected elements. In such a configuration, Figure 4, the elements are electrically coupled.

This section investigates the effect of the electrical coupling on the performance of piezoelectric energy harvesting generators. Two configurations are investigated, the electrical parallel connection as well as the electrical serial connection. Both cases are utilizing two elements as most basic version of a piezoelectric array. The model is based on the linear generator model in Equation 19, therefore a fore excitation is assumed. We further assume that the applied fore is equal on all elements, representing a common support. The resonance frequency tuning is made by an adoption of the modal mass of the elements (change of tip mass), all other parameters are assumed to be constant. The configurations including boundary conditions is given in Figure 4. Each element can be represented by the linear Equation 19. In the parallel configuration the two voltages are equal and $u_{P_1} = u_{P_2} =$

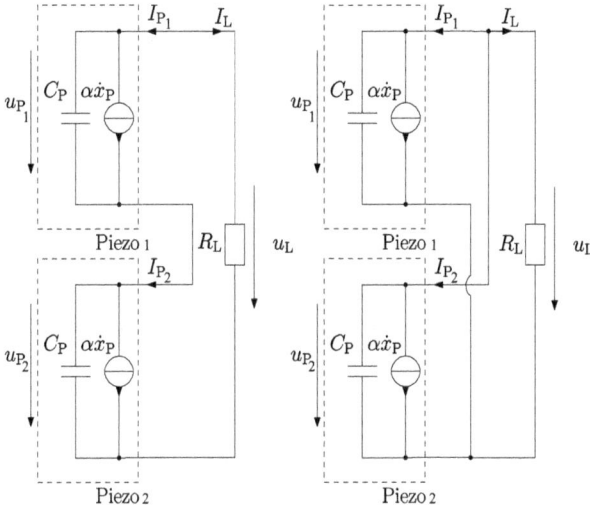

Figure 4. Schematic Circuit. Left: series configuration. Right: parallel configuration

$R_L \left(i_{P_1} + i_{P_2} \right)$ apply due to Kirchhoff's rules. Using both, the parallel configuration is described by

$$\begin{bmatrix} m_{\text{mech}} - \Delta m & 0 & 0 & 0 \\ 0 & m_{\text{mech}} + \Delta m & 0 & 0 \\ 0 & 0 & 0 & 0 \\ 0 & 0 & 0 & 0 \end{bmatrix} \begin{bmatrix} \ddot{q}_1 \\ \ddot{q}_2 \\ \ddot{Q}_{\text{p1}} \\ \ddot{Q}_{\text{p2}} \end{bmatrix} + \begin{bmatrix} d_{\text{mech}} & 0 & 0 & 0 \\ 0 & d_{\text{mech}} & 0 & 0 \\ 0 & 0 & R & R \\ 0 & 0 & R & R \end{bmatrix} \begin{bmatrix} \dot{q}_1 \\ \dot{q}_2 \\ \dot{Q}_{\text{p1}} \\ \dot{Q}_{\text{p2}} \end{bmatrix}$$

$$\dots + \begin{bmatrix} c_{\text{mech}} + \frac{\alpha^2}{C_\text{P}}\kappa^2 & 0 & \frac{\alpha}{C_\text{P}}\kappa & 0 \\ 0 & c_{\text{mech}} + \frac{\alpha^2}{C_\text{P}}\kappa^2 & 0 & \frac{\alpha}{C_\text{P}}\kappa \\ \frac{\alpha}{C_\text{P}}\kappa & 0 & \frac{1}{C_\text{P}} & 0 \\ 0 & \frac{\alpha}{C_\text{P}}\kappa & 0 & \frac{1}{C_\text{P}} \end{bmatrix} \begin{bmatrix} q_1 \\ q_2 \\ Q_{\text{p1}} \\ Q_{\text{p2}} \end{bmatrix} = \begin{bmatrix} F(t) \\ F(t) \\ 0 \\ 0 \end{bmatrix}. \tag{25}$$

The coupling between the elements is obviously seen in the damping matrix. In the series configuration, the current at both generators is equal. Therefore $u_{P_1} + u_{P_2} = R_L i_{P_1} = R_L i_{P_2}$ is applied to couple the two generators:

$$\begin{bmatrix} m_{\text{mech}} - \Delta m & 0 & 0 \\ 0 & m_{\text{mech}} + \Delta m & 0 \\ 0 & 0 & 0 \end{bmatrix} \begin{bmatrix} \ddot{q}_1 \\ \ddot{q}_2 \\ \ddot{Q}_\text{P} \end{bmatrix} + \begin{bmatrix} d_{\text{mech}} & 0 & 0 \\ 0 & d_{\text{mech}} & 0 \\ 0 & 0 & R \end{bmatrix} \begin{bmatrix} \dot{q}_1 \\ \dot{q}_2 \\ \dot{Q}_\text{P} \end{bmatrix} + \dots$$

$$\begin{bmatrix} c_{\text{mech}} + \frac{\alpha^2}{C_\text{P}}\kappa^2 & 0 & \frac{\alpha}{C_\text{P}}\kappa \\ 0 & c_{\text{mech}} + \frac{\alpha^2}{C_\text{P}}\kappa^2 & \frac{\alpha}{C_\text{P}}\kappa \\ \frac{\alpha}{C_\text{P}}\kappa & \frac{\alpha}{C_\text{P}}\kappa & \frac{2}{C_\text{P}} \end{bmatrix} \begin{bmatrix} q_1 \\ q_2 \\ Q_\text{P} \end{bmatrix} = \begin{bmatrix} F(t) \\ F(t) \\ 0 \end{bmatrix}. \tag{26}$$

Here the coupling is evidently in the stiffness matrix. Where all parameters correspond to the ones from the modeling section. The mass for frequency adoption is $\Delta m = 0.5g$. For the steady state the system can be solved and the transfer functions can be determined analog to the single system. The total dissipated energy at the resistor is again given by

$$E_{h,stat} = \int_0^{\frac{2\pi}{\Omega}} u(t)i(t)\mathrm{d}t. \tag{27}$$

Evaluating this equation for voltage and current over the load resistor gives the gained energy for the application, the result is depicted in Figure 5 for serial configuration and in Figure 6 for parallel configuration. For the in series connected generators the bandwidth is widened at high impedance loads and it is not significantly changed for low impedances. For the parallel configuration, a bandwidth expanded at low impedance loads and also not changed for high impedances.

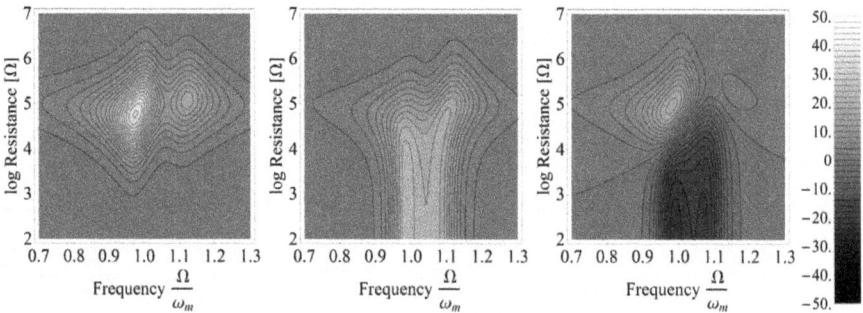

Figure 5. Energy per period in serial configuration in mJ/N. Left: gained useable energy. Middle: element 1. Right: element2.

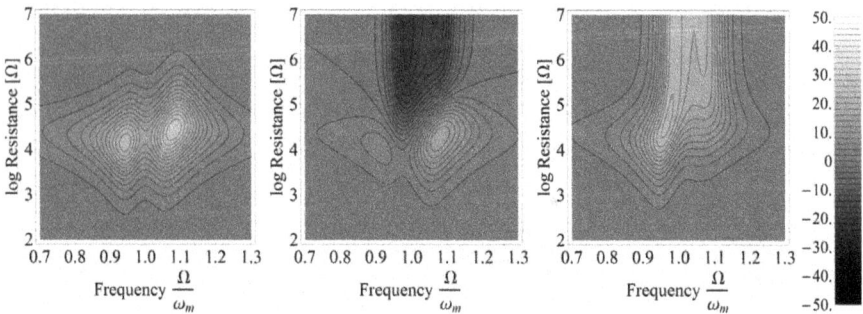

Figure 6. Energy per period in parallel configuration in mJ/N. Left: gained useable energy. Middle: element 1. Right: element2.

To explain why there is no widening of the bandwidth in serial connection at low impedances the Equation 27 is evaluated for both elements, using the individual voltage and the common current. Figure 5 show that the second element works as energy sink for low impedances, with the consequence, that the energy generated by element one is used to actuate the other one. Figure 7 shows this effect. Even with the overall maximum displacement of element two

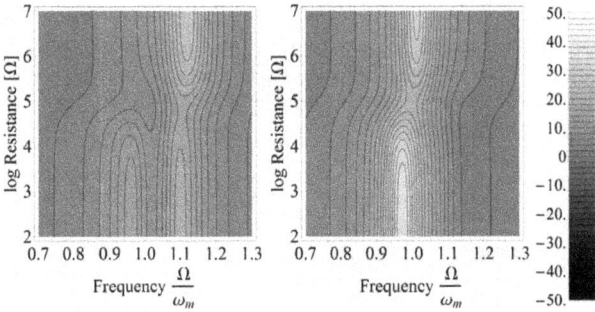

Figure 7. Tip displacement of both elements in serial configuration in mm/N. Left: element 1. Right: element2.

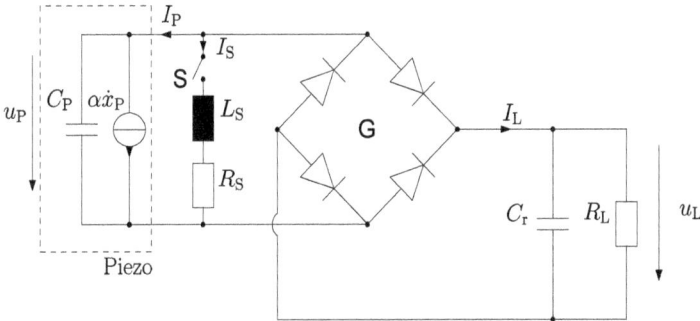

Figure 8. Schematic of SSHI circuit.

at low impedances the gained useable energy is low. The same effect but for high impedances is shown in Fig. 6 for the parallel configuration. In this case element one is the energy sink.

Concluding, the utilization of arrays with serials or parallel electrical coupling has only a major positiv effect on the bandwidth for a matched impedance, for unmatched impedances the coupling can be a drawback. The bandwidth can be enlarged with any further element, the mean energy output over bandwidth in general is higher if the mistuning of the resonance frequencies is smaller. For power and bandwidth comparison of generator arrays it is reasonable to keep the volume of active material constant.

4. Switching networks (SSHI)

An important technique to enhance the energy extraction is to use nonlinear switching networks. In detail, the 'Synchronized Switch Harvesting on Inductor' (SSHI) technique is studied. Such networks are an active field of research [3, 4, 7]. The corresponding network is shown in Figure 8. This nonlinear electric circuit consists of a switching LR-branch, a rectifier and load capacitor C_r. The load is again described as a resistor R_L. Assuming a sinusoidal mechanical deformation of the piezoceramics, the switch is briefly closed on minima and maxima of the deformation. During these times, an oscillating electric circuit is formed, as the capacitive piezoelectric transducer and the inductance are connected. During this electrical semi-period the switch is kept close and the voltage is inverted. As this electrical period time is generally much shorter than the mechanical one, this occurs nearly instantaneously. After inversion, the switch is opened again until the next deformation extremum. Consequently, the resulting voltage signal at the piezoelectrodes is nearly rectangular-shaped. Previous publications have proven the enhanced performance of SSHI circuits especially for systems with low piezoelectrical coupling.

The modeling and optimization of such networks is not straight forward, as the overall system is nonlinear. In the following we will present a modeling technique that is based on the harmonic balance method.

4.1. Period response for harmonic excitation

Firstly, the periodic response if the SSHI is studied. In order to simplify the results, the following approximations are defined,

$$L_s \to 0, \quad C_r \to \infty. \tag{28}$$

and the electrical losses remain constant. In practical realizations, all approximations are appropriate. A small inductance value results in a fast inversion, and a large storage capacitor means that the voltage at the load is nearly constant. Both situations are typically wanted. Further on, the nonlinear system can be treated as a piecewise linear system.

In order to obtain the stationary voltage signal it is necessary to study one semi-period of the system and consider the stationarity condition, which means the signal repeats after each period. We define the time axis in such a way that for $t = 0$ the voltage was just inverted and the switch is opened. With the approximation $C_r \to \infty$ the voltage u_L at the load capacitor is constant at u_0. This value has to be calculated yet. Because of the loss resistances - described by the electrical damping ratio ζ - the voltage changes from $\pm u_0$ to $\mp u_0 e^{-\pi\zeta}$. That means the absolute value of the voltage u_p at the piezoelectrodes after inversion is smaller than the voltage at the load u_0. Therefore the recifier blocks, and the piezovoltage changes linearly with the piezodeformation q,

$$u_p(t) = -\frac{\alpha\kappa}{C_p}q(t) + C_0, \tag{29}$$

with an integration constant C_0. According to the definitions, the piezodeformation must have an extremum at $t = 0$, so that the time signal reads

$$q(t) = \hat{q}\cos(\Omega t), \tag{30}$$

and for the voltage signal it follows

$$u_p(t) = -\frac{\alpha\kappa}{C_p}\hat{q}\cos(\Omega t) + u_0 e^{-\pi\zeta} + \frac{\alpha\kappa}{C_p}\hat{q}. \tag{31}$$

The voltage amplitude u_0 is still unknown, but at least the time t_1 can be calculated, at which the voltage at the piezoelectrodes equals the voltage at the load, $u_p(t = t_1) = u_0$:

$$t_1 = \text{acos}\left[\frac{u_0}{\frac{\alpha\kappa}{C_p}\hat{q}}\left(e^{-\pi\zeta} - 1\right) + 1\right]/\Omega. \tag{32}$$

At this time t_1 the rectifier changes from 'blocking' to 'conducting'. Practically this means that the piezovoltage remains constant at u_0 from t_1 until the end of the semi-period $T/2$.

The piezovoltage u_0 can be calculated based on the energy balance. Therefore, the transferred energy E_t, the harvested energy E_h and the energy dissipated in the switching branch E_s must be obtained. They read, respectively,

$$E_t = -2 \int_0^{T/2} F_p(t) \dot{q}(t) dt = 2 \alpha \kappa \Omega \hat{q} \left[\int_0^{t_1} u_p(t) \sin(\Omega t) dt + \int_{t_1}^{T/2} u_0 \sin(\Omega t) dt \right],$$

$$E_h = 2 \int_0^{T/2} \frac{u_0^2}{R} dt = 2 \frac{\pi}{\Omega} \frac{u_0^2}{R},$$

$$E_s = 2 \frac{1}{2} C_p u_0^2 \left[1 - \left(e^{-\pi \zeta} \right)^2 \right]. \tag{33}$$

The transferred energy corresponds to the total energy that is shifted from the mechanical system into the piezoelectric system, while the other terms are the energies that are dissipated within the load resistor and the resistor of the switching branch. In stationary situation, the equality

$$E_t = E_h + E_s \tag{34}$$

holds. With this equation, finally the stationary voltage amplitude can be recalculated as

$$u_0 = 2 \frac{\alpha \kappa}{C_p \left(1 - e^{-\pi \zeta} \right) + \frac{\pi}{\Omega R}} \hat{q}. \tag{35}$$

Figure 9 shows the time signal of the voltage u_p. Inserting the stationary voltage amplitude into the energies in Equation 33 gives us the stationary energies,

$$E_{t,stat} = 4 \frac{\alpha^2 \kappa^2}{C_p} \frac{1 - e^{-2\pi\zeta} + \frac{2\pi}{C_p \Omega R}}{\left(1 - e^{-\pi\zeta} + \frac{\pi}{C_p \Omega R} \right)^2} \hat{q}^2,$$

$$E_{h,stat} = 4 \frac{\alpha^2 \kappa^2}{C_p} \frac{\frac{2\pi}{C_p \Omega R}}{\left(1 - e^{-\pi\zeta} + \frac{\pi}{C_p \Omega R} \right)^2} \hat{q}^2,$$

$$E_{s,stat} = 4 \frac{\alpha^2 \kappa^2}{C_p} \frac{1 - e^{-2\pi\zeta}}{\left(1 - e^{-\pi\zeta} + \frac{\pi}{C_p \Omega R} \right)^2} \hat{q}^2. \tag{36}$$

All these energy terms have quadratic dependency with the force factor α, which should be maximized for a high energy conversion. Also the electrical damping ratio ζ in the switching branch should be low. They also grow quadratically with the vibration amplitude $\hat{q}$ of the oscillator.

However, for the force excited vibrations, the vibration amplitude is influenced by the harvesting device. The transferred energy $E_{t,stat}$ yields a damping effect upon the oscillator, which reduces the vibration amplitudes. In order to determine the vibration amplitud

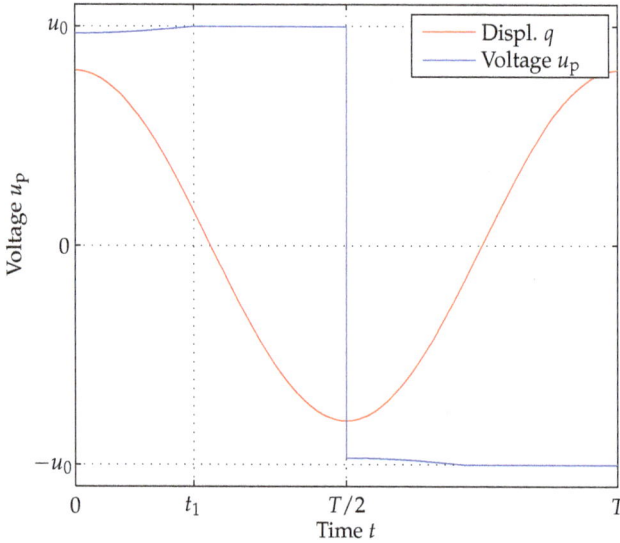

Figure 9. Time signals of piezodeformation and voltage at the electrodes.

the harmonic balance method is applied in the following. In this technique the shunted piezoceramics is replaced by a spring - damper combination. Therefore the period - but not harmonic - force response $F_p(t) = -\alpha u_p(t)$ of the shunted piezoceramics is expressed by its Fourier-series,

$$u_p(t) = \frac{1}{2}a_0 + \sum_{i=1}^{\infty} (a_i \cos(i\Omega t) + b_i \sin(i\Omega t)). \tag{37}$$

The Fourier-coefficients a_i, b_i are obtained by the periodic timesignal $u_p(t)$,

$$a_i = \frac{2}{T} \int_c^{c+T} u_p(t) \cos(i\Omega t)\mathrm{d}t; \quad b_i = \frac{2}{T} \int_c^{c+T} u_p(t) \sin(i\Omega t)\mathrm{d}t. \tag{38}$$

The idea of the proposed linearization techniques is to approximate the periodic voltage signal by its main harmonics,

$$u_p(t) \approx a_1 \cos(\Omega t) + b_1 \sin(\Omega t). \tag{39}$$

This harmonic force signal is also produced by a spring - damper combination with the following parameters,

$$c^* = \frac{a_1}{\hat{q}}, \quad d^* = -\frac{b_1}{\Omega \hat{q}}. \tag{40}$$

In general, these replacement parameters c^*, d^* are frequency dependent. With these results, the stationary vibration amplitudes of the oscillator with shunted piezoceramics can be recalculated,

$$\hat{q} = \frac{\hat{F}_p}{|-m\Omega^2 + j(d + d^*)\Omega + c + c^*|} \tag{41}$$

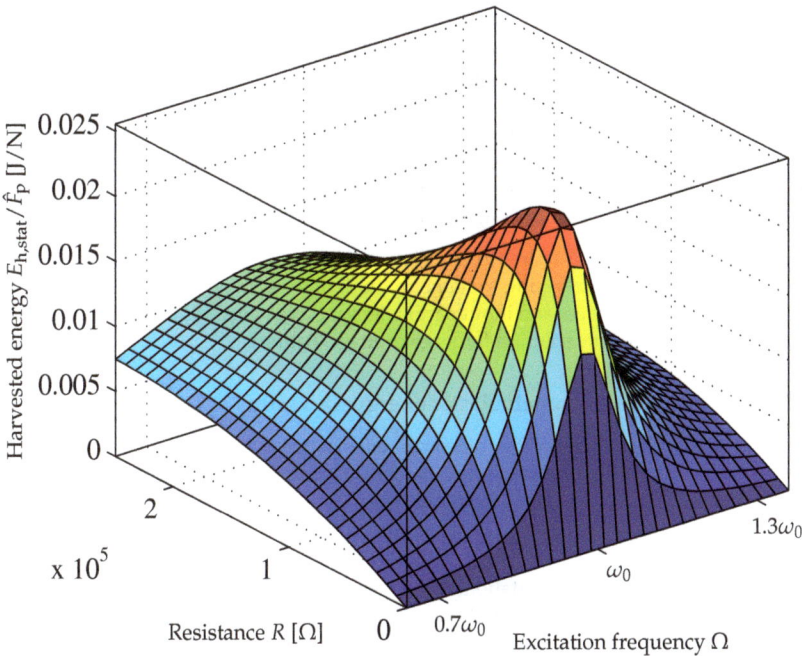

Figure 10. Harvested energy $E_{h,stat}$ versus normalized excitation frequency Ω and load resistance R.

Finally this stationary vibration amplitude $\hat{q}$ can be inserted into Equation 36 for the harvested energy $E_{h,stat}$. The resulting energy is shown in Figure 10 versus the excitation frequency Ω and the load resistance R. This figure can be compared with the linear resistance case in Figure 3. It again shows that most energy is harvested at the resonance frequency ω_0 of the oscillator, because the vibration amplitudes are highest in this case. But also the load resistor must be tuned correctly in order to achieve the maximum energy. Compared to the standard case the maximum amount of harvested energy is similar, while the frequency bandwidth with SSHI tends to be larger. However, the voltage at the load resistance with SSHI circuit is nearly constant which is wanted in most practical cases, while it is a harmonics oscillations with the linear resistance. Additionally the damping effect upon the mechanical structure is higher, because of the additional energy that is dissipated within the resistance of the switching branch.

5. Piezomagnetoelastic energy harvesting

One major drawback for energy harvesting systems is that conventional generators produce the maximum energy when the system is excited at its resonance frequency. If the excitation frequency shifts the output power is drastically reduced. To overcome this disadvantage multiple researchers work on different broadband techniques to widen the operational frequency range. The focus in this section is on piezomagnetoelastic energy harvesting strategies. The equations of motion (EOM) are derived in the previous section. The generator is based on a model with lumped parameters. The broadband response is achieved by using nonlinear magnetic forces. Piezomagnetoelastic generators are studied in a bunch of multiple research activities [1, 2, 13, 14]. In many approaches the system is modeled as Duffing oscillator. Usually the system parameters in the model are adjusted manually to match the amplitude or power response of the experiment. The aim in this work is to investigate an analytically approach to derive the duffing parameters out of the system parameters.

This section is organized as followed. The mechanical EOM is derived for the piezomagnetoelastic energy harvesting system. In the following the duffing parameters are derived with respect to the system and input parameters and the system dynamic is discussed. The last part shows the analytic solution for large orbit oscillations.

5.1. Modeling of the piezomagnetoelastic energy harvesting system

Figure 11 gives a schematic view of the piezomagnetoelastic system. The energy harvester consist of a cantilever with two piezoelectric patches mounted on each side of an inactive substructure. The system is excited by a harmonic force

$$F(t) = \hat{F}\sin(\omega t) \tag{42}$$

where $\hat{F}$ is the amplitude of the excitation and ω is the excitation frequency. A magnetic tip mass is attached to the free end of the beam. Another permanent magnet is stationary mounted near the free end. The magnets are oppositely poled so they exhibit a repulsive force. The nonlinear magnetic force leads to two stable equilibrium positions. Figure 11 shows both symmetric stable equilibrium positions. The tip displacement is given with q and the magnet spacing is s. The coupled mechanical and electrical differential equations are derived in the previous section

$$m\ddot{q} + d\dot{q} + cq - \frac{\alpha}{C_P}Q_P + \frac{dU_{\mathrm{mag}}}{dq} = F(t) \tag{43}$$

and

$$R\dot{Q}_P + \frac{1}{C_P}Q_P - \frac{\alpha}{C_P}q = 0. \tag{44}$$

Equation 43 and 44 are similar to 19 where $m = m_{\mathrm{mech}}$ is the modal mass. The modal damping is d and $c = c_{\mathrm{mech}} + \frac{\alpha^2}{C_p}\kappa^2$ is the total mechanical stiffness. Additionally to the linear differential equations the nonlinear magnetic force leads to

$$U_{\mathrm{mag}} = \frac{\mu_0}{4\pi}\boldsymbol{\mu}_A \nabla \frac{\mathbf{r}_{AB}}{\|\mathbf{r}_{AB}\|_2^3}\boldsymbol{\mu}_B \tag{45}$$

where μ_0 is the permeability of free space, $\boldsymbol{\mu}_A$ and $\boldsymbol{\mu}_B$ are the magnetic dipole moment vectors. $\mathbf{r}_{AB}$ is the vector from the source of magnet B to magnet A. $\|\cdot\|_2$ is the EUKLIDEAN norm and ∇ is the vector gradient defined with

$$\nabla \frac{1}{r^n} = \begin{bmatrix} \partial/\partial x \\ \partial/\partial y \\ \partial/\partial z \end{bmatrix}\frac{1}{r^n} = -\frac{n}{r^{n+1}}\begin{bmatrix} x/r \\ y/r \\ z/r \end{bmatrix} = -\frac{n\mathbf{r}}{r^{n+2}}. \tag{46}$$

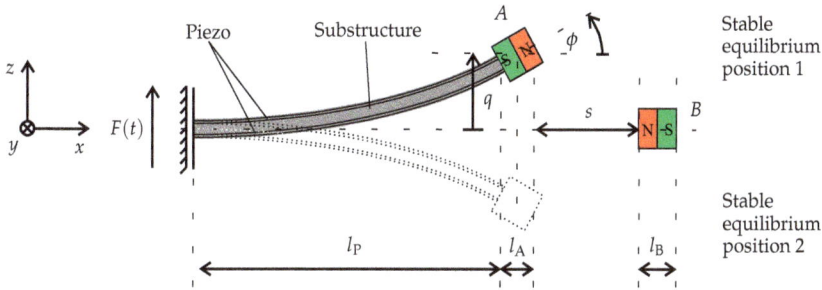

Figure 11. Schematic view of the piezomagnetoelastic energy harvesting system

The magnetic dipole moment vectors are written as

$$\mu_A = M_A V_A \begin{bmatrix} \cos\phi \\ 0 \\ \sin\phi \end{bmatrix}, \tag{47a}$$

$$\mu_B = M_B V_B \begin{bmatrix} -1 \\ 0 \\ 0 \end{bmatrix}, \tag{47b}$$

where M_A and M_B represents the vector sum of all microscopic magnetic moments within a ferromagnetic material and V_A and V_B are the volumes of the magnets. Details of the force between magnetic dipoles can be found in [16]. ϕ is the rotation angle at the magnet A.

The vector from the source of magnet B to magnet A is

$$\mathbf{r}_{AB} = \begin{bmatrix} -\left(s + \dfrac{l_A}{2} + \dfrac{l_B}{2} + \left(l_P + \dfrac{l_A}{2}\right)(1 - \cos\phi)\right) \\ 0 \\ q \end{bmatrix} \tag{48}$$

where l_A and l_B are the length of the magnets and l_P is the length of the beam in x-direction.

Figure 12 a) presents the potential energy

$$U = U_{mech} + U_{mag} = \frac{1}{2}cq^2 + \frac{\mu_0}{4\pi}\left(\frac{\mu_A \cdot \mu_B}{r_{AB}^3} - 3\frac{(\mu_A \cdot \mathbf{r}_{AB})(\mu_B \cdot \mathbf{r}_{AB})}{r_{AB}^5}\right) \tag{49}$$

where U_{mech} is the mechanical potential energy and c is the equivalent stiffness of the beam. U is normalized to the potential energy U_0 for a magnet distance of $s = 0.79s_0$. The tip displacement q is normalized to the tip displacement q_0 for $s = 0$. The derivation of the mechanical potential energy expression is shown in [15] and is proportional to q^2 for the first bending mode. U_{mag} is given in Equation 45. The Figure shows the potential energy for different magnet distances with respect to s_0 which is the critical magnet distance where the potential energy change from two to one stable equilibrium position. The nonlinear magnetic

a)   b)

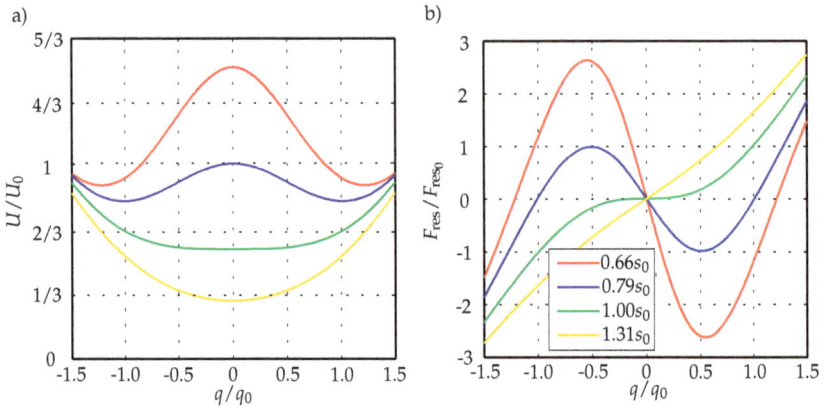

Figure 12. Potential energy and restoring force

force leads to an energy hump at zero tip displacement. A sharp peak results for a small spacing.

The system exhibits two stable equilibrium positions which are the local minimum positions in the potential energy and one unstable equilibrium for zero displacement ($s < s_0$). The hump disappears for large s and there is only one stable equilibrium for zero displacement ($s > s_0$).

The normalized restoring force

$$F_{res} = \frac{dU}{dq} \tag{50}$$

is additionally presented in Figure 12 b). F_{res} is normalized to the restoring force for a magnet spacing of $s = 0.79s_0$. Equation 50 is a nonlinear function of the system parameters in particular a function of s.

The benefit of the magnet force is the nonlinearity in the system response. Due to the restoring force the system exhibits overhanging resonance curves which strongly depends on the parameters. For $s < s_0$ and low excitation energy that the system only oscillates around one equilibrium the EOM has a hardening stiffness so the resonance curve overhang to the left. A softening response is given for $s > s_0$ with a distorted peak to the right. Specially for a hardening stiffness the system exhibit two equilibrium positions and the system bounces between both positions if the energy of the excitation is high enough. These large orbit deflections generate the most energy and are most important for designing the system setup.

5.2. Approximation with the Duffing oscillator

The potential energy shown in Figure 12 a) is approximated as fourth order polynomial function

$$U_{Duff} = \frac{1}{4}\alpha q^4 + \frac{1}{2}\beta q^2 \tag{51}$$

in a bunch of different research activities. The nonlinear restoring force shown in Figure 12 b) is than approximated as the Duffing equation

$$F_{\text{res,Duff}} = \frac{dU_{\text{Duff}}}{dq} = \alpha q^3 + \beta q \tag{52}$$

with the Duffing parameters α and β. The approach as Duffing equation is not only limited to piezomagnetoelastic energy harvesting techniques. [5] used this equation to model an electrostatic energy harvesting system and [8] approximate an electromagnetic energy generator. In the following α is the Duffing parameter before the cubic term instead of the coupling factor in Equation 43 to be conform to the most publications concerning the duffing equation. The uncoupled equation of motion (EOM) for the piezomagnetoelastic system in terms of the duffing oscillator is written as

$$m\ddot{q} + d\dot{q} + \alpha q^3 + \beta q = F(t). \tag{53}$$

The Duffing parameters are complicated functions of the beam, the piezo, the magnet parameters and in particular the magnet distance. It can be recognized in Figure 12 b) that the cubic part αq^3 is always positive for repulsive magnets because the function is always monotonically increasing for large q values. Only the linear part βq can be either positive or negative with respect to the magnet distance. For $s < s_0$ the parameter β is negative and the system exhibits two stable and one unstable equilibrium position. For $s \geq s_0$ the system has only one stable equilibrium position and β is positive. s_0 is the critical magnet spacing.

The nonlinear restoring force F_{res} shown in Figure 12 b) can be approximated as a third order Taylor serious given with

$$p(q) = k_1 q^3 + k_2 q^2 + k_3 q + k_4 \tag{54}$$

where k_{1-4} are constants derived from the system parameters. By comparing Equation 54 and Equation 52 note that $k_1 = \alpha$ and $k_3 = \beta$. To calculate the constant parameters in Equation 54 it takes four constrains.

1. The restoring force is point symmetric:

$$p(0) = 0, \tag{55}$$

$$\frac{d^2 p(0)}{dq^2} = 0. \tag{56}$$

2. The Taylor series must exhibit the same equilibrium positions:

$$p(q_{\text{eq}}) = 0. \tag{57}$$

3. The oscillation must be suitable for small magnet distances s around one equilibrium position

$$\frac{dp(q_{\text{eq}})}{dq} = a, \tag{58}$$

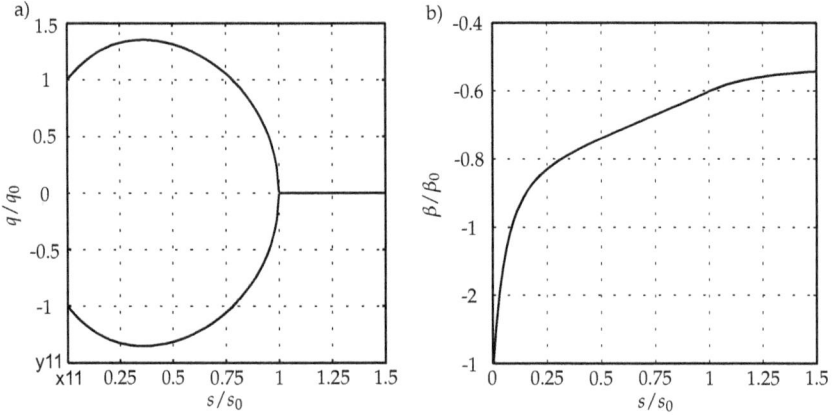

a)

b)

Figure 13. Equilibrium positions and β with respect to the magnet distance s.

where q_{eq} is the equilibrium position and a is the slope at q_{eq}. a and q_{eq} can be calculated from the original system. The solution for k_{1-4} is

$$k_1 = \alpha = \frac{a}{2q_{eq}^2},\tag{59a}$$

$$k_2 = 0,\tag{59b}$$

$$k_3 = \beta = -\frac{a}{2},\tag{59c}$$

$$k_4 = 0.\tag{59d}$$

Equation 53 becomes

$$m\ddot{q} + d\dot{q} + \frac{a}{2q_{eq}^2}q^3 - \frac{a}{2}q = F(t).\tag{60}$$

Equation 60 is usable if the potential energy exhibits two equilibrium positions. If the magnet distance is larger than s_0 the Duffing parameters α and β are simply the liner and the cubic part of the third order Taylor series of Equation 50 and α and β are positive.

This behavior can be recognized in Figure 13 a) and b). The graphs are calculated by using the potential energy of two point dipols in Equation 49. Figure a) shows the equilibrium position over the magnet distance. The equilibrium position q_{eq} is normalized to the known q_0. The magnet distance is normalized to the critical magnet distance s_0. Note that the distance between the two stable q_{eq} becomes smaller for very small s because of the rotation angle ϕ of magnet A. The Figure 13 b) gives the modification of the Duffing parameter β with respect to the normalized known magnet distance. β is normalized to β_0 the linear Duffing parameter

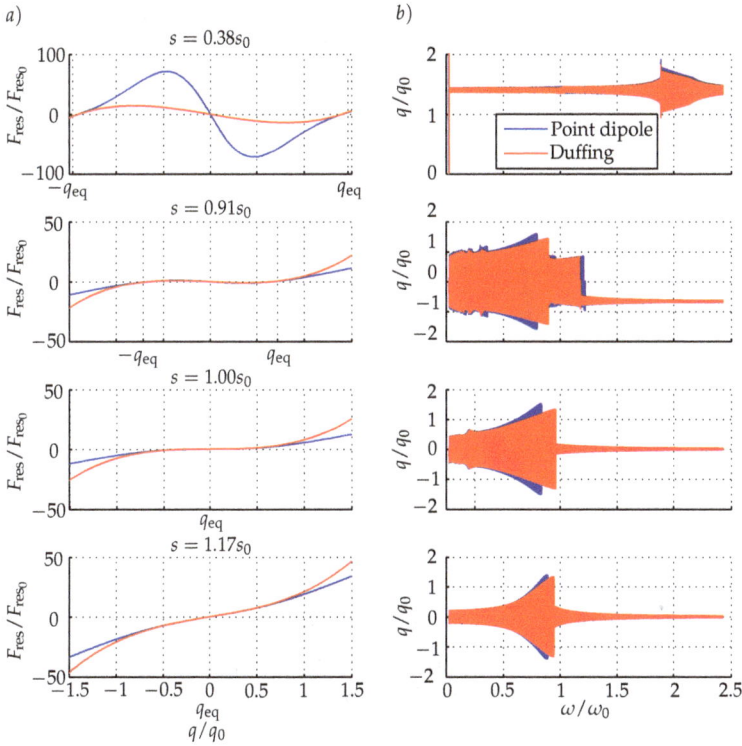

Figure 14. Duffing approximation of the restoring force and displacement

for $s = 0$. For $s < s_0$ the system exhibit a negative linear restoring force due to the negative parameter β. β is positive for $s > s_0$.

Figure 14 shows the approximation of Equation 50 with Equation 52 and the Duffing parameter known from Equation 59. The pictures a) show the restoring force derived from the force between two point dipoles and the Duffing equation for different magnet distances. The pictures b) gives the corresponding time variant tip displacement. It can be recognized that the Duffing equation approximate the restoring force very well for the different ranges of magnet spacing. The Duffing equation shows the hardening response for small magnet distances ($s = 0.38s_0$) and it approximates the influence of the different attractors ($s = 0.91s_0$). For $s > s_0$ the approximation is also very good.

5.3. Solution for large orbit oscillations

In Section 5.1 and Section 5.2 the piezomagnetoelastic energy harvesting system is presented and an approximation with the Duffing oscillator is given. In the following Equation 60 is solved for large orbit oscillations so that system bounces between both symmetric equilibrium positions or around the only one for $s > s_0$. The EOM can be written as

$$\ddot{q} + 2D\omega_0\dot{q} + \epsilon q^3 + \text{sgn}(\beta)\omega_0^2 q = f_0\cos(\omega t) \tag{61}$$

where $\text{sgn}(\beta)$ is either positive or negative for $\beta > 0$ or $\beta < 0$ with

$$D = \frac{d}{2\omega_0 m}, \tag{62a}$$

$$\omega_0^2 = \frac{|\beta|}{m}, \tag{62b}$$

$$\epsilon = \frac{\alpha}{m}, \tag{62c}$$

$$f_0 = \frac{\hat{F}}{m}. \tag{62d}$$

The Equation 61 can be solved by applying the harmonic balance method. The harmonic balance method is well known and details can be found in [12]. The amplitude response is assumed as harmonic with the frequency of the excitation

$$q = \hat{q}\cos(\omega t - \varphi) \tag{63}$$

where $\hat{q}$ is the amplitude and φ is the phase of the tip displacement. Insert Equation 63 in Equation 61 and only consider terms with the excitation frequency leads to

$$\sqrt{\left[\left(\text{sgn}(\beta)\omega_0^2 - \omega^2\right)\hat{q} + \frac{3}{4}\epsilon\hat{q}^3\right]^2 + 4D^2\omega_0^2\omega^2\hat{q}^2}\cos\left(\omega t - \ldots\right.$$

$$\left.\ldots - \varphi + \arctan\left(\frac{2D\omega_0\omega\hat{q}}{(\text{sgn}(\beta)\omega_0^2 - \omega^2)\hat{q} + \frac{3}{4}\epsilon\hat{q}^3}\right)\right) = f_0\cos(\omega t). \tag{64}$$

The Equation 61 is valid if the amplitude and the phase in Equation 63 solves the equations. The solution gives the frequency response with respect to the amplitude $\hat{q}$

$$\omega_{1/2}^2 = \text{sgn}(\beta)\omega_0^2 - 2D\omega_0^2 + \frac{3}{4}\epsilon\hat{q}^2 \pm \sqrt{\frac{f_0^2}{\hat{q}^2} + 4D^2\omega_0^2\left(D^2\omega_0^2 - \text{sgn}(\beta)\omega_0^2 - \frac{3}{4}\epsilon\hat{q}^2\right)}. \tag{65}$$

Figure 15 shows the analytical amplitude response given in Equation 65 and the numerical solution of the Duffing oscillator. The graph shows the solution for three different magnet distances. The frequency was slowly increased so the system remains a steady state response. One can recognize that the harmonic balance is well suited if the Duffing equation has a positive linear restoring force ($\beta > 0$). If the Duffing oscillator exhibits a negative restoring force and the system excitation delivers enough energy that the energy harvester bounces between both stable equilibrium positions than the harmonic balance predicts the influence of

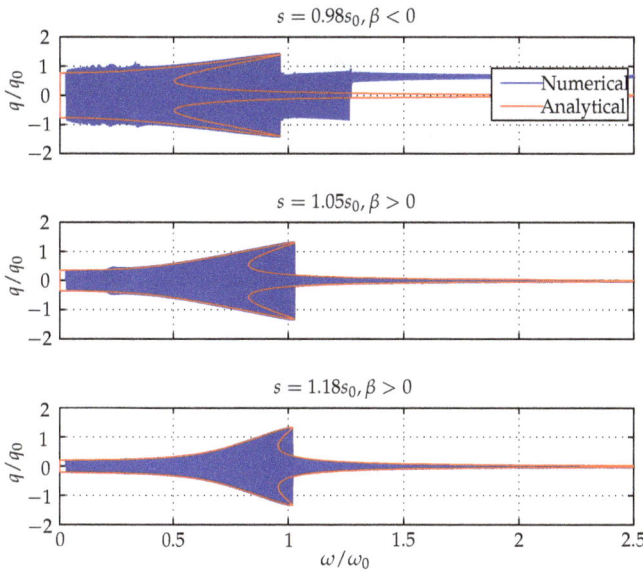

Figure 15. Numerical and analytical results for the large orbit duffing equation

the attractors till the first jump. The system behavior after the first jump can not be predicted with this harmonic approximation.

6. Conclusions

This chapter presents different techniques to enlarge the frequency bandwidth of piezoelectric energy harvester. A precise modeling of piezomechanical structures is given, and the linear harvesting system is given as the reference. In detail a nonlinear switching SSHI circuit, nonlinear magnet forces and an array configuration of several bimorphs are discussed. For all cases, appropriate modeling techniques are presented that allow an efficient yet precise analysis. The nonlinear techniques alter the system dynamics especially for off-resonance vibrations. The magnet forces generate a bistable system, in which the bimorph oscillates between both equilibria. The SSHI circuit increase the energy conversion rate by producing a rectangular shaped voltage signal, while the array configuration tunes the individual bimorphs to slightly different frequencies, so that the energy conversion is distributed to a broader frequency range.

Author details

Marcus Neubauer, Jens Twiefel, Henrik Westermann and Jörg Wallaschek
Leibniz University Hannover, Institute of Dynamics and Vibration Research, Appelstrasse 11, 30167 Hannover, Germany

7. References

[1] Erturk, A., Hoffmann, J. & Inman, D. [2009]. A piezomagnetoelastic structure for broadband vibration energy harvesting, *Applied Physics Letters* .

[2] Ferrari, M., Ferrari, V., Guizzetti, M., Andò, B., Baglio, S. & C.Trigona [2010]. Improved energy harvesting from wideband vibrations by nonlinear piezoelectric converters, *Sensors and Actuators A: Physical* .

[3] Guyomar, D., Badel, A., Lefeuvre, E. & Richard, C. [2005]. Toward energy harvesting using active materials and conversion improvement by nonlinear processing, *Ultrasonics, Ferroelectrics and Frequency Control, IEEE Transactions on* 52(4): 584–595.

[4] Guyomar, D., Magnet, C., Lefeuvre, E. & Richard, C. [2006]. Power capability enhancement of a piezoelectric transformer, *Smart Materials and Structures* 15(2): 571–580. URL: *http://stacks.iop.org/0964-1726/15/571*

[5] Halvorsen, E., Blystad, L.-C., Husa, S. & Westby, E. [2007]. Simulation of electromechanical systems driven by large random vibrations, *MEMSTECH*.

[6] *IEEE standard on piezoelectricity* [29 Jan 1988]. *ANSI/IEEE Std 176-1987* .

[7] Lallart, M., Guyomar, D., Richard, C. & Petit, L. [2010]. Nonlinear optimization of acoustic energy harvesting using piezoelectric devices, *Journal of the Acoustical Society of America* 128: 2739–2748.

[8] Mann, B. & Sims, N. [2009]. Energy harvesting from the nonlinear oscillations of magnetic levitation, *Journal of Sound and Vibration* 319: 515–530.

[9] Neubauer, M., Schwarzendahl, S. & Wallaschek, J. [2012]. A new solution for the determination of the generalized coupling coefficient for piezoelectric systems, *Journal of Vibroengineering* 14(1): 105–110.

[10] Neubauer, M. & Wallaschek, J. [2011]. Optimized geometry of bimorphs for piezoelectric shunt damping and energy harvesting, *Proceedings of Int. Conference of Engineering Against Fracture (ICEAF)*.

[11] Priya, S. & Inman, D. (eds) [2009]. *Energy Harvesting Technologies*, Springer.

[12] Schmidt, G. & Tondl, A. [1986]. *Non-Linear Vibrations*, Akademie-Verlag Berlin.

[13] Sebald, G., Kuwano, H., Guyomar, D. & Ducharne, B. [2011]. Experimental duffing oscillator for broadband piezoelectric energy harvesting, *Smart Materials and Structures* 20.

[14] Stanton, S., McGehee, C. & Mann, B. [2009]. Reversible hysteresis for broadband magnetopiezoelastic energy reversible hysteresis for broadband magnetopiezoelastic energy harvesting, *Applied Physics Letters* 95.

[15] Stanton, S., McGehee, C. & Mann, B. [2010]. Nonlinear dynamics for broadband energy harvesting: Investigation of a bistable piezoelectric inertial generator, *Physica D: Nonlinear Phenomena* .

[16] Yung, K., Landecker, P. & Villani, D. [1998]. An analytic solution for the force between two magnetic dipoles, *Magnetic and Electrical Separation* 9: 39–52.

Permissions

The contributors of this book come from diverse backgrounds, making this book a truly international effort. This book will bring forth new frontiers with its revolutionizing research information and detailed analysis of the nascent developments around the world.

We would like to thank Dr. Mickaël Lallart, for lending his expertise to make the book truly unique. He has played a crucial role in the development of this book. Without his invaluable contribution this book wouldn't have been possible. He has made vital efforts to compile up to date information on the varied aspects of this subject to make this book a valuable addition to the collection of many professionals and students.

This book was conceptualized with the vision of imparting up-to-date information and advanced data in this field. To ensure the same, a matchless editorial board was set up. Every individual on the board went through rigorous rounds of assessment to prove their worth. After which they invested a large part of their time researching and compiling the most relevant data for our readers. Conferences and sessions were held from time to time between the editorial board and the contributing authors to present the data in the most comprehensible form. The editorial team has worked tirelessly to provide valuable and valid information to help people across the globe.

Every chapter published in this book has been scrutinized by our experts. Their significance has been extensively debated. The topics covered herein carry significant findings which will fuel the growth of the discipline. They may even be implemented as practical applications or may be referred to as a beginning point for another development. Chapters in this book were first published by InTech; hereby published with permission under the Creative Commons Attribution License or equivalent.

The editorial board has been involved in producing this book since its inception. They have spent rigorous hours researching and exploring the diverse topics which have resulted in the successful publishing of this book. They have passed on their knowledge of decades through this book. To expedite this challenging task, the publisher supported the team at every step. A small team of assistant editors was also appointed to further simplify the editing procedure and attain best results for the readers.

Our editorial team has been hand-picked from every corner of the world. Their multi-ethnicity adds dynamic inputs to the discussions which result in innovative outcomes. These outcomes are then further discussed with the researchers and contributors who give their valuable feedback and opinion regarding the same. The feedback is then collaborated with the researches and they are edited in a comprehensive manner to aid the understanding of the subject.

Apart from the editorial board, the designing team has also invested a significant amount of their time in understanding the subject and creating the most relevant covers. They scrutinized every image to scout for the most suitable representation of the subject and create an appropriate cover for the book.

The publishing team has been involved in this book since its early stages. They were actively engaged in every process, be it collecting the data, connecting with the contributors or procuring relevant information. The team has been an ardent support to the editorial, designing and production team. Their endless efforts to recruit the best for this project, has resulted in the accomplishment of this book. They are a veteran in the field of academics and their pool of knowledge is as vast as their experience in printing. Their expertise and guidance has proved useful at every step. Their uncompromising quality standards have made this book an exceptional effort. Their encouragement from time to time has been an inspiration for everyone.

The publisher and the editorial board hope that this book will prove to be a valuable piece of knowledge for researchers, students, practitioners and scholars across the globe.

List of Contributors

Kai Ren and Yong X. Gan
Department of Mechanical, Industrial and Manufacturing Engineering, College of Engineering, University of Toledo, Toledo, OH, USA

Hongying Zhu, Pierre-Jean Cottinet and Daniel Guyomar
INSA de Lyon, LGEF Laboratoire de Génie Electrique et Ferroélectricité EA 682, Université de Lyon, Bâtiment Gustave FERRIE, 8 rue de la Physique, F-69621 Villeurbanne Cedex, France

Sébastien Pruvost
Université de Lyon, INSA-Lyon, Ingénierie des Matériaux Polymères (IMP) UMR CNRS 5223, 69621 Villeurbanne Cedex, France

Chris Gould and Noel Shammas
Faculty of Computing, Engineering and Technology, Staffordshire University, Stafford, United Kingdom

Igor L. Baginsky and Edward G. Kostsov
Institute of Automation and Electrometry, Russian Academy of Sciences, Russia

S. Boisseau, G. Despesse and B. Ahmed Seddik
LETI, CEA, Minatec Campus, Grenoble, France

Mickaël Lallart, Pierre-Jean Cottinet, Jean-Fabien Capsal, Laurent Lebrun and Daniel Guyomar
Université de Lyon, INSA-Lyon, LGEF EA 682, F-69621, Villeurbanne, France

Wen Jong Wu and Bor Shiun Lee
Department of Engineering Science and Ocean Engineering, National Taiwan University, Taipei, Taiwan

Yu-Jen Wang, Sheng-Chih Shen and Chung-De Chen
National Formosa University, National Cheng Kung University, Industrial Technology Research Institute, Taiwan

B. Ahmed Seddik, G. Despesse, S. Boisseau and E. Defay
LETI, CEA, Minatec Campus, Grenoble, France

Adam Wickenheiser
George Washington University, United States

Cuong Phu Le and Einar Halvorsen
Department of Micro and Nano Systems Technology, Faculty of Technology and Maritime Sciences, Vestfold University College, Tønsberg, Norway

Ji-Tzuoh Lin and Bruce William Alphenaar
Department of Electrical and Computer Engineering, University of Louisville, Louisville, KY, USA

Barclay Lee
Department of Bioengineering, California Institute of Technology, Pasadena, CA, USA

Yuan-Ping Liu and Dejan Vasic
SATIE, ENS Cachan, France

Marcus Neubauer, Jens Twiefel, Henrik Westermann and Jörg Wallaschek
Leibniz University Hannover, Institute of Dynamics and Vibration Research, Appelstrasse 11, 30167 Hannover, Germany

www.ingramcontent.com/pod-product-compliance
Lightning Source LLC
Chambersburg PA
CBHW070718190326
41458CB00004B/1015